# Designing Small Weapons

# Designing Small Weapons

Jose Martin Herrera-Ramirez
Luis Adrian Zuñiga-Aviles

CRC Press is an imprint of the
Taylor & Francis Group, an **informa** business

First edition published 2022
by CRC Press
6000 Broken Sound Parkway NW, Suite 300, Boca Raton, FL 33487-2742

and by CRC Press
4 Park Square, Milton Park, Abingdon, Oxon, OX14 4RN

*CRC Press is an imprint of Taylor & Francis Group, LLC*



*Library of Congress Cataloging-in-Publication Data*
Names: Herrera-Ramirez, Jose M., author. | Zuñiga-Aviles, Luis Adrian, author.
Title: Designing small weapons / Jose Martin Herrera-Ramirez, Luis Adrian Zuñiga-Aviles.
Description: First edition. | Boca Raton : CRC Press, 2022. |
Includes bibliographical references and index.
Identifiers: LCCN 2021059478 (print) | LCCN 2021059479 (ebook) |
ISBN 9781032052663 (hbk) | ISBN 9781032052694 (pbk) | ISBN 9781003196808 (ebk)
Subjects: LCSH: Firearms—Design and construction.
Classification: LCC TS535 .H47 2022 (print) | LCC TS535 (ebook) |
DDC 683.4—dc23/eng/20220223
LC record available at https://lccn.loc.gov/2021059478
LC ebook record available at https://lccn.loc.gov/2021059479

ISBN: 978-1-032-05266-3 (hbk)
ISBN: 978-1-032-05269-4 (pbk)
ISBN: 978-1-003-19680-8 (ebk)

DOI: 10.1201/9781003196808

Typeset in Times
by codeMantra

*To my wife Ana Maria and my son Martin Alejandro, without whose partnership and comprehension this book would not have been possible. It is our collective achievement! I love you.*

**Dr. Jose Martin Herrera-Ramirez**

*To my wife Sonia, my sons Irving and Kevin, and my daughter Katia, their support and love are my life balance to continue on this fantastic road, collaborating, learning, sharing, and dreaming together to achieve new goals, new ideas, new books.*

**Dr. Luis Adrian Zuñiga-Aviles**

# Contents


List of Abbreviations....xiii
Preface....xix
Acknowledgments....xxi
Authors....xxiii

**Chapter 1** History of the Design of Small Weapons....1

1.1 Introduction....1
1.1.1 The First Period....1
1.1.2 The Second Period....1
1.1.3 The Third Period....2
1.1.4 The Fourth Period....3
1.2 Evolution of Firearms....3
1.2.1 The Hand Cannon....4
1.2.2 The Matchlock System....4
1.2.3 The Wheellock System....4
1.2.4 The Snaphaunce System....4
1.2.5 The Flintlock System....4
1.2.6 The Percussion System....5
1.2.7 The Dreyse Needle System....5
1.2.8 The Pinfire System....5
1.2.9 The Rimfire Ammunition....6
1.2.10 The Centerfire Ammunition....6
1.2.11 The Rifling System....6
1.2.12 The Revolver....6
1.2.13 Self-Loading Firearms....6
1.3 Classification of Firearms....7
1.3.1 Classification by the Level of Harm....7
1.3.1.1 Lethal Firearms....7
1.3.1.2 Non-lethal Firearms....7
1.3.2 Classification by the Traditional Structure....7
1.3.2.1 Firearms....8
1.3.2.2 Conventional Weapons....8
1.3.2.3 Non-conventional Weapons....8
1.3.3 Classification by the Portability....8
1.3.3.1 Small Arms....8
1.3.3.2 Light Weapons....8
1.3.3.3 Heavy Weapons....8
1.3.4 Classification by the Physical Characteristics, Size, and Support....8
1.3.4.1 Short....8

1.3.4.2 Long ..... 8
1.3.5 Classification by the Weapon Action ..... 9
1.3.5.1 Single-Shot Action ..... 9
1.3.5.2 Repeating Action ..... 9
1.3.5.3 Semi-automatic Action ..... 9
1.3.5.4 Burst Action ..... 9
1.3.5.5 (Fully) Automatic Action ..... 9
1.3.6 Classification by the Type of Firearm ..... 9
1.3.6.1 Revolver ..... 9
1.3.6.2 Pistol ..... 10
1.3.6.3 Shotgun ..... 10
1.3.6.4 Rifle or Carbine ..... 10
1.3.6.5 Assault Rifle ..... 10
1.3.6.6 Sub-machine Gun ..... 10
1.3.6.7 Machine Gun ..... 10
1.3.6.8 Other Type of Firearms ..... 10
1.4 Evolution of Weapon Materials ..... 11
1.4.1 Stone, Wood, and Bone ..... 11
1.4.2 Metals ..... 11
1.4.2.1 Bronze ..... 11
1.4.2.2 Iron ..... 12
1.4.2.3 Steel ..... 13
1.4.2.4 Aluminum ..... 13
1.4.3 Polymers ..... 13
1.4.4 Composites ..... 14
1.4.5 Ceramics ..... 14
1.5 Evolution of Firearm Manufacturing Processes ..... 15
1.6 Evolution of Design Tools ..... 15
1.6.1 The First Period ..... 15
1.6.2 The Second Period ..... 16
1.6.3 The Third Period ..... 16
References ..... 18

**Chapter 2** Beginning the Product Design ..... 21

2.1 Introduction ..... 21
2.2 Product Lifecycle Management ..... 21
2.3 Design Methodologies ..... 24
2.3.1 Expectation for Innovation Using the Market Pull ..... 24
2.3.2 Expectation for Innovation Using the Technology Push ..... 25
2.3.3 Design Criteria ..... 26
2.3.4 Design Attributes ..... 26
2.3.5 Requirements ..... 27

2.3.6 Functional Requirements ... 27
2.3.7 Design Parameters ... 28
2.3.8 Constraints ... 28
2.3.9 Commonality Index ... 28
2.3.10 DFA Index ... 29
2.3.11 Modeling and Simulation ... 29
2.3.12 PDMs Frameworks ... 29
2.4 A Case Study Based on PDMs Toolkit ... 33
2.4.1 Preliminary Topics ... 33
2.4.2 Market Pull Analysis ... 33
2.4.3 Design Criteria ... 35
2.4.4 Requirements ... 38
2.4.5 Functional Requirements ... 38
2.4.6 Design Parameters ... 39
2.4.7 Constraints ... 39
2.4.8 Commonality Index ... 39
2.4.9 DFA Index ... 40
2.4.10 Product Portfolio ... 41
2.5 Closing Remarks and Perspectives ... 43
References ... 44

**Chapter 3** Custom and Functional Requirements ... 47

3.1 Introduction ... 47
3.2 Requirements ... 47
3.3 Requirements to Reach Readiness of a System ... 48
3.4 Requirement Identification ... 49
3.5 Determination of FR and Assessment of Its Difficulty ... 51
3.6 Determination of DPs Using Axiomatic Design ... 54
3.7 Relationship DP to FR ... 56
3.8 Relationship Critical Design Parameter (CDP) to Test Bench Feature (TBF) ... 56
3.9 Relationship Process Variable (PV) to DP ... 58
3.10 Determination of Instruments to Technology Transfer ... 59
3.11 Closing Remarks and Perspectives ... 59
References ... 60

**Chapter 4** CAD Modeling and CAE Simulation ... 63

4.1 Introduction ... 63
4.2 Modeling and Simulation ... 63
4.3 CAD Modeling ... 65
4.3.1 Digital Model Obtaining ... 65
4.3.2 Materials Database Storage ... 67
4.3.3 Bill of Material Property Manager ... 67

4.3.4 CAD Animation .................... 67
4.3.5 Interference Detection .................... 67
4.3.6 Tolerance Stack-up Analysis .................... 70
4.3.7 CAD Drawings .................... 74
4.4 CAE Simulation .................... 74
4.4.1 CAD Model Treatment .................... 79
4.4.2 CAE Analysis by FEA .................... 79
4.4.3 CAE Motion .................... 81
4.4.4 Flow CFD .................... 83
4.4.5 CAE Multibody Dynamics .................... 96
4.4.6 CAE Co-Simulation .................... 100
4.4.7 CAE Multi-Domain .................... 100
4.4.8 CAE Bullet Penetration and Perforation by Explicit Dynamics .................... 100
4.4.9 Thermal Simulation by FEA, CFD, and FSI .................... 105
4.4.10 CAE Emulation .................... 105
4.5 CAD-CAE Documentation and Report .................... 107
4.6 Closing Remarks and Perspectives .................... 107
References .................... 108

**Chapter 5** CAM Assessment and Rapid Prototyping .................... 111

5.1 Introduction .................... 111
5.2 Technical Preliminaries .................... 111
5.3 CAM Assessment .................... 112
5.4 CNC Machining .................... 115
5.5 Laser Cutting .................... 116
5.6 3D Printing .................... 117
5.7 Prototyping Workshop .................... 120
5.8 Rapid Prototyping .................... 120
5.9 Rapid Tooling and Manufacturing Devices .................... 121
5.10 Industry 4.0 .................... 122
5.11 Closing Remarks and Perspectives .................... 124
References .................... 124

**Chapter 6** Experimental Physical Models, Test Benches, and Prototypes .................... 129

6.1 Introduction .................... 129
6.2 Tests Protocols and Product Validation Process .................... 129
6.3 Firearm Usability and UX .................... 132
6.4 Firearm EPMs and Demonstration Prototypes .................... 132
6.5 Test Benches and Standards to Firearm Performance .................... 133
6.6 Firearm Prototypes in Real Environment .................... 136
6.7 Polymer Firearms and 3D Printed Prototypes .................... 137

6.8 Firearm Performance Case Studies ... 138
6.9 Closing Remarks and Perspectives ... 140
References ... 141

**Chapter 7** Materials Used in the Production of Small Weapons ... 145

7.1 Introduction ... 145
7.2 Classification of Materials ... 146
7.2.1 Metals ... 146
7.2.2 Polymers ... 147
7.2.3 Ceramics ... 147
7.2.4 Composites ... 147
7.3 Binary Phase Diagrams ... 148
7.4 The Fe-C Phase Diagram ... 149
7.5 Aluminum Alloys Phase Diagrams ... 151
7.6 Mechanical Properties of Materials ... 151
7.7 Steels ... 154
7.7.1 Classification of Steels ... 154
7.7.2 Influence of Alloying and Residual Elements on Steel Properties ... 156
7.7.3 Designation of Steels ... 157
7.8 Aluminum Alloys ... 162
7.8.1 Classification of Aluminum Alloys ... 164
7.8.2 Designation of Wrought Aluminum Alloys ... 165
7.8.3 Influence of Alloying Elements and Impurities on Aluminum Alloy Properties ... 165
7.9 Titanium Alloys ... 167
7.10 Synthetic Polymers ... 168
7.10.1 Glass Transition Temperature ... 168
7.10.2 Nomenclature of Synthetic Polymers ... 169
7.10.3 Classification of Polymers by Properties ... 169
7.11 Composites ... 170
7.11.1 Matrices ... 171
7.11.2 Reinforcements ... 172
7.12 Ceramics ... 174
7.13 Closing Remarks ... 175
References ... 175

**Chapter 8** Heat Treatments and Surface Hardening of Small Weapon Components ... 179

8.1 Introduction ... 179
8.2 Heat Treatments of Steels ... 179
8.2.1 The TTT Diagram ... 181

8.2.2 Annealing ........ 182
8.2.2.1 Full Annealing ........ 182
8.2.2.2 Process Annealing ........ 183
8.2.2.3 Spheroidizing Annealing ........ 183
8.2.3 Normalizing ........ 183
8.2.4 Hardening ........ 184
8.2.5 Tempering ........ 185
8.2.6 Martempering ........ 185
8.2.7 Austempering ........ 185
8.3 Surface Hardening of Steels ........ 187
8.3.1 Carburizing ........ 190
8.3.2 Nitriding ........ 190
8.3.3 Carbonitriding ........ 190
8.3.4 Induction Hardening ........ 190
8.3.5 Flame Hardening ........ 190
8.4 Heat Treatments of Aluminum Alloys ........ 191
8.5 Anodizing of Aluminum Alloys ........ 195
8.6 A Case Study Based on Heat Treatments of Steels ........ 195
8.7 Closing Remarks ........ 202
References ........ 202

**Chapter 9** Manufacturing Processes for Small Weapon Components ........ 205
9.1 Introduction ........ 205
9.2 Casting ........ 205
9.3 Forming ........ 208
9.3.1 Forging ........ 208
9.3.2 Extrusion ........ 210
9.3.3 Stamping ........ 210
9.4 Polymer Processes ........ 211
9.5 Powder Metallurgy ........ 211
9.6 Material Removal Processes ........ 212
9.7 Additive Manufacturing Processes ........ 215
9.8 Finishing, Assembly, and Testing ........ 218
9.8.1 Finishing ........ 218
9.8.2 Assembly ........ 218
9.8.3 Testing ........ 219
9.9 Closing Remarks ........ 219
References ........ 220

**Index** ........ 221

# List of Abbreviations

| | |
|---|---|
| **2PP** | two-photon polymerization |
| **3DGP** | 3D gel printing |
| **A2LA** | American Association for Laboratory Accreditation |
| **AA** | Aluminum Association |
| **ABC** | atomic, biological, and chemical weapons |
| **ABS** | acrylonitrile butadiene styrene |
| **ACP** | automatic colt pistol |
| **AEP** | Allied Engineering Publication |
| **AFNOR** | French Standardization Association |
| **AFSD** | additive friction stir deposition |
| **AI** | artificial intelligence |
| **AISI** | The American Iron and Steel Institute |
| **AJ** | aerosol jetting |
| **AM** | additive manufacturing |
| **API** | advanced primer ignition |
| **AQAP** | Allied Quality Assurance Publications |
| **AR** | augmented reality |
| **ASA** | acrylic styrene acrylonitrile |
| **ASTM** | American Society for Testing and Materials |
| **ATF** | Bureau of Alcohol, Tobacco, and Firearms |
| **BAAM** | big area additive manufacturing |
| **BC** | boundary condition |
| **BCC** | body-centered cubic crystal structure |
| **BCL** | binary cutter language |
| **BCT** | body-centered tetragonal crystal structure |
| **BDT** | bullet dwell time |
| **BFS** | back-face signature |
| **BI** | business intelligence |
| **BIS** | Bureau of Indian Standards |
| **BJ3DP** | binder jet three-dimensional printing |
| **BOM** | bill of materials |
| **BSI** | British Standards Institution |
| **CAD** | computer-aided design |
| **CAE** | computer-aided engineering |
| **CAGR** | compound annual growth rate |
| **CAM** | computer-aided manufacturing |
| **CAST** | Centre for Applied Science and Technology |
| **CDP** | critical design parameter |
| **CEM** | composite extrusion modeling |
| **CES** | European Committee for Standardization |
| **CFD** | computational fluid dynamic |

| | |
|---|---|
| **CFR** | Code of Federal Regulations |
| **CI** | communality index |
| **CIP** | Permanent International Commission (French abbreviation) |
| **CJM** | customer journey mapping |
| **CL** | cutter location |
| **CLF** | ceramic laser fusion |
| **CLIP** | continuous liquid interface production |
| **CMC** | ceramic-matrix composite |
| **CMM** | coordinate measuring machine |
| **CNC** | computer numerical control |
| **CRM** | customer relationship management |
| **CSAM** | cold spray additive manufacturing |
| **CX** | customer experience |
| **D** | difficulty |
| **DACs** | data acquisition cards |
| **DC** | design component |
| **DFA** | design for assembly |
| **DFSS** | design for Six Sigma |
| **DFT** | design for testability |
| **DFX** | design for excellence |
| **DIN** | German Institute for Standardization |
| **DLP** | digital light processing |
| **DMLS** | direct metal laser sintering |
| **DNC** | direct numerical control |
| **DoC** | declaration of conformity |
| **DOD** | U.S. Department of Defense |
| **DOF** | degree of freedom |
| **DOP** | depth of penetration |
| **DP** | design parameters |
| **DSM** | design structure matrix |
| **DT** | digital twin |
| **EBAM** | electron beam additive manufacturing |
| **EBM** | electron beam melting |
| **ECAM** | electrochemical additive manufacturing |
| **ECM** | electrochemical machining |
| **EDM** | electrical discharge machining |
| **EDS** | energy-dispersive spectroscopy |
| **EIA** | electronic industries alliance |
| **EPM** | experimental physical model |
| **EPVAT** | electronic pressure velocity and action time |
| **ERP** | enterprise resource planning |
| **ERP** | engineering resources product software |
| **FCC** | face-centered cubic crystal structure |
| **FDM** | fused deposition modeling |

| | |
|---|---|
| **FDM** | fused deposition modeling |
| **FEA** | finite element analysis |
| **FMEA** | failure modes and effects analysis |
| **FPCS** | firearm performance case studies |
| **FPM** | fused pellet modeling |
| **FR** | functional requirements |
| **FSC** | Firearm Safety Certificate |
| **FSI** | fluid-structure interaction |
| **GD&T** | geometric dimensioning and tolerancing |
| **GOST** | Governmental Standard (Russian acronym) |
| **GPU** | graphics processing unit |
| **Gtols** | geometric tolerances |
| **GUI** | graphical user interface |
| **HBW** | Brinell hardness |
| **HMI** | human machine interface |
| **HOQ** | house of quality |
| **HOSDB** | Home Office Scientific Development Branch |
| **HPC** | high-performance computing |
| **HR** | high resolution X-radiography |
| **HRB** | hardness Rockwell B |
| **HRC** | hardness Rockwell C |
| **HSS** | high speed sintering |
| **HTG** | homogeneous transformation graphical |
| **IADS** | the international alloy designation system |
| **IIoT** | industrial internet of things |
| **IJP** | ink jet printing |
| **IoT** | internet of things |
| **IRL** | investment readiness level |
| **ISO** | International Organization for Standardization |
| **IUPAC** | International Union of Pure and Applied Chemistry |
| **IW** | internet wave |
| **JIS** | Japanese Industrial Standards |
| **JIT** | just-in-time manufacturing |
| **JTBD** | Jobs to Be Done |
| **LCM** | lithography-based ceramic manufacturing |
| **LENS** | laser engineered net shaping |
| **LMHAM** | localized microwave heating-based additive manufacturing |
| **LOM** | laminated object manufacturing |
| **LT** | lock time |
| **MAPS** | microheater array powder sintering |
| **MBD** | multibody dynamics |
| **MC** | machining center |
| **MCM** | Monte Carlo method |
| **MDDM** | micro droplet deposition manufacturing |

**MIL-STD** Military Standard
**MIT** Massachusetts Institute of Technology
**MM** multi-operation machine
**MMC** metal-matrix composite
**MRL** manufacturing readiness level
**NASA** National Aeronautics and Space Administration
**NATA** National Association of Testing Authorities
**NBR** nitrile-butadiene rubber
**NC** numerical control
**NCAGE** NATO Commercial and Government Entity
**NDT** non-destructive testing
**NFA** National Firearms Act
**NIJ** National Institute of Justice
**NOM** Official Mexican Standards
**NP** new priority
**NPD** new product development
**NRA** National Rifle Association
**NRBC** nuclear, radiological, biological, chemical weapons
**NURBS** non-uniform rational B-spline
**NVH** noise, vibration, and harshness
**NVLAP** National Voluntary Laboratory Accreditation Program
**OAC** open architecture control
**PAD** plasma arc additive manufacturing
**PAM** product approval methodologies
**PC** polycarbonate
**PCB** printed circuit board
**PDEs** partial differential equations
**PDM** product data management
**PDMs** product design methods
**PET** poly(ethylene-terephthalate)
**PETG** polyethylene terephthalate glycol
**PJ** photopolymer jetting
**PLA** polylactic acid
**PLC** programmable logic controller
**PLM** product lifecycle management
**PMC** polymer-matrix composite
**PME** powder melt extrusion
**PTP** point to point
**PV** process variable
**QFD** quality function deploy
**RDOP** residual depth of penetration
**RFP** rapid freeze prototyping
**RP** rapid prototype
**RPD** rapid plasma deposition

**RSS** root sum squared
**RT** rapid tooling
**SAAMI** Sporting Arms and Ammunition Manufacturers' Institute
**SAE** The Society of Automotive Engineers
**SAM** serviceable available market
**SBR** styrene-butadiene rubber
**SCC** stress-corrosion cracking
**SCM** supply chain management
**SDK** software development kit
**SEM** scanning electron microscopy
**SHS** selective heat sintering
**SIS** selective inhibition sintering
**SLA** stereolithography
**SLM** selective laser melting
**SLS** selective laser sintering
**SMG** sub-machine gun
**STANAG** Standardization Agreement
**STL** Standard Triangle Language or Standard Tessellation Language
**SUS** system usability scale
**T3DP** thermoplastic 3D printing
**TAM** total available market
**TB** test bench
**TBF** test bench features
**TDTs** Toyota design techniques
**TIPS** theory of inventive problem solving
**TIR** tactical importance rating
**TPA** thermoplastic polyether block amide
**TPC** thermoplastic copolyester elastomer
**TPE** thermoplastic elastomer
**TPO** thermoplastic polyolefin elastomer
**TPS** thermoplastic styrenic block copolymer elastomer
**TPS** Toyota production system
**TPU** thermoplastic polyurethane elastomer
**TPV** thermoplastic vulcanizate elastomer
**TR** technical guidelines (German abbreviation)
**TRIZ** TIPS in Russian
**TRL** technology readiness level
**TTT** time-temperature-transformation diagram
**UCD** user-centered design
**UHS** ultra high speed
**UI** user interface
**UNODC** United Nations Office on Drugs and Crime
**UPS** uninterruptible power supply
**UT** usability testing

| | |
|---|---|
| **UX** | user experience |
| **VMP** | valuable minimum product |
| **VOC** | voice of the customer |
| **VPAM** | anti-attack materials and constructions (German abbreviation) |
| **VR** | virtual reality |
| **WAAM** | wire arc additive manufacturing |
| **WWII** | World War II |

# Preface

The idea and motivation of writing this book emanated from a friendship meeting. The authors reflected that at present books and courses cover either mechanical design—including that of weapons, materials, or manufacturing processes, but not one that encompasses all three aspects as a whole. Considering that both authors are military engineers with master's and doctorate degrees, the content of this book on the design of small weapons reflects the experience they have acquired in weapon development and materials science in both industrial and academic research.

This book is aimed at designers, workers in research and development, engineering and design students, students at military colleges, sportsmen, hunters, and those interested in firearms. The intent is to give the reader knowledge about the development of small weapons using modern design tools, involving the use of design methodologies such as Product Lifecycle Management, Quality Function Deploy, Axiomatic Design, Design for Assembly, and Platforms Products; the maturity of the technology (TRL), manufacture (MRL), and business (IRL) are taken into account. CAD/CAE/CAM software, rapid prototyping, test benches, materials, heat and surface treatments, and manufacturing processes are addressed in this book. Several case studies are presented to provide detailed considerations on developing specific topics.

An introductory chapter provides a review of the history of weapons from ancient to modern times, describing how firearms have progressed until reaching contemporary firearms; the evolution of materials, manufacturing processes, and design tools used for developing firearms along the history is told.

This book continues with the beginning of the product design using several design methodologies, understanding custom, and functional requirements. Afterward, it gets into the collaborative design and solver capabilities through modern software (CAD and CAE simulation), in addition to the CAM assessment until reaching an experimental physical model and a test bench.

This book progresses in the description of the current and emerging materials used in the production of firearms and heat treatments and surface treatments for enhancing the properties of firearm components. A review of the main conventional and unconventional manufacturing processes used for the production of firearm components is made.

The subject of small weapon design is multidisciplinary, so the various topics in this book are presented in as accessible a manner as possible, with the individual chapters being able to be consulted as they meet the reader's needs. It is expected that this book will encourage interest and learning in small weapon development using modern design tools and emerging materials and manufacturing processes.

# Acknowledgments

**Dr. Jose Martin Herrera-Ramirez** thanks the Research Center for Advanced Materials (CIMAV) and National Council of Science and Technology (CONACYT). He greatly thanks Dr. Caleb Carreño Gallardo for the preparation and analysis of samples; without his support and contribution, the case study on heat treatments of steels would not have been possible. Special thanks to MSc Jose Ernesto Ledezma Sillas for designing part of the 3D model images, as well as specimen machining and preparation. He is also grateful to the following people for the technical support provided during the analysis of samples: MSc Karla Campos Venegas, MSc Claudia Patricia Peregrino Ibarra, BSc Flor Griselda Nevarez Vargas, BSc Alejandro Benavides Montoya, and MSc Ruben Castañeda Balderas.

**Dr. Luis Adrian Zuñiga-Aviles** thanks the Faculty of Medicine and Faculty of Engineering from Autonomous University of Mexico State, Military School for Engineers and Catedras Program from the National Council of Science and Technology (CONACYT). He greatly thanks Dr. Usiel Sandino Silva Rivera for the simulation of internal ballistic of polygonal and grooved barrels.

Both authors thank the CRC Press/Taylor & Francis Group team for trusting them and accepting the proposal of this book. A special acknowledgment goes out to Nicola Sharpe and Nishan Bhagat, who promptly followed up on all administrative tasks concerning this book.

# Authors

**Dr. Jose Martin Herrera-Ramirez** is a military engineer with experience in the field of weapon and ammunition development. After receiving his PhD in Materials Science and Engineering from the Paris School of Mines in France, he was the head of the Applied Research Center and Technology Development for the Mexican Military Industry (CIADTIM). He now researches the development of metallic alloys and composites at the Research Center for Advanced Materials (CIMAV) in Chihuahua, Mexico.

**Dr. Luis Adrian Zuñiga-Aviles** is a military engineer with wide experience in the field of weapon and ammunition development. He was head of the prototypes and simulation departments at the Applied Research Center and Technology Development for the Mexican Military Industry (CIADTIM) and head of engineering of the Production directorate. He received his PhD in Science and Technology on Mechatronics from the Center for Engineering and Industrial Development (CIDESI) in Queretaro, Mexico. He now researches the new product design and development for military application, machinery, robotics, and medical devices in the Faculty of Medicine at the Autonomous University of Mexico State (UAEMex) and the Faculty of Engineering at UAEMex as part of the Researchers for Mexico program CONACYT.

# 1 History of the Design of Small Weapons

## 1.1 INTRODUCTION

The history of land armament has several periods, which are well defined:

### 1.1.1 THE FIRST PERIOD

The first period comprises the use of the combatant's own strength as a driving force. Whether he was using the spear, the ax, the sword, the lance, the mace, the pike, or the saber [1], he has to draw on his physical energy to use these weapons. The same happened when used weapons that store this energy to restore it in a single blow, with greater power and brutality, as is the case with bows [2], crossbows [3], ballista, or catapults [4–6]. In front of the offensive armament of antiquity, there is its corresponding defensive armament, such as leather or wood shields up to the use of metallic armor.

### 1.1.2 THE SECOND PERIOD

The second period arises with the use of gunpowder [7], which provided man with a new source of energy a thousand times greater than that provided by muscle. It is interesting to affirm that the first use of chemical reaction energy to create movement had military purposes, taking several centuries for the same phenomenon to be applied to peaceful uses. In this sense, centuries later, a new source of energy much more powerful than chemical energy was applied for the first time to military purposes; we refer to nuclear energy.

If the appearance of gunpowder on the battlefields revealed wide horizons in the art of war, the development of firearms was not rapid due to the slow progress of the mechanical and chemical arts. For this reason, firearms originally offered almost the same danger for the user as for the enemy. Effective performance of firearms had to be achieved to make them safe, mobile, and accurate. The transformation took place over the centuries. Progress was so slow that even during World War I (1914–1918), the main combatants started the conflict with cavalry forces, equipped with sabers and helmets [8].

DOI: 10.1201/9781003196808-1

### 1.1.3 The Third Period

At the end of the second period, the pace of progress began to accelerate by virtue of the advantages produced as a consequence of the industrial revolution at the beginning of the 20th century. The third period started from this time, which marked the great development of science. The armament progressed in a few years what it could not achieve in centuries of the previous period. The rate of fire of the artillery was greatly increased; the use of black powder, which generated those smoke clouds characteristic in the battle paintings of the French Empire and Revolution, disappeared thanks to the invention of the smokeless powder [9]; the power of the projectiles was greatly increased because of the use of new substances. Finally, the organization, power, range, and mobility of artillery materials changed their qualities and appearance to such a degree that the difference with those used some 10 years earlier is radical.

In small-caliber firearms, the transformation was much more remarkable. Thus, the old one-shot rifle is replaced by the repeating rifle, while automatic weapons made their appearance. One year of a world war is enough to consider these weapons as essential for the infantry, rapidly developing an extensive series of models: automatic pistol, automatic rifle, machine gun, and submachine gun.

Simultaneously with the improvement of traditional weapons, others were born. The internal combustion engine left the laboratories and was immediately applied to motor vehicles and aircraft, which were soon used by the military to install weapons on them, thus achieving great mobility of firearms [10]. On the other hand, the advances of the steel industry and the ease of transporting thick armor plates on engine-powered chassis opened new paths to the struggle between the projectile and the armor.

The characteristic of this period is the appearance of a considerable number of auxiliary firearm accessories. Power, ammunition consumption, and range are high and, therefore, the guarantee of maximum performance is required. For this, it is necessary to accurately determine the targets to be destroyed, to carry out a good preparation of the shot, a target quick fix, and a maximum fire concentration; new techniques derived from optics, acoustics, electromagnetism, electronics, infrared, and ultrasound, among others, allowed to solve these issues satisfactorily. Thus, sighting, fire control, reference and detection, transmission, and remote control accessories arose. Without these accessories, the firearms would only be like a blind monster, ineffective in combat.

Finally, this third period is differentiated from the previous ones by the speed with which the firearms and accessories expire. Progress is so fast that a firearm that has just entered in service is obsolete due to the increase in power of the armament in front of it: the small-caliber anti-tank guns have been eliminated due to the increased protection of armored units, or simply replaced by a more powerful one.

### 1.1.4 The Fourth Period

We should distinguish an important fourth period in the history of weapons, which began on August 6, 1945, with the dropping of the first atomic bomb over Hiroshima [11]. Although this date marked an event of capital significance, this chapter will deal with the third period. The goal is to comprehend the fundamental laws of contemporary weapons to take them as a basis in modern weapon design, as well as in the choice of optimal materials and manufacturing processes.

## 1.2 EVOLUTION OF FIREARMS

The appearance of firearms is uncertain, but several references report their first uses in the early 13th century [7,12–14] (Figure 1.1). The models began to emerge imprecisely but with enough force to be classified into infantry armaments and artillery material.

The first firearms were heavy, unmanageable, and inaccurate. The gunpowder for the priming charge had to be placed at the last moment. Thus, the shooter was unable to move with his weapon when he had already loaded it. He quickly needed a wick to light the bait, from which the slowness and imprecision of the shot are deduced.

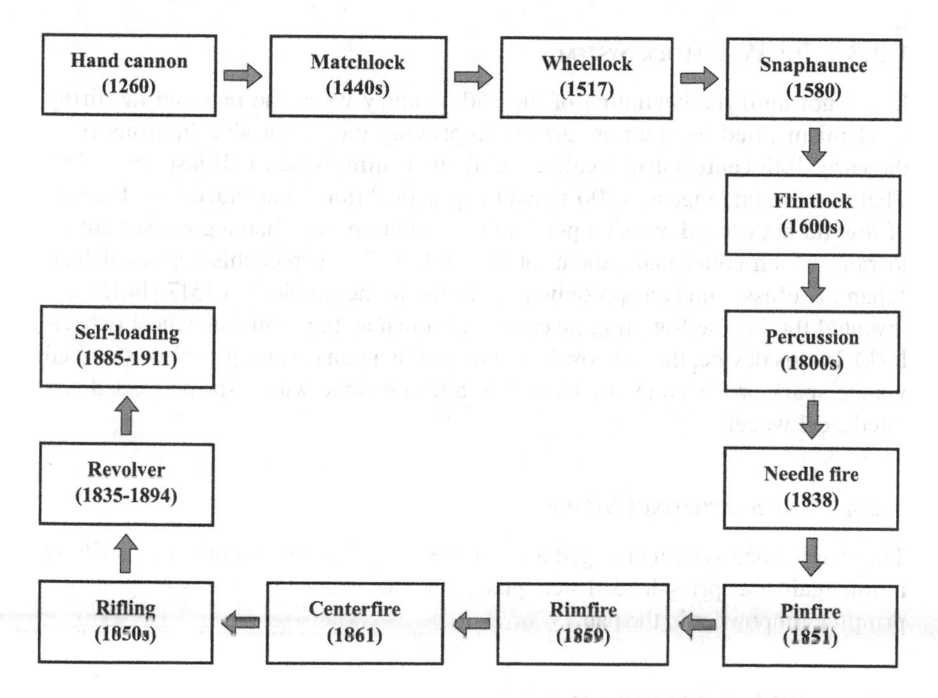


FIGURE 1.1 Evolution of firearms.

### 1.2.1 The Hand Cannon

The hand cannon [12] is considered the oldest type of small arms, being the most simplistic form of metal barrel firearms, mechanically speaking [13]. Its first recorded use was during the Battle of Ain Jalut in 1260. It was made of bronze or wrought iron, fixed to a frame or stock with metal or leather straps. It could be used by infantry or cavalry. This weapon did not have any type of firing mechanism; instead, it was manually ignited through a touch hole at the breech end of the barrel.

### 1.2.2 The Matchlock System

The matchlock system was introduced by the Janissary corps of the Ottoman army in 1440 [14]. It consisted of a slow-burning cord (match) placed in a clamp at the end of a small serpentine-shaped lever, a kind of early trigger. Similar to the hand cannon, the matchlock was loaded through the muzzle, with the difference that the priming gunpowder was ignited by the cord, which was introduced into the touch hole to ignite the main charge. Despite these advances, firearms were still trouble to operate: accurately aiming was difficult and users had to go through the complicated reloading process after every shot.

### 1.2.3 The Wheellock System

It was not until the beginning of the 16th century when the pan and the firing mechanism fitted to firearms, greatly improving them. Portable firearms from the early 16th century had a caliber of about 18 mm, which will last until 1850. Their maximum range was 200 m, but the practical range was much less. The rate of fire did not exceed one shot per 2 min. In addition, they had a great sensitivity to rain, which could make them totally useless. To correct this serious defect, Johann Kiefuss found an approximate solution to the problem in 1517 [14,15]. He invented the first self-igniting mechanism known as the arquebus wheel (wheellock). In this device, the gunpowder contained in a pan (priming pan) was ignited via the spark produced by the friction of a pyrite stone with a spring-loaded serrated steel wheel.

### 1.2.4 The Snaphaunce System

The snaphaunce system emerged around 1580 [14,16], which worked by striking a flint against a spring-loaded steel plate, generating a set of sparks to ignite the priming gunpowder in the pan.

### 1.2.5 The Flintlock System

The flintlock mechanism was introduced in the early 17th century [14,17]. It is said that Marin le Bourgeoys designed the first true flintlock mechanism.

The flintlock consisted of a piece of flint pressed between two jaws at the end of a cock beak-shaped hammer. The hammer was released by the trigger, causing the flint to hit the pan placed below a steel piece named frizzen. This blow generated a shower of sparks that ignited the priming gunpowder charge, which subsequently ignited the main charge through the touch hole. The flintlock made it possible to accelerate the shot to reach a rate of two shots per minute and to greatly reduce failures and sensitivity to rain, in addition to being cheaper.

### 1.2.6 The Percussion System

At the beginning of the 19th century, Edward Charles Howard discovered mercury fulminate [18], which allowed Alexander John Forsyth to patent the percussion ignition. The early percussion system worked almost identically to flintlock firearms, but using a priming made of mercury fulminate, which was ignited by the hammer impact. Such an impact generated a flash to ignite the main gunpowder charge. However, the system was enhanced by Joshua Shaw, who designed and patented the percussion cap in 1822. The cap was made of copper and filled with mercury fulminate. The whole was placed in a cone or nipple, where the hammer hit to initiate the mercury fulminate, whose detonation ignited the main charge.

### 1.2.7 The Dreyse Needle System

In 1808, Jean Samuel Pauly created the first self-contained ammunition made of a copper base filled with mercury fulminate as a priming charge, a brass or paper case, and a round bullet [19]. Then, he patented the first breechloading firearm in 1812. Pauly's inventions served as the basis for Nicolaus von Dreyse to invent the Dreyse needle gun, which was adopted by the Prussian army in 1838 [19]. The cartridge of this firearm was similar to that of Pauly, with the difference that it included a primer cap, which was percussed by a long, thin needle-like firing pin. The Dreyse gun was the first to use the bolt-action lock to open and close the chamber. The rate of fire of this weapon was seven shots per minute. Although imperfect, e.g., the needle corroded by the effect of the gunpowder gases and broke, the Dreyse gun was one of the most sensational events in the evolution of modern firearms.

### 1.2.8 The Pinfire System

The pinfire system was patented by Casimir Lefaucheux in 1835 [20] and exhibited in 1851 [21]. The invention included a cartridge containing all its components: a brass case, a percussion cap with the priming charge, the main charge, and the bullet embedded in the case. Further, the cartridge had a small pin aligned to the percussion cap, which protruded radially from the base of the case. When the weapon's hammer struck the pin, this hit the priming charge, producing its detonation and subsequent ignition of the main charge; the accumulation of gases caused the bullet to come out of the case, causing it to travel through the barrel. The protruding pin also served to extract the used case from the weapon. Nevertheless, it could cause accidents if the cartridges were handled incorrectly.

### 1.2.9 The Rimfire Ammunition

The rimfire ammunition was invented by Louis-Nicolas Flobert in 1845, but Smith & Wesson produced the first revolver to shoot rimfire cartridges in 1859 [14]. This ammunition was similar to that of the pinfire system, differing in that the priming charge was contained in the base of a thin-walled case. The firing pin struck the base's rim to ignite the priming charge, which expands to ignite the main charge.

### 1.2.10 The Centerfire Ammunition

The centerfire ammunition was introduced by George H. Daw in 1861 [22]. It consisted of a thicker metal case manufactured in a single piece, with a central cavity at its base where an independent percussion cap was inserted. The firing pin struck and crushed only the percussion cap. With some improvements, this is the prevailing firing system today for both infantry and artillery weapons.

### 1.2.11 The Rifling System

The term rifling refers to a helical grooving that is made inside the barrel of a firearm, in order to give the bullet a rotational movement (spin) around its longitudinal axis. This spin stabilizes the bullet as it travels through the air, increasing the bullet velocity and improving the shooting accuracy. The rifling is attributed to August Kotter in 1520, but it was not until the second half of the 19th century when it was put into practice.

### 1.2.12 The Revolver

A revolver is defined as a single-barrel firearm with a multi-chambered cylinder, which is successively lined up with the barrel and fired with the same hammer. Several primitive revolvers were developed as early as the mid-16th century. However, true revolvers appeared later, being the snaphaunce revolver one of the most recognized around 1700 [23], and the flintlock revolver developed by Elisha H. Collier in 1814. Samuel Colt patented the first revolver with the cylinder revolved and locked by cocking the hammer in 1835 [14], which was improved in other patents [24–26]. Other improvements followed, as those of Rollin White [27,28] and Smith & Wesson [29,30].

### 1.2.13 Self-Loading Firearms

A self-loading or semi-automatic firearm is a repeating firearm whose first round, coming from a magazine, is manually loaded into the chamber by the user.

The subsequent rounds are automatically loaded through part of the energy generated by the propellant charge of the cartridge; the user is required to actuate the trigger to discharge each shot. This system's operation principle had already been invented in the mid-17th century, but the lack of suitable cartridges, smokeless gunpowder, and materials prevented it from being used.

Some successful designs for self-loading firearms can be mentioned, such as those of Ferdinand Ritter von Mannlicher in 1885 [31], Hugo Borchardt in 1890–1893 [14], and Theodor Bergmann in 1893 [14]. Other outstanding designs were the Model 1900 Colt 0.38″ automatic, the Model 1902 Colt Military Automatic Pistol, which was used by the U.S. Navy [14], and the Model 1911 in 0.45″ ACP adopted by the U.S. military [14,32]. These designs continue to be used today, with perfections made from the point of view of new materials, manufacturing processes, and heat treatments.

## 1.3 CLASSIFICATION OF FIREARMS

The United Nations Office on Drugs and Crime (UNODC) classifies firearms based on different criteria [33], which are summarized in Figure 1.2 and explained below.

### 1.3.1 Classification by the Level of Harm

According to the level of harm they produce, firearms are classified into:

#### 1.3.1.1 Lethal Firearms

They are designed to kill the target.

#### 1.3.1.2 Non-lethal Firearms

They are designed to scare or incapacitate the target.

### 1.3.2 Classification by the Traditional Structure

Depending on the traditional structure, weapons are classified into the following categories:

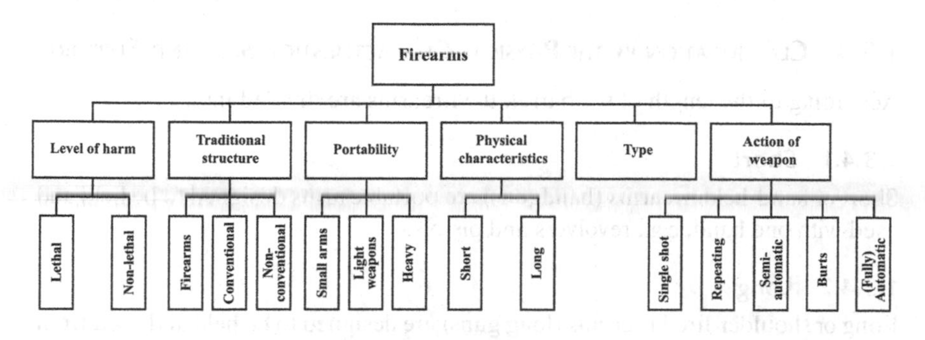


**FIGURE 1.2** General classification of firearms.

#### 1.3.2.1 Firearms

A firearm is a weapon from which a shot is discharged through the barrel by the action of gunpowder propulsion gases.

#### 1.3.2.2 Conventional Weapons

Conventional weapons are those designed for military uses. They include but are not limited to small arms and light weapons, battle tanks, armored combat vehicles and helicopters, combat aircraft, artillery systems, warships, cluster munitions, landmines, missiles, and missile launchers.

#### 1.3.2.3 Non-conventional Weapons

The non-conventional weapons are also known as ABC (atomic, biological, and chemical) or NRBC (nuclear, radiological, biological, chemical) weapons, which refer to weapons of mass destruction capable of killing and damaging numerous humans, or destructing natural or man-made structures, as well as the biosphere.

### 1.3.3 Classification by the Portability

According to the number of people to transport the weapons, they are classified into:

#### 1.3.3.1 Small Arms

They can be transported by one person. Examples of small weapons are pistols, revolvers, rifles, carbines, shotguns, and sub-machine guns.

#### 1.3.3.2 Light Weapons

They can be transported by more than one person. They include machine guns, anti-tank rifles, mortars, cannons, and howitzers.

#### 1.3.3.3 Heavy Weapons

They are systems that are cumbersome for foot transportation and must be transported by technical means as wheeled frames and/or vehicles, aircrafts, etc. They comprise mortars, artillery systems, and rocket launchers.

### 1.3.4 Classification by the Physical Characteristics, Size, and Support

According to the length of the barrel, the firearms are divided into:

#### 1.3.4.1 Short

Short or hand-held firearms (handguns) are portable arms designed to be held and used with one hand, e.g., revolvers and pistols.

#### 1.3.4.2 Long

Long or shoulder-fired firearms (long guns) are designed to be held and used from the hip or shoulder by both hands, e.g., rifles, shotguns, and carbines.

### 1.3.5 Classification by the Weapon Action

The action of a weapon is the firing mechanism through which cartridges are loaded, locked, fired, extracted, and ejected; this mechanism also controls the firing frequency when the trigger is actioned. Actions can be classified into five types:

#### 1.3.5.1 Single-Shot Action

In this system, a single cartridge is fired; the user performs four functions manually to fire each shot: bring the bolt back, load the cartridge into the barrel, close the bolt, and act on the trigger.

#### 1.3.5.2 Repeating Action

In this system, a single cartridge is fired; for each shot, the user manually loads a cartridge from a magazine into the barrel and acts on the trigger.

#### 1.3.5.3 Semi-automatic Action

In this system, the user manually loads the first cartridge from a magazine into the barrel and acts on the trigger. Then, the firearm uses part of the energy generated by the gases to eject the fired case and load a new cartridge from the magazine into the barrel, ready to complete another firing cycle. This process is repeated as long as there are cartridges in the magazine.

#### 1.3.5.4 Burst Action

In this system, the user is able to perform a predetermined number of shots for every depression on the trigger; the cycle automatically repeats each time the trigger is actuated until the magazine is empty. The burst action may be considered as an intermediate between semi-automatic and (fully) automatic actions.

#### 1.3.5.5 (Fully) Automatic Action

In this system, the firing cycle is automatically carried out until the trigger is released or the magazine is empty.

### 1.3.6 Classification by the Type of Firearm

The firearms are classified according to their technical-operational characteristics as follows:

#### 1.3.6.1 Revolver

A revolver is a single-barrel firearm with a multi-chambered cylinder, which is manually loaded with cartridges. The number of chambers can vary from five to nine. During its operation, such a cylinder is successively lined up with the barrel and fired with the same hammer. Fired cases are manually unloaded. A revolver is considered a repeating firearm and can be of single or double action.

#### 1.3.6.2 Pistol

A pistol is a hand-held firearm with a single chamber integrated into its barrel. The operation of a pistol can be single shot, repeating, semi-automatic, and automatic, being the semi-automatic one the most common, in which cartridges are loaded into the barrel chamber from a magazine. Once the first cartridge is shot, the fired case is automatically ejected and the next cartridge is loaded from the magazine, thanks to the energy generated by the propulsion gases. This cycle repeats each time the trigger is actuated until the magazine is empty.

#### 1.3.6.3 Shotgun

A shotgun is a shoulder-fired long firearm having one or two smooth barrels. Instead of a bullet, a shotgun is usually designed to shoot numerous small pellets called "shots," although it is possible to find those that shoot a single projectile called a "slug." Almost all shotguns are breech-loading and their action can be single shot, repeating, or semi-automatic.

#### 1.3.6.4 Rifle or Carbine

A rifle or carbine is a shoulder-fired long firearm whose barrel is rifled to impart a spin to the bullet. A carbine has a shorter barrel than a rifle, whose use was originally conceived for mounted troops. The action of these firearms may be single shot, repeating, semi-automatic, or fully automatic.

#### 1.3.6.5 Assault Rifle

An assault rifle can be considered a subcategory of a rifle, which uses an intermediate cartridge. The action of an assault rifle can be semi-automatic and/or fully automatic.

#### 1.3.6.6 Sub-machine Gun

A sub-machine gun (SMG) is a small firearm chambered for relatively low-energy pistol cartridges, which can be from the hand, hip, or shoulder. The action of an SMG can be semi-automatic or fully automatic. In the latter case, the weapon is said to be an automatic pistol or automatic SMG.

#### 1.3.6.7 Machine Gun

A machine gun is a long-barrelled firearm designed to fire rifle caliber cartridges, which continues to fire automatically until the pressure is released from the trigger. Machine guns can be fed by belts and magazines. It is possible to find relatively light machine guns, which can be operated by a single person, and heavy machine guns, which require to be operated by a crew. The action of a machine gun can be semi-automatic or fully automatic.

#### 1.3.6.8 Other Type of Firearms

UNODC considers a generic classification of firearms, the characteristics of which may overlap with those mentioned above. However, the production and possession of these firearms can represent a legal challenge. Such a classification is:

- Craft production and rudimentary arms
- 3D printed firearms
- Unlicensed copies
- Replica and imitation firearms
- Deactivated and converted firearms
- Modular firearms
- Concealable firearms
- Firearm kits
- Lethal autonomous weapons

More information about their characteristics can be found in Ref. [33].

## 1.4 EVOLUTION OF WEAPON MATERIALS

Materials have been of great importance to human's life in such a way that their development and predominant use have defined different periods, such as the Stone Age, the Bronze Age, and the Iron Age. The materials employed for manufacturing weapons have progressed over the years. As manufacturing processes have developed, different materials have been used in the manufacture of weapons. The materials range from stone, wood, and bone, through different metals to modern materials such as polymers, composites, and ceramics (Figure 1.3); some of their applications in weapons are summarized in Table 1.1. As stated in Section 1.2.13, firearm systems developed in the early 20th century prevail currently, being the new improvements focused on the development of lightweight materials.

### 1.4.1 Stone, Wood, and Bone

These materials were used to make different weapons such as tips, arrows, bows, spears, and axes, among others since prehistoric times [1,2,34].

### 1.4.2 Metals

#### 1.4.2.1 Bronze

Native copper was the first metal used by humans [35]. However, this metal in its pure form is fairly soft. Arsenical bronze, an apparently accidental alloy made of

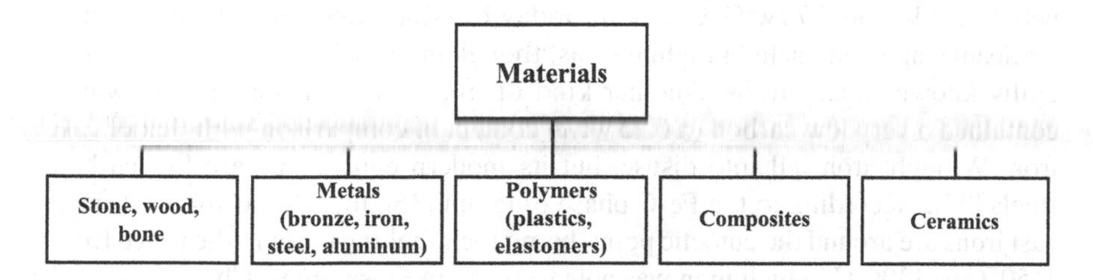


**FIGURE 1.3** Evolution of materials in the manufacture of weapons.

**TABLE 1.1**
**Materials and Applications for Weapons**

| Material | Applications |
|---|---|
| Stone, wood, and bone | Tips, arrows, bows, spears, axes |
| Bronze and iron | Spears, axes, arrows, bows, swords, knives, maces, cannons |
| Steels | Swords, barrels, sights, magazines, structural components |
| Aluminum | Handguards, frames, sights |
| Polymers | Frames, handguards, grips, stocks, forearms, Picatinny rails, trigger guards, head guards, magazines, components of the trigger mechanism, sights, recoil pads |
| Composites[a] | Barrels |
| Ceramics | Barrels (in study) |

[a] Most polymers are reinforced, becoming composites with the same applications.

copper (Cu) and arsenic (As), was used to obtain useful forms. Considering the Cu-As phase diagram and depending on the As content, the melting point of arsenical bronzes was above 600°C [36]. This means that in ancient times man was able to reach such temperatures to manufacture different weapons such as spears, axes, arrows, bows, swords, knives, and maces. Nowadays, bronze is an alloy of copper and other elements, including tin, aluminum, silicon, and nickel [37]. Considering a bronze made of copper (Cu) and tin (Sn), and according to the Cu-Sn phase diagram [36], the melting point of bronzes may vary from 830°C to 1020°C.

One of the first firearms built of bronze was the hand cannon. Bronze was subsequently used to make barrels, frames and slides for pistols and rifles, as well as cannons. Examples of bronze-made firearms are the Wuwei Bronze Cannon [38] and the Winchester Model 1866, the "yellow boy" [39].

#### 1.4.2.2 Iron

One of the disadvantages of ancient weapons made of bronze was that their edge did not hold for long. When man acquired knowledge to obtain and process iron, it gradually replaced bronze.

Ferrous alloys are the combination of iron (Fe) as the prime constituent, and they are classified in steels and cast irons, depending on the carbon (C) content [37]. The former will be explained in the next section. Cast irons may contain between 2.14 and 6.70 wt% C and are today classified into gray, ductile, white, malleable, and compacted graphite irons, though in ancient times they were generally known as cast irons. Another kind of iron was the wrought iron, which contained a very low carbon (< 0.25 wt%) content in comparison with that of cast iron. Wrought iron fell into disuse, but its modern equivalents are low-carbon steels [37]. According to the Fe-C phase diagram [36], the compositions of most cast irons are around the eutectic point, being their melting point in the range from 1150°C to 1300°C, which man was able to reach in those times. China produced cast iron as early as 200 BC, but it was until the 14th century when it was used for making cannons in Europe [13,35]. It is worth mentioning that some cannons were made from different pieces of wrought iron, joined by iron rings and longitudinal staves [40]. Despite their low cost, iron cannons were not initially very successful, since they were more prone to corrode by the effect of the gunpowder combustion gases and saline environments. Another of their problems were that, when failed, they exploded into fragments due to their brittleness. Therefore, in some places it was preferred to continue using bronze cannons.

#### 1.4.2.3 Steel

Steels are the other types of ferrous alloys, which are iron-carbon alloys (0.008–2.14 wt% C) containing appreciable concentrations of alloying elements, usually added to improve mechanical and corrosion-resistance properties [37]. The melting point of most steels is about 300°C higher than for cast irons [36].

The earliest proof of steel production dates back to 300 BC, when man discovered that iron became harder, stronger, and more durable if it was combined with charcoal [35]. There is evidence that the first high-carbon steel productions were carried out in India, with the steel known as wootz, which had 1.0–2.0 wt% C. It is believed that Damascus blades were forged directly from this steel during the 17th century [35,41]. Today, it is known that these blades contain carbon nanotubes—a carbon-based form—and cementite nanowires, which impart them enormous mechanical properties and an exceptionally sharp cutting edge [41]. The modern Age of Steel begins with Sir Henry Bessemer in 1856, who invented a pneumatic process for the production of steel, with the aim of improving gun construction and reducing costs [42].

Swords were the first weapons to be fabricated of steel, when blacksmiths realized that mixing iron with charcoal hardened it. The development of firearms was due to the advancement in steel metallurgy. At present, different steels are the primary materials in any firearm.

#### 1.4.2.4 Aluminum

Aluminum is the most abundant metal in the earth's crust. Among other properties, aluminum and its alloys are characterized by having a relatively low density (2.7 g/cm$^3$ compared to 7.9 g/cm$^3$ for ferrous alloys) and good resistance to atmospheric corrosion [37]. Thanks to these properties, aluminum alloys are important materials in the manufacture of firearm components. Examples of such components are handguards and frames or receivers. One of the disadvantages of aluminum alloys is their low mechanical resistance (yield strength, impact) compared to that of steels. The frames of the Colt Lightweight Commander pistol [43], the Colt M16 rifles [44], and the Beretta 92 series pistol [45] are made of aluminum.

### 1.4.3 Polymers

Polymers are classified into classes depending on their thermomechanical properties [46]. One of these classes is thermoplastics, commonly known as plastics. Plastics can be molded into numerous shapes through various manufacturing processes, such as injection molding and extrusion. Plastics have very low densities compared to steel and aluminum, which saves weight. For instance, the density of the polyamide 66 (Nylon 66), which is used in the manufacture of firearm components, is around 1.17 $g/cm^3$ (versus 2.7 $g/cm^3$ for aluminum alloys and 7.9 $g/cm^3$ for steels) [47]. It is worth mentioning that plastics are often blended with fillers, additives, and modifiers to improve their properties. Another class of polymers is elastomers, which are rubbery materials that can be stretched many times their original dimension and that recover their initial dimensions when the applied stress is released [46]. The fabrication of polymer components for firearms is usually cheaper and faster than that of metallic components. Further, they offer excellent corrosion and recoil impact resistances, they are thermal insulators, and their color can be customized, among other advantages. Different firearm components are produced from polymers: frames, handguards, grips, stocks, forearms, Picatinny rails, trigger guards, head guards, magazines, some components of the trigger mechanism, and recoil pads, among others.

### 1.4.4 Composites

A composite is a material produced from two or more materials to obtain the combination of properties, which are different from those of the original materials. Composites may be selected to achieve unusual combinations of stiffness, strength, weight, high-temperature performance, corrosion resistance, hardness, or conductivity [48]. According to the nature of the matrix, the composites are classified into metal-matrix, polymer-matrix, and ceramic-matrix composites. The reinforcing phase may be particulates, fibers, or laminates, which are normally stiffer, stronger, and/or harder than the matrix.

In the case of metal-matrix composites, there is research focused on gun barrels, where the metallic matrix is reinforced with both a metal [49] and a ceramic [50]. About polymer-matrix composites, the plastics used for manufacturing firearm components are habitually reinforced with particles or fibers [51]. In addition, there are investigations where hybrid materials (ceramic-matrix reinforced with fibers + metal matrix-material reinforced with fibers) are used [52].

### 1.4.5 Ceramics

A ceramic is an inorganic compound made up of metallic and non-metallic elements, whose crystal structure is generally more complex than that of metals [48]. Ceramics possess high hardness and stiffness, superior wear and heat resistance, high-temperature capability, and relatively low density, compared to metals. For example, the density of alumina ($Al_2O_3$) is around 3.8 $g/cm^3$ [53]. The main drawback of ceramic materials is their disposition to catastrophic brittle fracture; thus, their processing to finished product is normally slow, laborious, and costly. Due to these disadvantages, they are somewhat limited in applicability [37]. Notwithstanding, attempts are being made with ceramics for gun barrel applications [54–56], with the aim of providing a significant increase in the barrel life and a reduction in weight for small caliber systems. The main limitation has been the difficulty in introducing the rifling pattern inside the barrels. Attempts have also been made to produce hybrid ceramic/steel barrels [57].

Chapter 7 will be dedicated to a review of the materials used in the modern manufacture of firearms components.

## 1.5 EVOLUTION OF FIREARM MANUFACTURING PROCESSES

The development of firearms has been parallel to industrial technology progress (skills, machines, techniques). The history of manufacturing can be divided into two periods. First, the discovery and invention by humans of materials and processes to make things. Second, the development of production systems [58]. In the past, the manufacture of implements and weapons was done as handicrafts and trades, whereas at present most of the manufacturing is accomplished by automated and computer-controlled machinery.

Processes as forging, casting, machining, and stamping have been used since ancient times, with the difference that before they were done manually, while today they are done with the help of machinery. For example, at the beginning, forging ("hammering") was intended for hammering firearm small components by blacksmiths; today, modern forge machines can make larger parts such as pistol and rifle frames; casting techniques such as investment casting are used since olden times, but, currently, different improvements have been made to increase the dimensional accuracy of components; machining was previously done by hand, with the inherent low dimensional accuracy and no interchangeability of components, whereas today manual machine tools and/or computer numerical control machines are used; the long firearm stocks were made of wood, while now they are made of plastics.

Nowadays, the production of firearm components requires tight dimensional tolerances; hence, the selection of suitable technology and manufacturing process is necessary. There are multiple methods for producing such components. For metals, casting (sand, die injection molding, investment, lost foam), powder metallurgy, machining, welding, forging, stamping, rolling, and additive manufacturing processes, among others, are used; they may be complemented by heat and/or surface treatments. For polymers, injection molding, extrusion, and additive manufacturing processes are used.

Chapter 9 will be devoted to the revision of modern manufacturing processes for the production of firearm components.

## 1.6 EVOLUTION OF DESIGN TOOLS

From World War II (WWII), the evolution of design tools can be divided into the following three periods.

### 1.6.1 The First Period

The first period comprises the use of sequential design. The production of systems is realized by project; several mechanisms were created for the operation of weapons; however, the action systems for weapon operations were slowly changed. The action systems as bolt-action rifle, blowback, and gas-operated reloading were the traditional reloading systems.

The production in mass was gradually growing with the production of the automobile. A technique was Fordism, a manufacturing technology that served as the basis of modern economic and social systems in industrialized, standardized mass production and mass consumption. The concept is named for Henry Ford. It was used in social, economic, and management theory about production, working conditions, consumption, and related phenomena, especially regarding the 20th century [59].

### 1.6.2 The Second Period

The second period arises after WWII with the new design techniques. Overall, the design techniques that highlighted are the Toyota design techniques (TDTs) as Toyota production system (TPS), Kaizen (change for better—continuously improve), Ishikawa diagram (also called fishbone diagrams, herringbone diagrams, cause-and-effect diagrams, or Fishikawa), just in time (JIT), poka-joke (mistake-proofing), and waterfall Toyota.

Other design techniques also arose, such as theory of inventive problem solving (TIPS in English or TRIZ in Russian) in 1946, failure modes and effects analysis (FMEA) in 1949, design for test or design for testing or design for testability (DFT) in 1950, Supplier development in 1951, bill of materials (BOM) in the early 1960s, design structure matrix (DSM) in 1960, and quality function deploy (QFD) developed in Japan in the late 1960s.

### 1.6.3 The Third Period

The third period arises with the computer-aided design (CAD) in 1959. Unisurf software was a pioneering surface CAD and computer-aided manufacturing (CAM) system designed to assist with car body design and tooling; it was developed in 1968 [60]. Later, the Sketchpad system was developed in 1963 [61]. As computers became more affordable, the application of CAD gradually expanded into weapon development. Table 1.2 summarizes the evolution of CAD software.

The design techniques were consolidated and new techniques arose, such as technology readiness level (TRL) during the 1970s, user-centered design (UCD) in 1977, Design Thinking in 1978, product lifecycle management (PLM) in 1985, Six Sigma in 1986, Lean Manufacturing in 1988, Platforms (product families, modular architecture, commonality index) in 1990, enterprise resource planning (ERP) systems in 1990, Axiomatic Design in 1998, product data management (PDM) in 2001, Design for Inspection in 2014, and several design techniques developed in 1996, including the Design for X or Design for Excellence [62], as seen in Table 1.3.

Thanks to the aforementioned design techniques, skills for design, and CAD software, the new weapons increased the solution alternatives for enhancing

**TABLE 1.2**
**Evolution of CAD Software**

| | | |
|---|---|---|
| 1963: Sketchpad | 1986: Sla format | 2007: NX |
| 1968: Unisurf | 1987: Pro-engineer | 2011: LibreCAD |
| 1970s: Mouse device | 1989: FDM format | 2012: Autodesk 360 |
| 1971: Adam | 1994: Step format | 2013: 3D CAD apps |
| 1977: Catia | 1995: SolidWorks | 2014: Selective laser sintering |
| 1978: Unigraphics | 1995: Solidedge | 2015: Onshape |
| 1980: IGES format | 1996: Catia conferencing groupware | 2017: Virtual reality in CAD |
| 1981: Geomod | 1999: Autodesk inventor | 2019 Extended Reality in SolidWorks |
| 1982: Autodesk autoCAD (dxf & dwg file formats) | 2002: FreeCAD | |

**TABLE 1.3**
**Design for X**

| | | |
|---|---|---|
| Design for Six Sigma (DFSS) | Design to cost | Design for ergonomics |
| Design for reliability | Design for logistics | Design for aesthetics |
| Design for minimum risk | Design for user-friendliness | Design for serviceability |
| Design for environment | Design for repair-reuse-recyclability | Design for maintainability |

**TABLE 1.4**
**Weapon Action Systems**

| Category | Subcategory |
|---|---|
| **Blowback** | |
| Simple blowback | |
| Advanced primer ignition (API) blowback | |
| Delayed blowback | Roller-delayed, lever-delayed, gas-delayed, chamber-ring delayed, hesitation locked, flywheel delayed blowback, toggle-delayed, off-axis bolt travel, radial-delayed, screw-delayed |
| Other blowback systems | Floating chamber, primer actuated, case setback |
| Limited-utility designs | Blish lock, savage rotating barrel, headspace actuated unlocking, magnet delay |
| **Gas-operated reloading** | Long-stroke piston, short-stroke piston, direct impingement |

the target, firepower, versatility, accessories, ammunition, sight, sustainability, safety, and action systems, as well as the building of low cost and weight systems. Table 1.4 presents various action systems developed.

Chapter 2 will be devoted to revising modern design techniques for the generation of new firearm components.

## REFERENCES


1. DeVries, K. and R.D. Smith, *Medieval Weapons: An Illustrated History of Their Impact*. 2007, ABC-CLIO: Santa Barbara, CA.
2. Blitz, J.H., Adoption of the bow in prehistoric North America. *North American Archaeologist*, 1988. **9**(2): pp. 123–145.
3. Arnade, P. and L. Crombie, *Archery and Crossbow Guilds in Medieval Flanders, 1300–1500*. 2018, Oxford University Press: Oxford.
4. Gurstelle, W., *Art of the Catapult: Build Greek Ballistae, Roman Onagers, English Trebuchets, and More Ancient Artillery*. 2004, Chicago Review Press: Chicago, IL.
5. Campbell, D.B. and B. Delf, *Greek and Roman Artillery 399 BC-AD 363*. 2003, Osprey: Oxford.
6. Cuomo, S., The sinews of war: Ancient catapults. *Science*, 2004. **303**(5659): pp. 771–772.
7. Herbst, J., *The History of Weapons*. 2005, Lerner Publications: Minneapolis, MN.
8. Jarymowycz, R.J., *Cavalry from Hoof to Track*. 2008, Greenwood Publishing Group: Westport, CT.
9. Davis, T.L., *The Chemistry of Powder and Explosives*. 2016, Pickle Partners Publishing: Auckland.
10. Fletcher, D., D. Crow, and N.W. Duncan, *Armoured Fighting Vehicles of the World*. 1998, World War 2 Books and Video, Arco Publishing Co: Milton, MA.
11. Hersey, J., *Hiroshima*. 2015, Penguin UK: London.
12. Broughton, G. and D. Burris, War and medicine: A brief history of the military's contribution to wound care through World War I, in *Advances in Wound Care*, C.K. Sen, Editor. 2010, Mary Ann Liebert, Inc.: New Rochelle, NY, pp. 1–5.
13. Harding, D., *Weapons: An International Encyclopedia from 5000 BC to 2000 AD*. 1982, St. Martin's Press: New York.
14. Arnold, L., A firearms history timeline. 2011 [February 25, 2021]; Available from: http://www.talonsite.com/tlineframe.htm.
15. Blair, C., Further notes on the origins of the wheellock, in *Arms and Armour Annual*, R. Held, Editor. 1973, Digest Books, Inc: New York, pp. 28–47.
16. Godwin, B., English firearms from the late 1500s to the English Civil Wars. *Arms & Armour*, 2013. **10**(1): pp. 51–70.
17. Lenk, T., *The Flintlock: Its Origin, Development, and Use*. 2007: Skyhorse Publishing Inc: New York.
18. Howard, E., On a new fulminating mercury. *Philosophical Transactions of the Royal Society of London*, 1800. **90**(1): pp. 204–238.
19. Pauly, R.A. and R. Pauly, *Firearms: The Life Story of a Technology*. 2004: Greenwood Publishing Group: Westport, CT.
20. Lefaucheux, C., *Fusil se chargeant par la culasse, au moyen d'un mécanisme qui fait basculer le canon*. French patent No. 6348, 1835, France, p. 5.
21. Lautissier, G. and M. Renonciat, *Casimir Lefaucheux, arquebusier: 1802–1852*. 1999: Ed. du Portail, La Tour-du-Pin, France.
22. Warlow, T., *Firearms, the Law, and Forensic Ballistics*. 2nd ed. 2011, CRC Press: New York.
23. Snaphaunce revolver - Annely pattern. [March 6, 2021]; Available from: https://collections.royalarmouries.org/object/rac-object-15214.html.
24. Colt, S., Improvement in repeating fire-arms and the apparatus used there with, Patent No. 1304, U.S.P. Office. 1839, United States of America, p. 8.
25. Colt, S., Improvement in repeating fire-arms, Patent No. 7613, U.S.P. Office. 1850, United States of America, p. 4.

26. Colt, S., Improvement in revolving chambered fire-arms, Patent No. 7629, U.S.P. Office. 1850, United States of America, p. 3.
27. White, R., Improvement in repeating fire-arms, Patent No. 12648, U.S.P. Office. 1855, United States of America, p. 3.
28. White, R., Improvement in revolving fire-arms, Patent No. 93653, U.S.P. Office. 1869, United States of America, p. 3.
29. Bates, J. and M. Cumpston, *Percussion Pistols and Revolvers: History, Performance and Practical Use*. 2005, iUniverse, Inc: New York.
30. Wesson, D.B., Revolver-lock mechanism, Patent No. 520468, U.S.P. Office. 1894, United States of America, p. 5.
31. Smith, W.H.B. and V. Kromar, *Mannlicher Rifles and Pistols: Famous Sporting and Military Weapons*. 2013, Literary Licensing, LLC: Whitefish, MT.
32. Browning, J.M., Firearm, Patent No. 984,519, U.S.P.a.T. Office. 1911, United States of America, p. 15.
33. United Nations Office on Drugs and Crime (UNODC). Module 2: Basics on firearms and ammunition, [March 8, 2021]; Available from: https://www.unodc.org/e4j/en/firearms/module-2/key-issues/typology-and-classification-of-firearms.html.
34. Lombard, M. and L. Phillipson, Indications of bow and stone-tipped arrow use 64 000 years ago in KwaZulu-Natal, South Africa. *Antiquity*, 2010. **84**(325): pp. 635–648.
35. Srinivasan, S. and S. Ranganathan, *India's Legendary Wootz Steel: An Advanced Material of the Ancient World*. 2004, National Institute of Advanced Studies: Bengaluru.
36. Okamoto, H., M.E. Schlesinger, and E.M. Mueller, *ASM Handbook Volume 3: Alloy Phase Diagrams*. 2016, ASM International: Materials Park, OH.
37. Callister, W.D. and D.G. Rethwisch, *Materials Science and Engineering: An Introduction*. 10th ed., 2018, Wiley: New York.
38. Rogers, C.J., K. DeVries, and J. France, *Journal of Medieval Military History*. Vol. 11, 2013, Boydell & Brewer Ltd: Woodbridge.
39. Henshaw, T., *The History of Winchester Firearms 1866–1992*. 6th ed., 1993, Academic Learning Company LLC: El Monte, CA.
40. Simmons, J.J., Early modern wrought-iron artillery macroanalyses of instruments of enforcement. *Materials Characterization*, 1992. **29**(2): pp. 129–138.
41. Reibold, M., et al., Carbon nanotubes in an ancient Damascus sabre. *Nature*, 2006. **444**(7117): pp. 286–286.
42. Birat, J.-P., The relevance of Sir Henry Bessemer's ideas to the steel industry in the twenty-first century. *Ironmaking & Steelmaking*, 2004. **31**(3): pp. 183–189.
43. Lightweight Commander (45ACP). [March 25, 2021]; Available from: https://www.colt.com/detail-page/lw-commander-45acp.
44. Colt M16 Rifle. [March 25, 2021]; Available from: https://www.colt.com/search/M16.
45. 92FS Inox. [March 25, 2021]; Available from: https://www.beretta.com/en/92-fs-inox/.
46. Hamley, I.W., *Introduction to Soft Matter: Synthetic and Biological Self-Assembling Materials*. 2013, John Wiley & Sons: Hoboken, NJ.
47. Material property data. [March 25, 2021]; Available from: http://www.matweb.com/search/QuickText.aspx?SearchText=nylon%2066.
48. Askeland, D.R. and W.J. Wright, *The Science and Engineering of Materials*. 7th ed., 2015, Cengage Learning: Boston, MA.
49. Glisovic, A., S. Gravely, and J. Gravely, Enhanced metal-metal-matrix composite weapon barrels and ways of making the same. Patent No. US 2017 / 0261280 A1, 2017, Google Patents.

50. Pyka, D., et al. Concept of a gun barrel based on the layer composite reinforced with continuous filament. In *AIP Conference Proceedings*. 2019, AIP Publishing LLC.
51. Underwood, J.M. and L.C. Underwood, Polymer/composite firearms and a process for strengthening polymer/composite firearms. Patent No. US 9.297,599 B2, 2016, Google Patents: USA.
52. Katz, R., et al., Hybrid ceramic matrix/metal matrix composite gun barrels. *Materials and Manufacturing Processes*, 2006. **21**(6): pp. 579–583.
53. Material property data. [March 26, 2021]; Available from: http://www.matweb.com/search/QuickText.aspx?SearchText=al2o3.
54. Swab, J.J., et al., *Mechanical and Thermal Properties of Advanced Ceramics for Gun Barrel Applications*. 2005, Army Research Laboratory: Adelphi, MD.
55. Bose, A., R.J. Dowding, and J.J. Swab, Processing of ceramic rifled gun barrel. *Materials and Manufacturing Processes*, 2006. **21**(6): pp. 591–596.
56. Emerson, R., et al., Approaches for the design of ceramic gun barrels. 2006, U.S. Army Research Laboratory, Weapons & Materials Research Directorate.
57. Grujicic, M., J. Delong, and W. Derosset, Reliability analysis of hybrid ceramic/steel gun barrels. *Fatigue & Fracture of Engineering Materials & Structures*, 2003. **26**(5): pp. 405–420.
58. Groover, M.P., *Fundamentals of Modern Manufacturing: Materials, Processes, and Systems*. 7th ed., 2020 John Wiley & Sons: Hoboken, NJ.
59. Thompson, G.F. Fordism, post-Fordism and the flexible system of production. Center for Digital Discourse and Culture 2003 [April 19, 2021]; Available from: https://www.cddc.vt.edu/digitalfordism/fordism_materials/thompson.htm.
60. Gannon, M. Reverberating across the divide: Bridging virtual and physical contexts in digital design and fabrication. 2014 [April 19, 2021]; Available from: http://papers.cumincad.org/cgi-bin/works/paper/acadia14_357.
61. Bhoosan, S., Collaborative design: Combining computer-aided geometry design and building information modelling. *Architectural Design*, 2017. **87**(3): pp. 82–89.
62. Eastman, C.M., *Design for X: Concurrent Engineering Imperatives*. 2012, Springer Science & Business Media: Berlin/Heidelberg, Germany.

# 2 Beginning the Product Design

## 2.1 INTRODUCTION

Product design is the process that designers and engineers use to solve real problems and satisfy market needs. This process has evolved enormously in recent years with the occurrence of modern tools. The current trends in the development of products focus on multidisciplinary teams. The development process begins with the design strategy and methodology, which allow outlining the aspects and methods that will be included in the product design. This chapter addresses the beginning of the product design using several design methodologies, such as product lifecycle management (PLM), quality function deploy (QFD), axiomatic design, design for assembly (DFA), and Platforms products. A case study based on the implementation of a toolkit of product design methodologies is presented here, which concerns the design of a product family comprising a semi-automatic pistol, an assault rifle, and a bullpup firearm. Different parameters involved in the operating principle of a firearm are taken into account for such a product family.

## 2.2 PRODUCT LIFECYCLE MANAGEMENT

The design is understandable in a simple way as the activity of sketching forms of ideas; however, creating products in 2022 is more complex than only forms. The fast fabrication is the base of the new meaning for the design word, so reaching the design target requires the strategy and organization of the work team. The recent trends in the development of products focus on multidisciplinary teams.

The PLM is realized with multidisciplinary teams working based on concurrent engineering using technologies platforms including servers and clients who have access as authors or visualizers. From a collaborative environment, the PLM software interacts with developers and suppliers, as well as with developers and customers. Both interactions support the product development, so customers participate in apportioning the customer's requirements, rejected claims, and letters of user's satisfaction; the suppliers participate during design and industrial development [1]. The PLM software manages the collaboration with engineering resources product (ERP) software, the product data management (PDM) software, and the industrial internet of things (IIoT) software [2].

The PDM software and ERP software collaborate using the Bill of Materials (BOM); this BOM includes the materials stock of developers and the materials

DOI: 10.1201/9781003196808-2

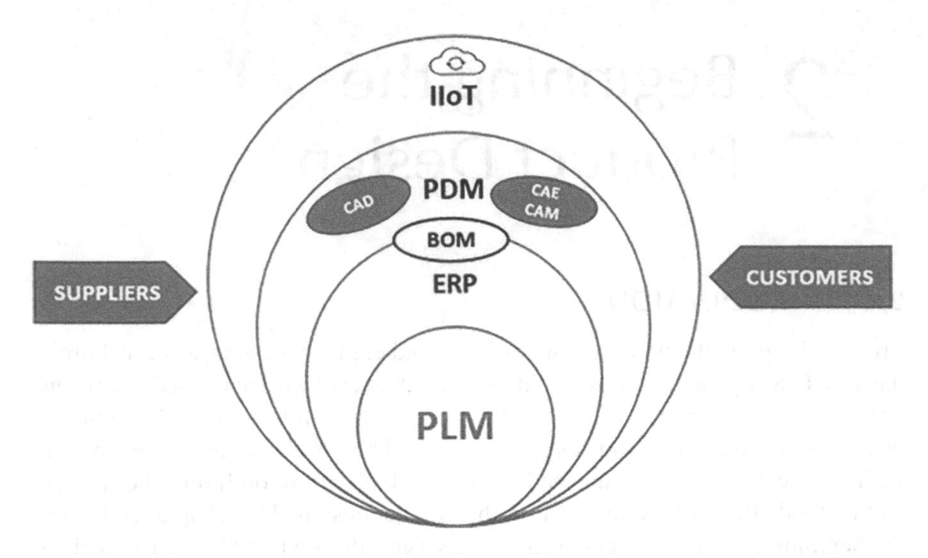


**FIGURE 2.1** Framework of PLM.

stock of suppliers; by means of PLM, the suppliers allow access to their material inventory for it to be used in the product development.

The IIoT software allows collaboration in the internet cloud, such as among the design workstations, machinery, quality control, production monitoring, other factories, developers of the test bench, and suppliers [3].

The PDM includes computer-aided design (CAD) software, computer-aided engineering (CAE) software, and computer-aided manufacturing (CAM) software. Currently, most of the CAD, CAE, and CAM include PDM connections. Figure 2.1 shows the interaction among the informatics technologies inclusive in the PLM.

The evolution of technology has motivated the update of definitions according to emergent approaches and techniques used to new product development (NPD); today, the stages for NPD are Research, Development, and Innovation [4].

The product lifecycle begins with the product's conception in the research stage, and it continues interacting with the development and later with the innovation. Hence, it reaches the product's first generation. Figure 2.2 presents the cycle where the product is induced to market, grows, reaches maturity at the top of the curve, and begins the generational replacement with the product of the next generation.

The design process requires the interaction in the cloud of technological tools that accelerate and control the development of a new product. Technological platforms manage the planning, lifecycle, development phases, prioritization of participation in the project, intellectual property, and product maturity until it is commercialized; then, they plan the next generation of the product. The second block describes tools that allow interaction among software, emphasizing the elements that make up modeling and prototyping. Prototyping is related to

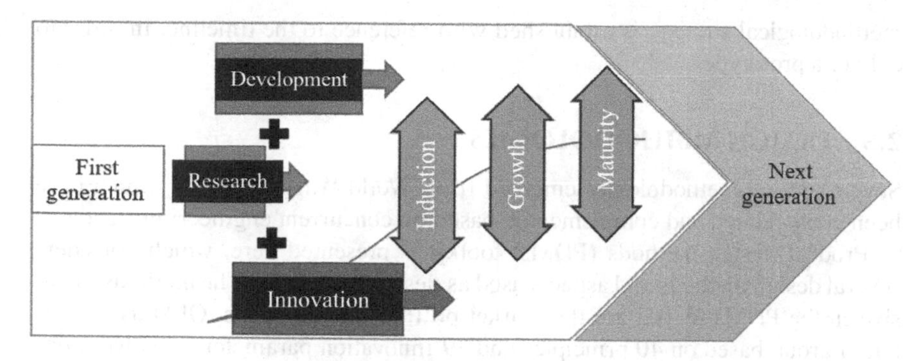


FIGURE 2.2 Cycle of the products development.

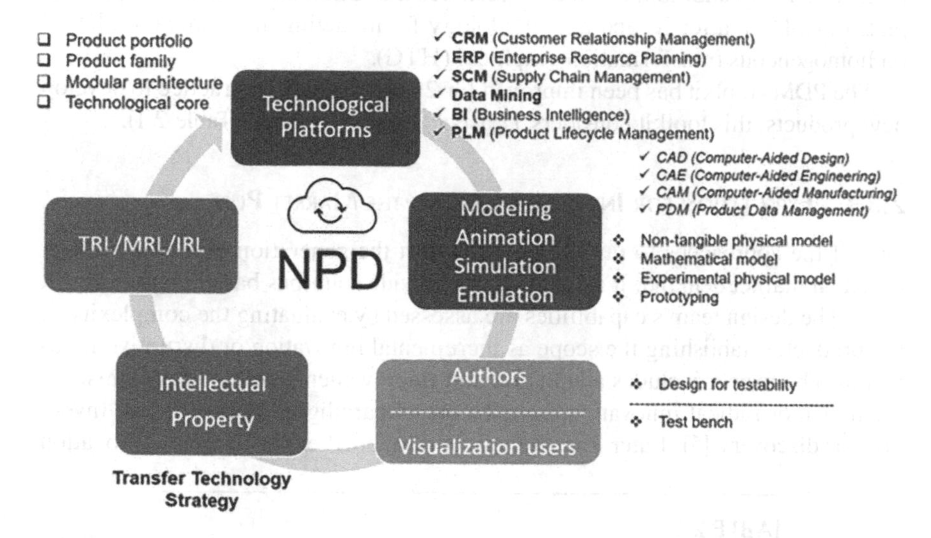


FIGURE 2.3 Interaction of key technological tools for new product development.

the emulation of systems on the test bench, i.e., in a laboratory environment with experiments conducted under controlled conditions. A hierarchy is used to facilitate the interaction of all the technological tools and give privileges to the authors, who generate information, and to the visualization users who cannot modify the models but can use the information to structure the tests. On the test bench, simulations are performed, and engineering drawings and process sheets are generated. Figure 2.3 shows the participation and interaction of key technological tools and highlights some important aspects, including the product portfolio, which today follows the development of products to satisfy different market segments.

In this book, the design process is limited to the phase of the product formulation, which starts from a creative process of idealization concerning a work environment, criteria (conditions, problem/opportunity, swot, budget, risk mitigation plan), and specifications (parameters and constraints), for which a methodological strategy is established with reference to the timeline, in order to deliver a prototype.

## 2.3 DESIGN METHODOLOGIES

Several design methodologies emerged from World War II, some of which have been consolidated and complemented based on concurrent engineering trends.

Product design methods (PDMs) toolkit is presented here, which considers several design methods and aspects used as design guidelines. The methods inclusive in the PDMs toolkit are the market pull, technology push, QFD as well as a tool group based on 40 principles and 39 innovation parameters to solve problems; this is the theory of inventive problem solving (TIPS), whose Russian acronym is TRIZ. In addition, this toolkit includes the Axiomatic Design, the product platforms, Design for X, and a methodology for modeling and simulation based on homogeneous transformation graphical (HTG).

The PDMs toolkit has been improved for 20 years of design practice to develop new products; this toolkit considers 11 aspects and 9 methods (Table 2.1).

### 2.3.1 EXPECTATION FOR INNOVATION USING THE MARKET PULL

One of the approaches to innovation is to start the generation of a new product based on market demand. It is expected to provide solutions based on the market need. The design team's capabilities are assessed by evaluating the complexity of the product, establishing the scope as incremental innovation or disruptive innovation. The former includes adaptation and improvements within paradigms; the disruptive or radical innovation includes out-of-paradigm improvements, invention, or discovery [5]. Later, the strategy is developed either by open innovation

**TABLE 2.1**
**Aspects and Methods Included in PDMs Toolkit**

| Item | Aspect | Method |
|---|---|---|
| 1 | Expectation for innovation | Market pull |
| | | Technology push |
| 2 | Design criteria | |
| 3 | Design attributes | |
| 4 | Requirements | QFD |
| 5 | Functional requirements | QFD and TRIZ |
| 6 | Design parameters | QFD and Axiomatic design |
| 7 | Constraints | |
| 8 | Communality index | Product platforms |
| 9 | DFA index | Design for X |
| 10 | Modeling and simulation | HTG methodology |
| 11 | Implementation | PDMs frameworks |

(collaboration among universities, startups, and others) or closed innovation (typically when the development process is confidential); so, the design, manufacturing, marketing, and sales begin [6].

In principle, one might think that the pull of the market is open innovation; however, companies such as Google, Microsoft, and Apple prevent other companies from benefiting from their products, developing confidentially [7]. Different companies use open innovation collaborating to determine market needs. Bayer, through startups, studies the need for medical systems by integrating the stakeholders involved in the rehabilitation process (doctor, physiotherapist, and patient); the market needs can be better feedbacked through networking with the hospitals [8].

### 2.3.2 Expectation for Innovation Using the Technology Push

Another approach to innovation is to start the generation of new products based on the evolution of technology. It is expected to give solutions usually as improvements to existing operating principles or from the results of basic science, such as the case of increasing the mechanical properties of some material. Subsequently, the technology implementation is evaluated, establishing the scope as incremental innovation or disruptive innovation.

The technology push generally contrasts with the pull of the market; it is how technology comes out and creates value in society [9]. Someone has devised a technology that often worked from a scientific perspective, e.g., some development of a system that moves air in larger quantities with less power; however, there is no guarantee that it will fit in the market, who will be the client, how it will really add value in society, etc. Thus, it is necessary to make a lot of effort to bring technology to the community ("if I had asked people what they wanted, they would have said faster horses"—Henry Ford quote). Other examples are the car, the camera, the internet, the touch screen, and the cell phone; all these things that nobody asked for, but the developers came up with based on new technology, and it could be pushed to the market, as well as assimilate and adopt it [10].

Ideas are often brought to market and tailored based on feedback. Therefore, there is a combination of technology push and market pull back and forth to create a long-term product reaching be a minimally viable product. But it must be brought to market, quickly learned from customers, and improved. As the technology push approach relates to patent/technology licensing, it is done with a way of thinking of open innovation. It is negotiated in a market; if it does not fit, it moves to other markets or is licensed to those who can use the technology in other ways. This technology licensing is why intellectual assets are used to transfer rights in multiple contexts at the same time, so each intellectual asset is a portfolio of possibilities, both for one market and another, as long as they do not overlap. It is a relatively easy value proposition; licensees want to collaborate to implement the technology, but some do not want to pay a lot for it [11]. The capabilities needed for a successful technology push approach are understanding the technology deeply using the intellectual active management framework where the licensees can capture the different technological elements that allow the operation of the solution. Finally, a comparison must be made to analyze the position in the market and work toward success [12]. There are some others that resort to closed innovation, sensing the market. Another example of technology push with open innovation is BMW® with an idea platform, where any user could contribute ideas, concepts, and patents on new technologies, where 3% are a reality. Others such as Metalsa®, Enel®, AstraZeneca®, and Ford®, through the inocentive.com platform, propose their innovation challenges together with an economic incentive [13]. The National Aeronautics and Space Administration (NASA) is investigating the ability of technology to fold wings during flight. The solution proposes the technologies to solve problems without focusing on demand but on the technological advance reached after the improvements have been implemented [14].

### 2.3.3 Design Criteria

Each application environment is different, so that the product design must be congruent with the market segment, the application sector, and the development area. Thus, the solutions must be pertinent concerning references to the application environment; then, said reference serves as a design guideline and involves the design trend, the sector of use, standards that regulate its use, its tests, and its validation. The design criteria are conditions that must be accomplished for design approval; an example of these criteria is the congruence criterion, where two triangles are congruent if two of their indicated angles have the same value [15]. Suppose the use of axiomatic design as a design tool. In that case, the axiom of independence, among others, should be fulfilled; the axiomatic design is defined as the process of seeking the uncoupled solution in the relationship between functional requirement (FR) and design parameter (DP). Then, the independence axiom will be fulfilled for a given key under the axiomatic design approach, which is a design criterion [16]. As has been observed, the design criteria are an obligatory step when starting the design of a product.

### 2.3.4 Design Attributes

The term attributes refers to qualities that make a product different from those known in the market through a value offered. This contribution can be an adaptation, an improvement within paradigms, an improvement outside paradigms, or an invention. Price and quality are often the most used at the beginning of the search for different aspects of the products. However, it is not the most recommended when starting the design of a new product. When the operating principle and the minimum viable product are reached, it is recommended to work on the quality strategy to meet customer expectations through certifications as well as comply with standards that occur at a level of technological maturity TRL8, which represents the product as a reliable option. With this level of technological maturity, the pricing strategy can be generated, sometimes through a high price and high investment in promotion; subsequently, its cost will be reduced to reach the rest of the consumers, which is known as a skimming strategy. Other times, a penetration strategy is used, where a reduced initial price is introduced to attract the maximum number of consumers and win the market. This strategy is usually used in products with fewer attributes or different aspects and as a test method to publicize the product in its initial stages [17].

As explained, the price and quality can be confusing when starting the product design, so it is preferable to focus on other attributes (e.g., factors like modularity, scalability, robustness, portability, wearable, gadgets, styling, and customizing) at the beginning so that the value of the product is sustained. For this, it is important to perform a benchmarking to identify and differentiate the product from that of competitors and define the client's requirements or clients in the displayed market segments.

### 2.3.5 Requirements

The user requirements, also called the voice of the customer (VOC), consolidate needs and improvements expected in the new product. Today, there are multiple methodologies to understand the requirements that must be met in the product, such as design thinking, where empathizing with the user is essential, in addition to defining the needs and perceptions of users; another methodology is the user-centered design, which seeks to find the definition of the problem and, from this problem, build a value offer [18].

However, first, the user and the customer are defined, understanding that there may be different types of users, but only one customer, who is the one who pays for the product. For this reason, the house of quality (HOQ) is used to build the list of requirements. HOQ is part of the QFD methodology, where user requirements are listed and prioritized in conjunction with improvements to products that are already in the market, as own as others. Then, a stakeholder survey is conducted aimed at stakeholders who are part of a select group of users, such as suppliers, distributors, and users with proven experience.

### 2.3.6 Functional Requirements

User requirements are analyzed and, using the first house of quality (HOQ1), technical solutions are proposed to meet those requirements; these solutions are called FRs. For instance, suppose that a VOC is to have hot water when washing dishes that are self-regulating so as not to have to compensate between hot and cold water, then the FR would self-regulate the water temperature at 35°C±3°C. In the QFD methodology, the user requirements are called WHATS, while the FR are called HOWS. Every FR can meet one or more VOCs, although sometimes the FRs contradict each other. Take as an example, high speed versus low weight. The TRIZ methodology is used to solve this contradiction, which includes 39 parameters and 40 innovation principles. When the parameters come into contradiction, where one improves and the other worsens, they are resolved by the matrix of contradictions with some inventive principle; then the relationship of each FR with each VOC is evaluated, comparing the proposed technical solutions with the solutions of the competitors for each VOC [19].

### 2.3.7 Design Parameters

The FRs help us to consolidate the design target in a numerical way. In the second house of quality (HOQ2), the FRs are taken as input; now, the WHATS are the FRs, and the HOWS are the DPs. First, the FRs are analyzed in numerical form; following the example of the previous section, suppose that an FR is to have a water temperature of 35°C±3°C. The DP would use a thermostat with digital temperature control with an error of ±1°C. In HOQ2, the QFD Methodology is used, where each DP corresponds to an FR. Once the DPs are numerically defined, the relationships of each DP with each FR are evaluated concerning the equation of the axiomatic design, specifically the equation of the axiom of independence (Equation 2.1).

$$\left\{ FR_1 \vdots FR_n \right\} = \left( X \cdots 0 \vdots \ddots \vdots 0 \cdots X \right) \left\{ DP_1 \vdots DP_n \right\} \tag{2.1}$$

where the list of FRs is used as a vector that is equal to the product of the design matrix by the vector of DPs. There are three possible design matrices; one of them is a coupled design matrix, where there are multiple relationships between the DPs and the FRs, which is not desirable since it means that while one FR is fulfilled, another FR is damaged. Another matrix is a decoupled design matrix, where the design improves by decreasing the relationships; this option usually presents a triangular matrix. Finally, the ideal design concerning the axiomatic design is represented by an uncoupled matrix, a matrix with minimal or null relationships, which presents a diagonal matrix [20].

### 2.3.8 Constraints

The technical specifications of the product (Tech Specs) include the DPs and the design restrictions; these restrictions can be external or internal. External constraints consider the operating environment and the coupling of subsystems in the product or the coupling of the product in a system that incorporates or integrates it. The internal restrictions condition the operation of the product, i.e., the interactions in a mechanism, the assembly means, or the contacts between components [21].

### 2.3.9 Commonality Index

Product platforms include modular architectures and integral architectures. Modular architectures include modules and connectors or buses to be interchangeable and achieve modularity or scalability of the product. Integral architectures are typically dedicated to systems that do not allow the interchange of components, e.g., an electronic card, a control system, a mechanism. The commonality index (CI) is used as a measure of the modularity of the product, which must be >60% to achieve an acceptable CI [22].

### 2.3.10 DFA Index

The design for X methodology, also known as design for excellence (DFX) is a group of methodologies where the X is a variable that can have one of many possible values [4]. The DFX includes the DFA, which is measured by an index that allows determining the assembly faster by simplifying the structure of the product, selecting essential and non-essential components. The acceptable DFA index must be >60%, which reduces the number of components required to be assembled [23].

### 2.3.11 Modeling and Simulation

Modeling and simulation are the central stages of the design and development of the product. After having generated the idealization and the design concept, the modeling phase is reached, which includes three kinds of models: the computational, iconic, digital model, virtual in CAD software; the mathematical model that is made based on the HTG; and the experimental physical model (EPM) [24]. The simulation is developed based on the modeling and includes three kinds: the simulation in CAE software for structural evaluation, computational fluid dynamics, and motion assessment; the numerical simulation using mathematical software as MATLAB®, using the multibody systems module in conjunction with the mathematical model; and the CAM simulation that is performed by taking the CAD components of the modeling phase, which are generated in the assessment for the manufacturing and are refined according to the machining times and the manufacturing process. The simulation in CAE software is usually validated with at least two solvers, obtaining a permissible correlation index; the numerical simulation is validated with the position in the workspace through a graphical user interface (GUI). The simulation is subsequently validated by comparing it with the emulation carried out on a test bench, for which a permissible error is taken. The validation of CAM simulation is done by a machining test. However, when machining complex workpieces (high accuracy), its waste is expensive and has a high possibility of breaking the tool by material features. It is necessary to use techniques based on kinematics and dynamics simulation [25].

### 2.3.12 PDMs Frameworks

The design methodologies included in the PDMs toolkit are implemented based on five frameworks. Figure 2.4 shows the first framework consisting of seven stages for product generation development. The design concept stage is where the solution is formulated. The animation stage is distinguished by the fact that there is no physical force feedback, but rather the system operation is proposed; animation software such as Unreal Engine®, Unity®, 3D Max®, Maya®, Character

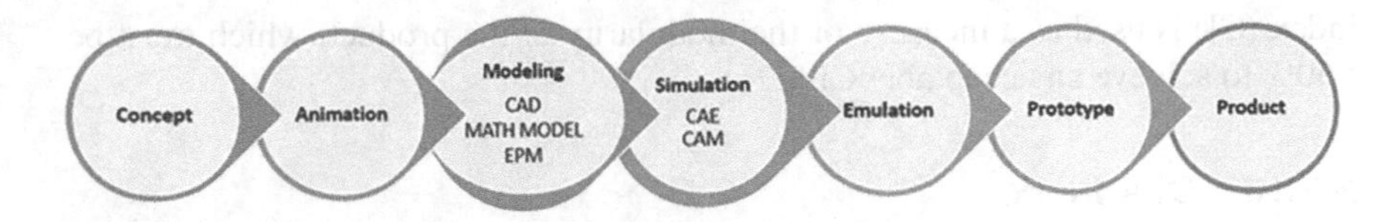


**FIGURE 2.4** Stages for the generation of new products.

Studio®, and Poser® are used for this task. The modeling stage includes three kinds of models, the first is the CAD model (non-tangible physical model), distinguished by the fact that CAD software is used to allow the feedback of physical force (SolidWorks®, Solid Edge®, and Inventor®, among others); the second is the Math model, which is referred to kinematics and dynamics model as: $M(q)\ddot{q}+V(q)\dot{q}+G(q)+\beta(q)=\tau$; the third model is the EPM, which is used to validate the operating principle of the system. The simulation stage is identified because CAE software is used to verify the system behavior in the computational environment. Simulation is validated, obtaining a correlation index among this behavior and its behavior of EPM in the test bench. The emulation stage is characterized by the use of instrumented test benches to perform emulating tests of the system behavior in a real environment. The prototyping stage is dedicated to obtaining a prototype, i.e., the first replicable type with which it is expected to have field tests; once the prototype is re-controlled and released, the consolidation stage of the product can be reached [26].

Figure 2.5 presents the second framework where the interaction among technology readiness level (TRL), manufacturing readiness level (MRL), and investment readiness level (IRL) is shown. As mentioned in the first framework, the design starts with the design concept, which corresponds to TRL1 and TRL2 levels. The EPM ALPHA, where the design concept is validated and the TRL3 level is reached, is related to the MRL1 and MRL2 levels, which correspond to initial materials. The TRL4 level is reached when there is an EPM validated in a laboratory environment, which is related to an IRL1 level where a preliminary market analysis is started (complete first-pass business model canvas or lean model canvas). TRL5 level is reached when there is a BETA prototype; TRL5 is related to MRL3, where an experimental lot is manufactured, as well as to IRL2 in which the market analysis is defined (market size: total available market (TAM), serviceable available market (SAM), and Target market). TRL6 level is reached when there is a prototype operating in relevant environment testing; it is related to MRL4 level, where the manufacturing technologies are determined, as well as to IRL3, where there is the solution validation. TRL7 is reached when there is a prototype operating in real testing; it is related to MRL5, where most of the materials and tooling are refined, as well as to IRL4, where there is a valuable minimum product (VMP) of low fidelity. TRL8 level is reached when there is a certified product. It is related to MRL6 and MRL7, where all materials are defined and tested in the pilot production line, and IRL7, a VMP of high fidelity. TRL9 level is reached when there is a commercial product. So, in MRL8 is

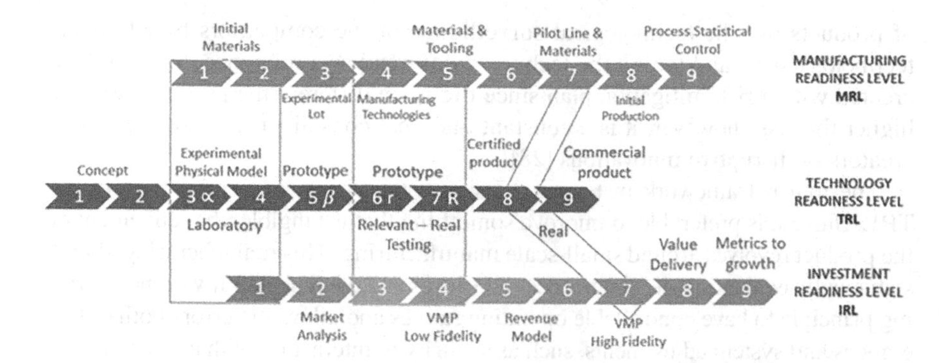


**FIGURE 2.5** Interaction among TRL, MRL, and IRL.

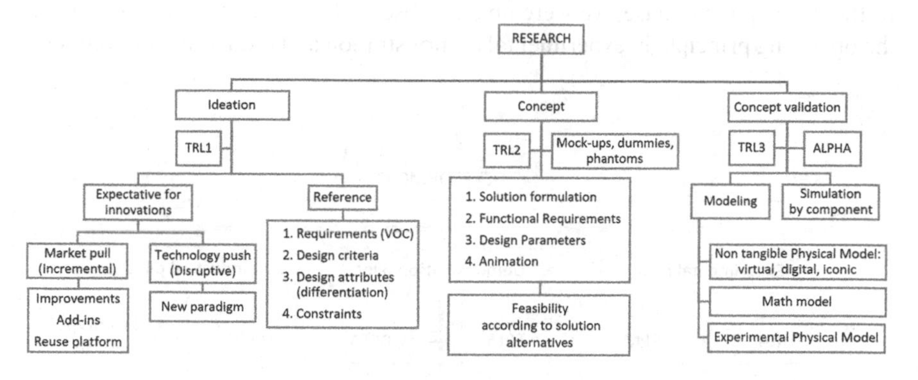


**FIGURE 2.6** Phases of research in the TRL.

released the initial production and in the MRL9 is carried out the process statistical control. In IRL8, there is a value delivery, and IRL9 is focused on metrics-based growth [27].

Figure 2.6 shows the third framework corresponding to the phases of research in the TRL. The premise is to create a value offer based on a clear differential concerning competitors. However, this task is not so easy. Suppose the market demand (users) is analyzed and based on this. In that case, a solution that is expected fast is proposed; it is most likely to think of it as an incremental innovation at best, in which case it will be an enhancement to a reuse platform plugin or app. Since a solution that implies a greater differential is based on new technology, this is not so trivial, which requires more time for scientific research, since the uncertainty is higher and then the approach would initially be to use the technology push and later make the pertinent modifications to adapt the product to an application that is in demand in the market; therefore, a mixture of the two approaches will be made. It should be considered that this procedure is recommended in designs made by universities and entrepreneurs with limited experience in the industrialization of products to obtain disruptive innovations. The companies or entrepreneurs with extensive experience in the industrialization of products use the technological surveillance of the competitors based on the technology push and taking the technological trend as a reference. Solutions are created with a risk mitigation plan since the more different the products are, the higher the risk; however, it is a constant task that consolidates the reputation as creators of disruptive innovations [28].

The fourth framework in Figure 2.7 shows the phases of development in the TRL. Since it is preferable to improve something that is tangible, the refinement of the product revolves around small-scale manufacturing. This manufacturing allows statistically evaluating the occurrence of failures and the repeatability of the operating principle to have controllable operating ranges and allowable errors both of tolerances and system adjustments, such as security in interaction with users. For this, usability metrics are used, which will ensure the performance of the product [29].

Figure 2.8 shows the fifth framework in the phases of deployment in the TRL. In the development phase, we were able to observe how important it is to repeat the operating principle in experimental demonstration and industrial environment

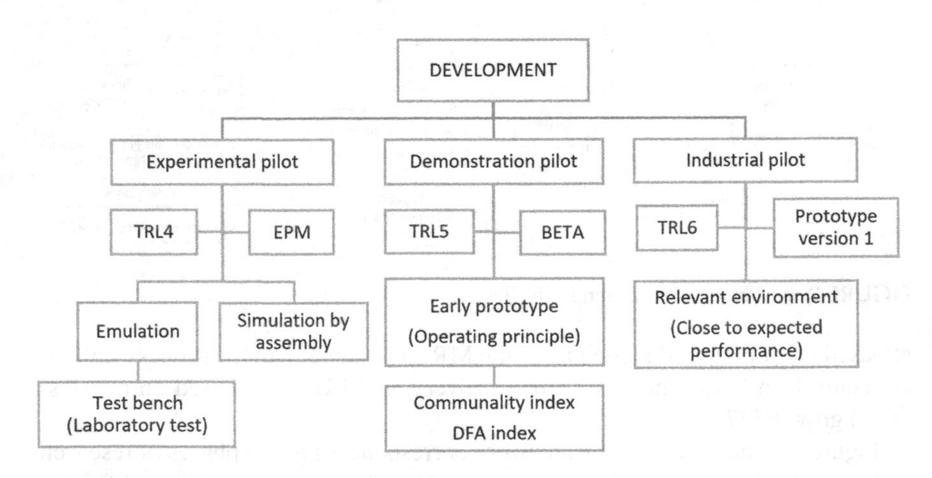


**FIGURE 2.7** Phases of development in the TRL.

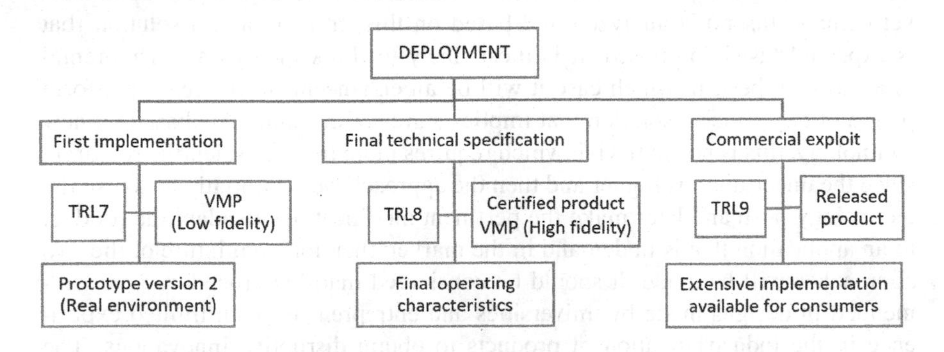


**FIGURE 2.8** Phases of deployment in the TRL.

in order to reach the necessary maturity of the product so that it can be implemented and commercially exploited based on the tech specs validated both in test benches as with users [30].

## 2.4 A CASE STUDY BASED ON PDMs TOOLKIT

### 2.4.1 Preliminary Topics

A firearm shot is a mechanical, chemical, and ballistic phenomenon that produces marks that cannot be repeated with another weapon; these traces remain on the bullets and caps [31]. Although the shooting phenomenon has the characteristics described above, this phenomenon has particularities according to each type of weapon, ammunition, chamber, and barrel. Next, four types of firearms will be explained, which represent, in general, the most used systems today.

A semi-automatic pistol is a type of single-chamber repeating pistol that automatically cycles its action to insert the next cartridge into the chamber but requires manual activation of the trigger to release the next shot [32].

An assault rifle is a military firearm that is chambered for small ammunition or propellant charge and has the ability to switch between semi-automatic and fully automatic fire. Although these weapons are lightweight and portable, they can still deliver a large volume of fire with reasonable accuracy at modern combat ranges of 1000 to 1600 feet (300–500 m) [33].

A sniper rifle is a high-precision, long-range rifle. Requirements include precision, reliability, mobility, concealment, and optics for military sniper surveillance and anti-personnel uses. The modern sniper rifle is a shoulder-fired portable weapon system with a choice between bolt action or semi-automatic action, equipped with a telescopic sight for extreme precision and chambered for a high-performance ballistic centerfire cartridge [34].

A bullpup firearm is one whose action is placed behind the trigger. This firearm creates an overall more compact and maneuverable weapon compared to conventional firearm designs with the same barrel length and maintains the advantages of a long barrel, such as better muzzle velocity and accuracy while reducing the size and the overall weight [35].

Although the semi-automatic pistol, the assault rifle, the sniper rifle, and the bullpup firearm have their own technical specifications such as pressure, range, speed, and caliber, among others, they all share the same shooting phases, which are depicted in Figure 2.9.

### 2.4.2 Market Pull Analysis

Briefly, an analysis of the TAM for the firearms is carried out; for this, the sport gun market size was selected. The global sport gun market size was valued at $2.36 billion in 2019 and is forecast to reach $3.46 billion by 2027 at a compound annual growth rate (CAGR) of 4.8%. A sporting gun refers to a non-lethal weapon that can include a rifle, a pistol, and a shotgun. It is intended for competitive

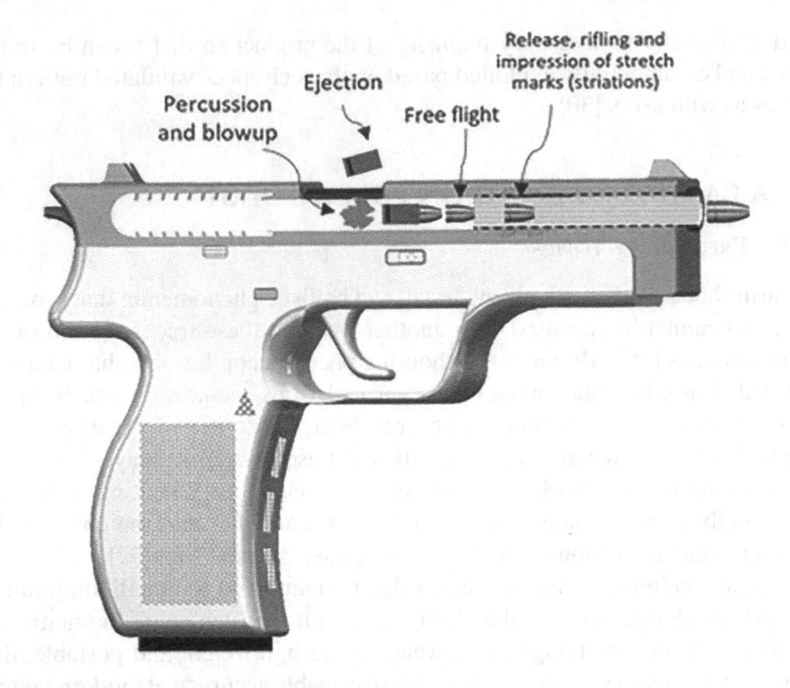


**FIGURE 2.9** Phases of firearm shooting.

shooting and recreational purposes [36]. These firearms differ from each other according to their technology, projectile, caliber, and others. Sport guns are lighter and more comfortable to use as compared to lethal firearms.

Increasing participation in various shooting sports acts as a key factor driving the sport gun market growth. Sports include Target Shooting, High Power Rifle, Popinjay, Sporting Clays, Skeet Shooting, Cowboy Action Shooting, Skirmish, and others. For instance, military pentathlon is a shooting discipline that consists of sighting shots and competition shooting [37].

People are increasingly participating in shooting sports owing to various benefits associated since they enhance physical discipline, focus, and sharpen eyesight. Furthermore, the growing participation of school and college kids/students in national-level shooting sports is expected to increase the growth of the market.

The global market landscape of sport guns is expected to remain in a very competitive and highly fragmented landscape consisting of several small startups, medium enterprises, and large conglomerates. During the projected era, increasing demand for technological development and higher diversification in offered products ensure the enormous potential for innovative players. The segment of materials used consists of steel, aluminum, polymer, and others. The polymer sub-segment is growing with the fastest CAGR due to its higher manufacturing efficiency with 3D printing techniques [38].

Further development is expected in the sport industry in North America, especially in the United States, owing to its leading investment in the defense

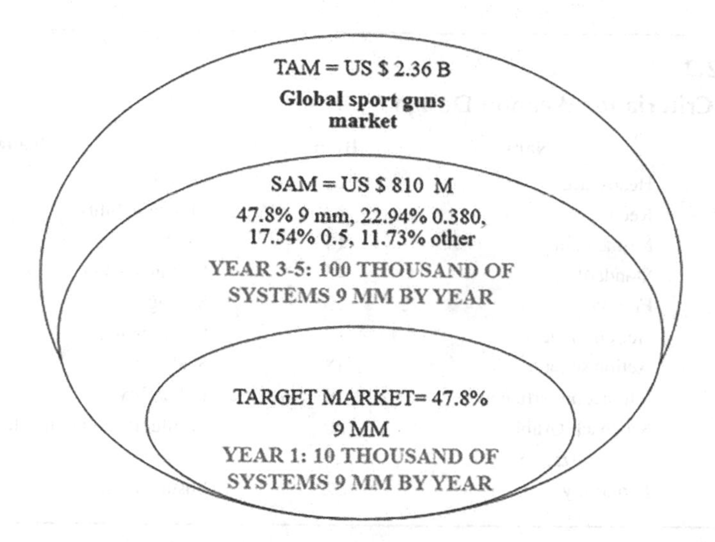


**FIGURE 2.10** Market analysis of sport guns for North America.

and special taskforce budgets. Regional statistics indicate that North America (Canada, the United States, Mexico) is a saturated market. Europe is a stable market, Latin America is a futuristic market, and Asia-Pacific (China, Japan, India, Australia, rest of Asia-Pacific) accounted for the fastest growth with a CAGR of 7%. Despite these data, North America is the region that acquires the most arms annually; therefore, it is the market chosen to compete. Figure 2.10 shows the market analysis for the North America area.

### 2.4.3 Design Criteria

Most of these criteria are considered to define a robust weapon design. The standard is used to evaluate the design concerning norms such as Military Standard, Sporting Arms and Ammunition Manufacturers' Institute (SAAMI), and Standardization Agreement (STANAG). The standard includes the ballistic standard, while standardization is used as a technique whose aim is to reduce the number of different parts within a product. In other words, the standardization is a part of having more common parts than unique ones, which is represented by the CI. Table 2.2 shows the design criteria taken into account for weapon design.

The headspace is the distance measured in the barrel chamber between the fulcrum of the cartridge housed and the frontal plane of the closure. Each kind of firearm has its headspace concerning its chamber and ammunition. Figure 2.11 schematizes different examples of headspace.

Recoil is the rearward motion of the firearm as a result of firing. It considers the forces acting on a fired firearm, which can cause the handgrip to gently rotate in the hand or the buttstock's heel to push against the shooter's shoulder. So, the recoil is the reaction force that is absorbed through the shooter's hand or

**TABLE 2.2**
**Design Criteria for Weapon Design**

| Item | Name | Item | Name |
|---|---|---|---|
| 1 | Headspace | 12 | Usability |
| 2 | Recoil | 13 | Maintainability |
| 3 | Momentum | 14 | Safety |
| 4 | Standard | 15 | Elegant design |
| 5 | Fire power | 16 | Styling |
| 6 | Action trigger | 17 | Customizing |
| 7 | Action shoot | 18 | Scalability |
| 8 | Interaction virtual axis | 19 | Reliability |
| 9 | Kutzbach-Grubler | 20 | Profitability-viability-feasibility |
| 10 | Standardization | 21 | Quality |
| 11 | Portability | 22 | Sustainability |

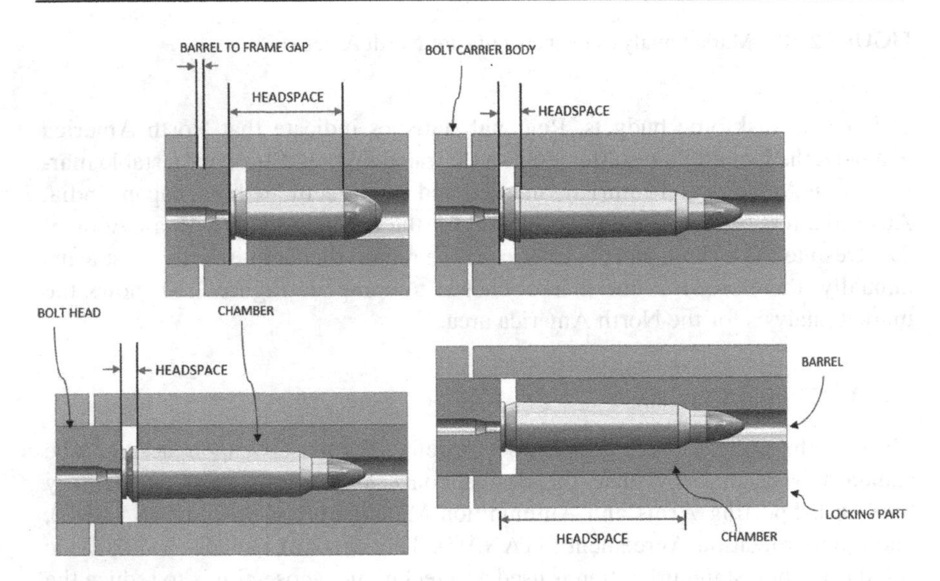


**FIGURE 2.11** Headspace in several kinds of breech.

shoulder, caused by the force that releases the explosion in ammunition and sends the bullet through the barrel, according to Newton's third law of motion. Figure 2.12 shows the components involved in the recoil [39].

The locking part resists the firing effect by keeping the chamber closed with the bolt, then the bolt head is unblocked through the recoil system (gas or mass), so the body of the bolt carrier is braked by both the recoil spring as the backplate with the recoil buffer. Newton's third law, known as conservation of momentum, explains that the application of forces and accelerations causes changes in the motion of a mass. Furthermore, if all the masses and velocities involved are

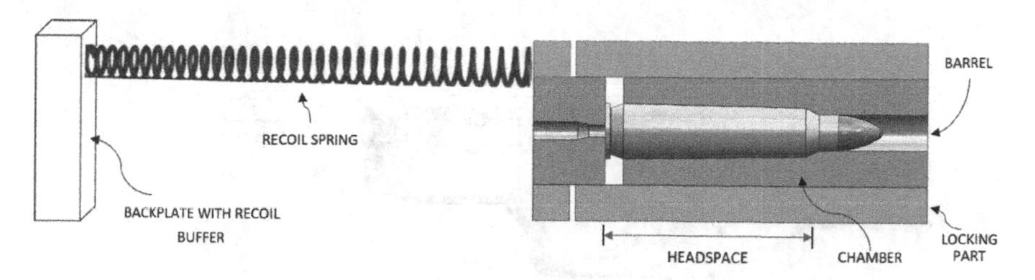


**FIGURE 2.12** Components involved in the recoil.

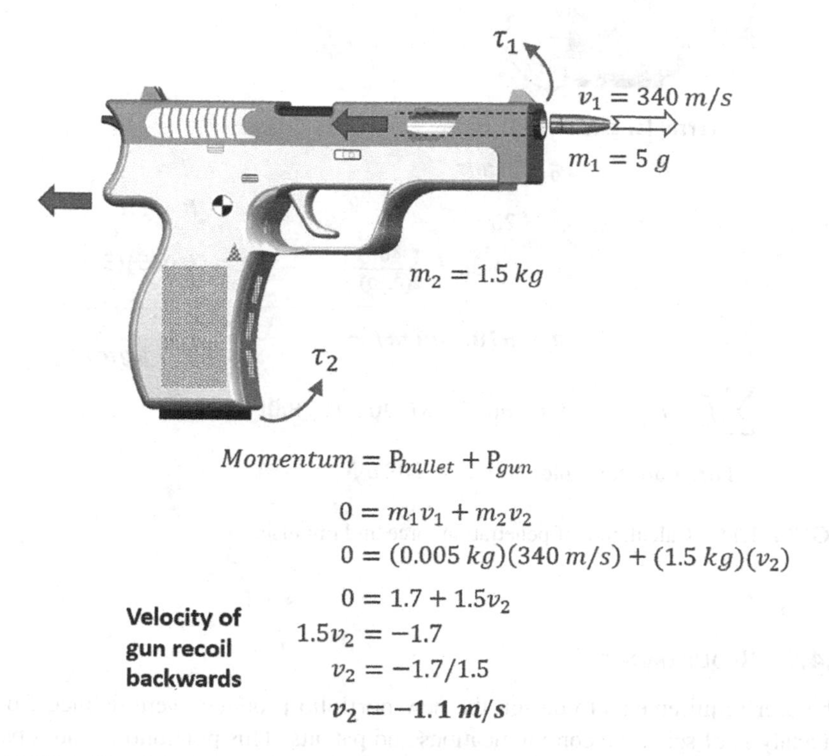

$\tau_1$
$v_1 = 340\ m/s$
$m_1 = 5\ g$
$m_2 = 1.5\ kg$
$\tau_2$
$Momentum = P_{bullet} + P_{gun}$
$0 = m_1 v_1 + m_2 v_2$
$0 = (0.005\ kg)(340\ m/s) + (1.5\ kg)(v_2)$
$0 = 1.7 + 1.5v_2$
$1.5v_2 = -1.7$
$v_2 = -1.7/1.5$
$\mathbf{v_2 = -1.1\ m/s}$
Velocity of gun recoil backwards


**FIGURE 2.13** Calculation of recoil velocity.

accounted for, the system's momentum is conserved [40]. This conservation of momentum is why gun recoil occurs in the opposite direction of the bullet projection; the mass multiplied by the projectile's velocity in the positive direction equals the mass multiplied by the gun's velocity in the negative direction.

Once a prototype firearm is manufactured, the projectile, gun recoil energy, and momentum can be directly measured using a ballistic pendulum with a ballistic chronograph [41]. Recoil buffering systems are adequately designed to safely dissipate that momentum and energy, based on estimates of the projectile velocity coming out of the barrel. Figure 2.13 shows the calculation of recoil velocity, and Figure 2.14 shows the calculation of penetration force and impulse.

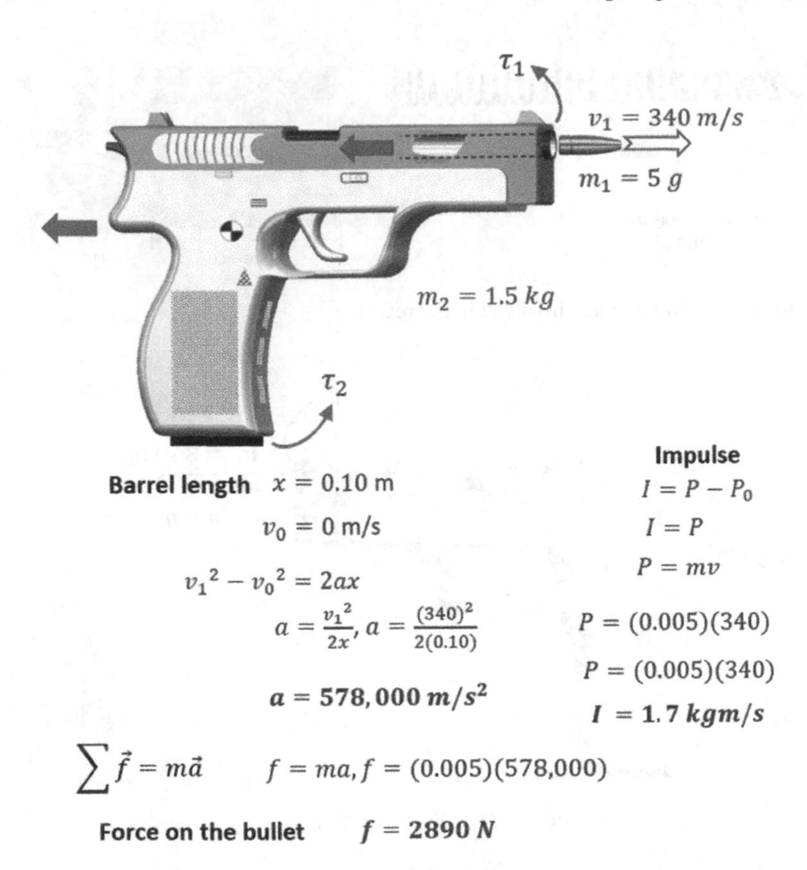

$\tau_1$
$v_1 = 340\ m/s$
$m_1 = 5\ g$
$m_2 = 1.5\ kg$
$\tau_2$
Impulse
$I = P - P_0$
$I = P$
$P = mv$
Barrel length $x = 0.10$ m
$v_0 = 0$ m/s
$v_1^2 - v_0^2 = 2ax$
$a = \frac{v_1^2}{2x}, a = \frac{(340)^2}{2(0.10)}$
$P = (0.005)(340)$
$P = (0.005)(340)$
$a = 578,000\ m/s^2$
$I = 1.7\ kgm/s$
$\sum \vec{f} = m\vec{a}$
$f = ma, f = (0.005)(578,000)$
Force on the bullet $f = 2890\ N$


**FIGURE 2.14** Calculation of penetration force and impulse.

### 2.4.4 Requirements

The user requirements to design the new portfolio products were defined from an analysis of scientific communications and patents. This portfolio includes two product families, a platform, and a modular architecture.

### 2.4.5 Functional Requirements

The FRs are detailed numerical solutions that are proposed to meet user requirements. When these FRs are in contradiction when using the TRIZ methodology, in this case supposing the integral architecture is an FR that is in contradiction with the CI, then the recommended principle is the modular architecture. Table 2.3 shows the relationship between user requirements (VOC) and FRs.

**TABLE 2.3**
**User Requirements and Functional Requirements**

| VOC | FR |
|---|---|
| Low weight | Polymers and sintered materials |
| Customization | Modular architecture |
| Scalability | CI>60% |
| Standardization | CI>60% |
| Elegant design | DFA index>60% |

**TABLE 2.4**
**Functional Requirements and Design Parameters**

| FR | DP |
|---|---|
| Polymers and sintered materials | Seventy percent of components include receiver, magazine, and parts of the trigger mechanism |
| Modular architecture | Methodology of product platforms, creating two product families |
| CI>60% | Methodology of product platforms, creating two product families |
| DFA index>60% | DFA algorithm to select the essential and non-essential parts |

### 2.4.6 Design Parameters

The DPs are detailed numerical solutions that are proposed to meet FRs, based on the axiom of independence from axiomatic design. Table 2.4 shows the relationship between FRs and DPs.

### 2.4.7 Constraints

Design constraints are conditions that delimit the formulation of solutions and the principle of operation in the design problem. They can be described in a mathematical way when related to the operating environment or the coupling with other systems; they are called external constraints. The internal constraints intervene in the relations of movement, position, force, moment, time, and others. An example of internal constraints is perpendicularity, concentrically, symmetry, collinearity, parallelism, and proportionality. Table 2.5 shows the external and internal constraints in the design of firearms.

### 2.4.8 Commonality Index

The example below shows a platform of products with two product families, A and B, each with common and unique components (Figure 2.15).

**TABLE 2.5**
**External and Internal Constraints**

| Constraints | External | Internal |
|---|---|---|
| Anthropometry and ergonomics | X | |
| Ergonomic | X | |
| The trigger mechanism of simple action | | X |
| Picatinny rail | X | |
| Overall tolerances of 0.1 mm | | X |

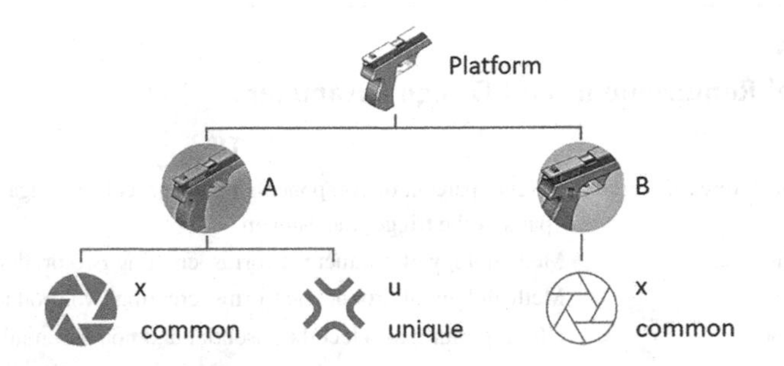


**FIGURE 2.15** Modular architecture of a product.

The calculation of the commonality index $C$ (Equation 2.2) shows that the more common components there are, the larger the commonality index.

$$C = \frac{100x}{x+u}$$

$$C = \frac{100(2)}{(2)+(1)} = 66.7\%$$

$$C = \frac{100(3)}{(3)+(1)} = 75\% \qquad (2.2)$$

### 2.4.9 DFA Index

The following example shows the original components of trigger mechanisms before simplification. Table 2.6 shows the minimum part criteria related to the essential parts, where the DFA index (Equation 2.3) is 53%.

$$\text{DFA} = \frac{100 N_m t_m}{T_P} \qquad (2.3)$$

**TABLE 2.6**
**Components of Trigger Mechanisms Before DFA Simplification**

| Item | Part | Quantity | Minimum Part Criteria |
|---|---|---|---|
| 1 | Bushing of harm | 1 | None |
| 2 | Body of harm | 1 | Movement |
| 3 | Body of trigger | 1 | Movement |
| 4 | Bushing of trigger | 1 | None |
| 5 | Base casing | 1 | Base |
| 6 | Bushing of selector | 1 | None |
| 8 | Right selector lever | 1 | Assembly |
| 9 | Left selector lever | 1 | Assembly |
| 10 | Pin | 3 | Assembly |
| 11 | Pin of selector | 2 | Fastener |
| 12 | Disconnector | 1 | Movement |
| 13 | Automaticity sear | 1 | Movement |
| 14 | Spring | 4 | None |
| Total | | 19 | Three meet theoretical minimum |

DFA: Design for Assembly Index
$N_m$ : Number of minimum part criteria
$t_m$ : Minimum assembly time by part
$T_P$ : Total of parts

$$DFA = \frac{100\ (3)(3)}{(19)} = 47\%$$

After DFA simplification using the minimum part criteria (Table 2.7), a DFA index of 60% is obtained:

$$DFA = \frac{100\ (3)(3)}{(15)} = 60\%$$

It can be seen that the smaller the number of components in relation to the minimum part criterion, the DFA index will be greater.

### 2.4.10 Product Portfolio

The following product portfolio is composed of a platform with modular architecture with two product families. One of them is a pistol, and the other is an assault rifle with two products, the first being the assault rifle itself and the second a bullpup. The design concept of the platform begins with the pistol; each subsystem does commonality and DFA indexes as modular architecture so that the product family can be scaled. Figure 2.16 shows the platform concept.

**TABLE 2.7**
**Components of Trigger Mechanisms after DFA Simplification**

| Item | Part | Quantity | Minimum Part Criteria |
|---|---|---|---|
| 1 | Harm | 1 | Movement |
| 2 | Trigger | 1 | Movement |
| 3 | Base casing | 1 | Base |
| 4 | Right selector lever | 1 | Assembly |
| 5 | Selector coupling | 1 | Assembly |
| 6 | Pin of selector | 1 | Fastener |
| 7 | Pin | 3 | Assembly |
| 8 | Disconnector | 1 | Movement |
| 9 | Automaticity sear | 1 | Movement |
| 10 | Spring | 4 | None |
| Total | | 15 | Three meet theoretical minimum |

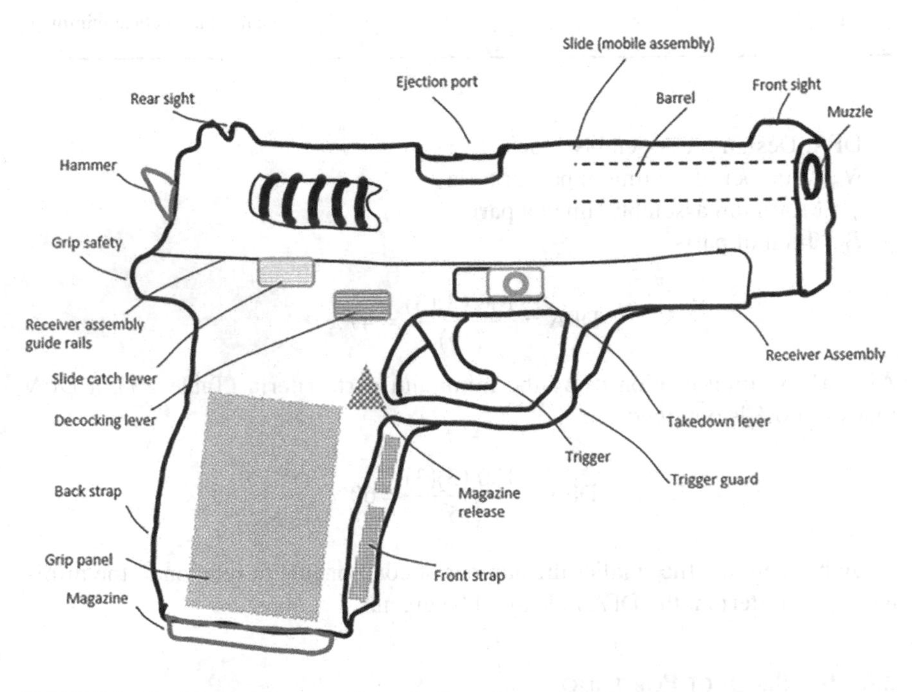


**FIGURE 2.16** Design concept of the platform for product families.

This product family is a pistol, and sometimes it is convenient to generate a brand per platform and a brand for each family product (Figure 2.17).

The following product family is an assault rifle with two products; different market segments are satisfied with these solutions (Figures 2.18 and 2.19).

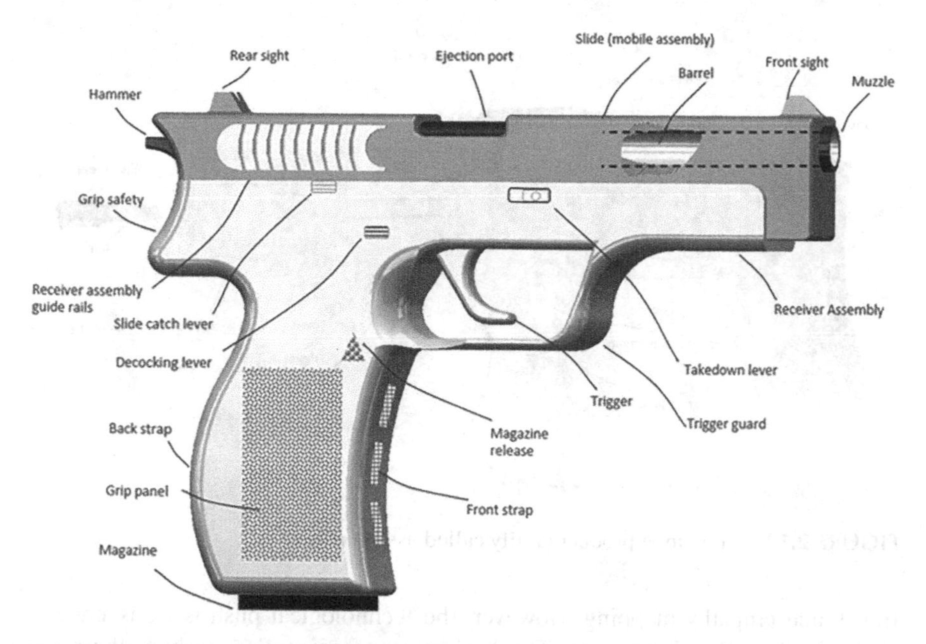


**FIGURE 2.17** Pistol of the product family.

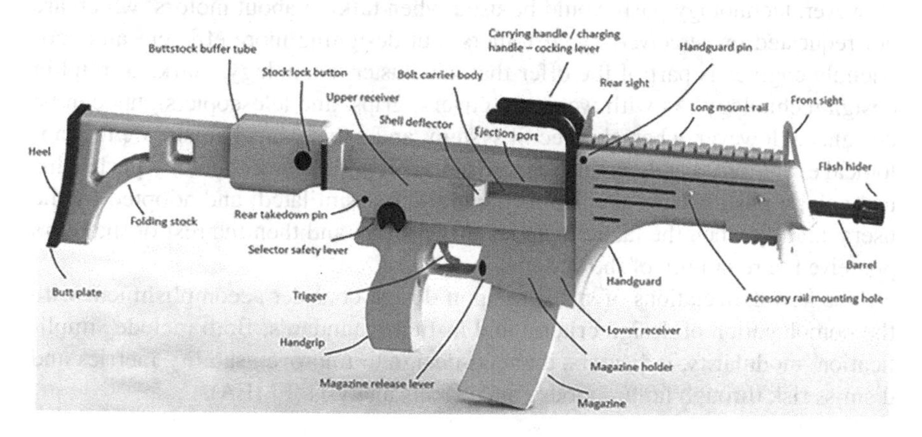


**FIGURE 2.18** Assault rifle of the product family.

## 2.5 CLOSING REMARKS AND PERSPECTIVES

The design strategy and methodology are the best way to start the development process. In this sense, both parts allow outlining the aspects and methods that will be included in the product design. Currently, innovation trends focus on the market pull: the user-centered design (UCD), which implies empathizing with the user and using tools to identify user preferences such as storytelling, mapping

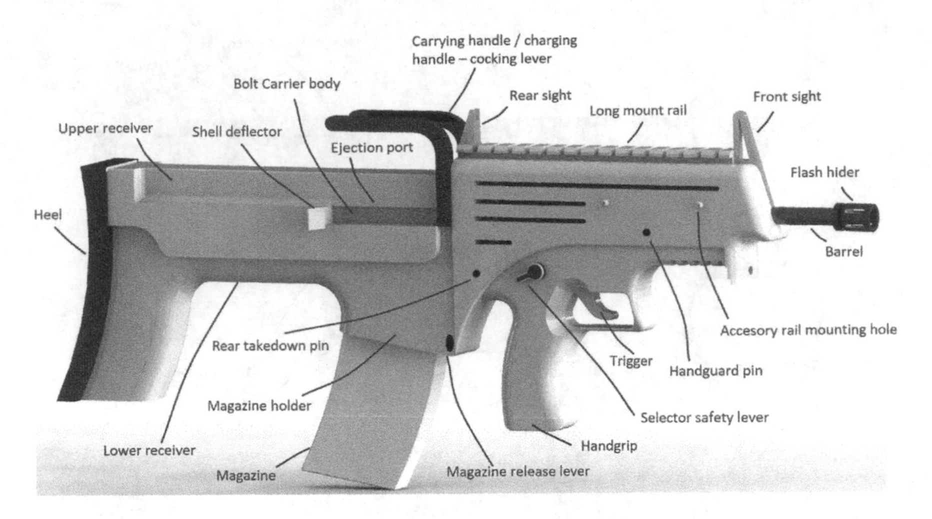


**FIGURE 2.19** Bullpup of product family called assault rifle.

travel, and empathy mapping. However, the technological push is transcendental in the disruptive innovation. Think of cars or motorcycles; perhaps the market pull could be used when talking about comfort, luxury, clothing, and color. However, technology push would be used when talking about motors, which are not requested or perceived by customers, but designing more efficient and eco-friendly engines is part of the offer that, as cluster technology, marks a trend in design technology. As with weapons, covers, grips, and telescopic sights can be designed. However, a barrel, a recoil system, and a firing mechanism are technological evolutions that do not arise from a market pull but a technology push. This technology push is later adapted to the market, assimilated, and adopted by the users, mainly when the tactic includes this firearm and then the rest of the users perceive the reliability of the systems.

The new expectations of small weapon design consider accomplishment with the homologation of design criteria and military standards. Both include simplification, modularity, indicators, and eco-design to improve usability metrics and dismiss risk through failure modes and effects analysis (FMEA).

## REFERENCES

1. Ullman, D.G., *The Mechanical Design Process*. Vol. 2, 1992, McGraw-Hill: New York.
2. Sodhro, A.H., et al., Convergence of IoT and product lifecycle management in medical health care. *Future Generation Computer Systems*, 2018. **86**: pp. 380–391.
3. Bowland, N.W., et al., A PDM and CAD-integrated assembly modelling environment for manufacturing planning. *Journal of Materials Processing Technology*, 2003. **138**(1–3): pp. 82–88.
4. Kahn, K.B., et al., An examination of new product development best practice. *Journal of Product Innovation Management*, 2012. **29**(2): pp. 180–192.

5. Brem, A. and K.I. Voigt, Integration of market pull and technology push in the corporate front end and innovation management: Insights from the German software industry. *Technovation*, 2009. **29**(5): pp. 351–367.
6. Chesbrough, H.W., et al., *Open Innovation: Researching a New Paradigm*. 2006, Oxford University Press on Demand: Oxford.
7. Cusumano, M.A., *Staying Power: Six Enduring Principles for Managing Strategy and Innovation in an Uncertain World (Lessons from Microsoft, Apple, Intel, Google, Toyota and More)*. 2010, Oxford University Press: Oxford.
8. Lazzarotti, V. and R. Manzini, Different modes of open innovation: A theoretical framework and an empirical study. *International Journal of Innovation Management*, 2009. **13**(04): pp. 615–636.
9. Di Stefano, G., A. Gambardella, and G. Verona, Technology push and demand pull perspectives in innovation studies: Current findings and future research directions. *Research Policy*, 2012. **41**(8): pp. 1283–1295.
10. Herstatt, C. and C. Lettl, Management of "technology push" development projects. *International Journal of Technology Management*, 2004. **27**(2–3): pp. 155–175.
11. Moogk, D.R., Minimum viable product and the importance of experimentation in technology startups. *Technology Innovation Management Review*, 2012. **2**(3): 23–26.
12. Nemet, G.F., Demand-pull, technology-push, and government-led incentives for non-incremental technical change. *Research Policy*, 2009. **38**(5): pp. 700–709.
13. Jovanović, T., et al. The crowdfunding idea contest of BMW. In *Proceedings of the 26th International Association for Management of Technology Conference*, Vienna, Austria, 2017.
14. Sadin, S.R., F.P. Povinelli, and R. Rosen, The NASA technology push towards future space mission systems, in *Space and Humanity*, L.G. Napolitano, Editor. 1989, Elsevier: Amsterdam, Netherlands, pp. 73–77.
15. Dávila-Vilchis, J.M., L.A. Zuñiga-Aviles, J.C. Ávila-Vilchis, and A.H. Vilchis-González, Design methodology for soft wearable devices: The MOSAR case. *Applied Sciences*, 2019. **9**(22): p. 4727.
16. Suh, N.P., Axiomatic design theory for systems. *Research in Engineering Design*, 1998. **10**(4): pp. 189–209.
17. Sauser, B., et al. From TRL to SRL: The concept of systems readiness levels. In *Conference on Systems Engineering Research*, Los Angeles, CA, 2006, Citeseer.
18. Consolvo, S., et al. Design requirements for technologies that encourage physical activity. In *Proceedings of the SIGCHI Conference on Human Factors in Computing Systems*, Montréal, Québec, Canada, 2006.
19. Thompson, M.K. A classification of procedural errors in the definition of functional requirements in Axiomatic Design theory. In *Seventh International Conference on Axiomatic Design (ICAD 2013)*, Worcester, MA, 2013.
20. Kulak, O., S. Cebi, and C. Kahraman, Applications of axiomatic design principles: A literature review. *Expert Systems with Applications*, 2010. **37**(9): pp. 6705–6717.
21. Thompson, M.K., et al., Design for additive manufacturing: Trends, opportunities, considerations, and constraints. *CIRP Annals*, 2016. **65**(2): pp. 737–760.
22. Kota, S., K. Sethuraman, and R. Miller, A metric for evaluating design commonality in product families. *Journal of Mechanical Design*, 2000. **122**(4): pp. 403–410.
23. Prakash, W.N., V. Sridhar, and K. Annamalai, New product development by DFMA and rapid prototyping. *ARPN: Journal of Engineering and Applied Sciences (JEAS)*, 2014. **9**(3): pp. 274–279.
24. Zúñiga-Avilés, L., et al., HTG-based kinematic modeling for positioning of a multi-articulated wheeled mobile manipulator. *Journal of Intelligent & Robotic Systems*, 2014. **76**(2): pp. 267–282.

25. Zivanovic, S., et al., Machining simulation and verification of tool path for CNC machine tools with serial and hybrid kinematics, in *Heavy Machinery- HM 2017*, M. Gašić, Editor. 2017, Faculty of Mechanical and Civil Engineering, Kraljevo: Zlatibor, Serbia. pp. 63–68.
26. Cruz Martinez, G.M. and L.A. Zuñiga Aviles, Methodology for the design of rehabilitation robots: Application in an exoskeleton for upper limb rehabilitation. *Applied Sciences*, 2020. **10**(5459). doi: 10.3390/app10165459.
27. Fernandez, J.A., *Contextual Role of TRLs and MRLs in Technology Management*. 2010, Office of Scientific and Technical Information (OSTI): California.
28. Ross, S., Application of system and integration readiness levels to department of defense research and development. *Defense Acquisition Research Journal: A Publication of the Defense Acquisition University*, 2016. **23**(3): pp. 248–273.
29. Engel, D.W., et al., *Development of Technology Readiness Level (TRL) Metrics and Risk Measures*. 2012, Pacific Northwest National Lab.(PNNL): Richland, WA.
30. Puig, L., A. Barton, and N. Rando, A review on large deployable structures for astrophysics missions. *Acta Astronautica*, 2010. **67**(1–2): pp. 12–26.
31. Rosenberg, Z. and E. Dekel, *Terminal Ballistics*. 2012: Springer: Berlin/Heidelberg.
32. Jenzen-Jones, N. and J. Ferguson, Weapons identification: Small arms. in *An Introductory Guide to the Identification of Small Arms, Light Weapons, and Associated Ammunition*, N.R. Jenzen-Jones and M. Schroeder, Editors. 2018, Small Arms Survey: Geneva, Switzerland.
33. Kuo, C.L., C.K. Yuan, and B.S. Liu, Using human-centered design to improve the assault rifle. *Applied Ergonomics*, 2012. **43**(6): pp. 1002–1007.
34. Von Wahlde, R., and D. Metz, *Sniper Weapon Fire Control Error Budget Analysis*. 1999, Army Research Lab: Aberdeen Proving Ground, MD.
35. Stone, R.T., B.F. Moeller, R.R. Mayer, B. Rosenquist, D. Van Ryswyk, and D. Eichorn, Biomechanical and performance implications of weapon design: comparison of bullpup and conventional configurations. *Human Factors*, 2014. **56**(4): pp. 684–695.
36. Koene, B., F. Id-Boufker, and A. Papy, Kinetic non-lethal weapons. Netherlands annual review of military studies, 2008: pp. 9–24.
37. Petrovič, P., et al., Impact of selected factors on performance in sporting shooting from air rifle in standing position. *Journal of Physical Education and Sport*, 2020. **20**(2): pp. 768–773.
38. Crouch, I. (Ed.), *The Science of Armor Materials*. 2016, Woodhead Publishing: Sawston.
39. Li, Z.C. and J. Wang, A gun recoil system employing a magnetorheological fluid damper. *Smart Materials and Structures*, 2012. **21**(10): p. 105003.
40. Burns, B.P., *Recoil Considerations for Shoulder-Fired Weapons*. 2012, Army Research Laboratory: Adelphi, MD
41. Lugini, C. and M. Romano, A ballistic-pendulum test stand to characterize small cold-gas thruster nozzles. *Acta Astronautica*, 2009. **64**(5–6): pp. 615–625.

# 3 Custom and Functional Requirements

## 3.1 INTRODUCTION

Today, more than ever, firearms should be safe and reliable. For this, their design requires tools that include state-of-the-art technological levels. Apart from those features, firearms should be accepted in the market, for which they should meet different requirements. The evolution of firearm design tools has driven the creation of new methodologies of defining user requirements, which should take into account the speed of their establishment in order not to affect firearm development times. This chapter addresses different tools to establish such requirements and reach the objectives in every phase and stage of the firearm design.

## 3.2 REQUIREMENTS

The success of a firearm is the acceptance of the product in the market, which can be an incremental or radical innovation when the users recognize the product. Incremental innovation is when users assimilate the technology quickly due to known operating paradigms. Radical innovation has slower assimilation due to the change in operating paradigms. Once users assimilate and adopt it, this innovation forms a new operating paradigm; an example of radical innovation is replacing the revolver with the pistol.

The acceptance of new products is related to the establishment of appropriate requirements, which allow understanding user needs through an exhaustive analysis of requirements. Various tools are used to establish requirements, such as the Kano model, storytelling, mapping journey, and affinity diagram.

With the advancement of technology, the classification of requirements depends on the design area. In the case of weapon design, the requirements are determined based on the type of use of the weapon. In general, requirements are established based on the user needs and the required performance of the weapon. For this, the design team proposes improvements in various areas, such as weight reduction, material selection, coatings, manufacturing processes, manufacturing cost reduction, ease of manufacture, interchangeability, refurbishment, maintenance, compatibility between accessories (sights, bipod, grenade launchers, lamps), customization, styling, security, ballistic parameters (velocity, momentum, trigger pull force, headspace, barrel dimensions), and part control.

The satisfaction of user requirements is fundamental; for this reason, understanding the technology tendency according to the requirements of every market

DOI: 10.1201/9781003196808-3

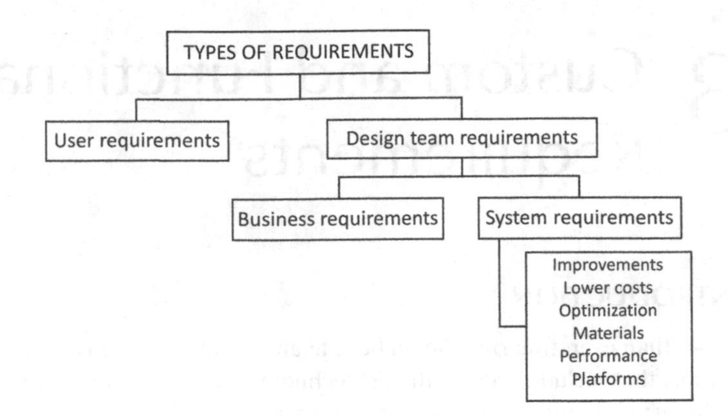


**FIGURE 3.1** Types of requirements for firearm design.

segment allows the product to be accepted, assimilated, and adopted. Overall, the types of requirements are divided into user and design team requirements; the latter are divided into business and system requirements. The business requirements are dedicated to linkage the market opportunity with the firearm design [1]. The system requirements are divided into improvements, lower costs, optimization, materials, performance, and platforms, as shown in Figure 3.1.

## 3.3 REQUIREMENTS TO REACH READINESS OF A SYSTEM

There are several requirements by phase of readiness during firearm development, such as the technology readiness level (TRL), manufacturing readiness level (MRL), and investment readiness level (IRL) [2]. Figure 3.2 shows the requirements to reach the readiness of technology, manufacturing, and investment levels as well as how they are related to each other according to the research, development, and deployment stages.

The research stage has three sections, the first and second sections include requirements to reach TRL1 and TRL2, respectively, and the third section includes requirements to reach TRL3, which is at the laboratory level. At this level, the operating evaluation is done on a test bench using controlled parameters [3]. Laboratory level includes requirements to TRL3 and MRL1 of the research stage, as well as requirements TRL4, MRL2, and IRL1 of the development stage.

The demonstration level is where the conditions that emulate the real operating environment are implemented. The prototype word is used to describe the system; this level includes TRL5, MRL3, and IRL2, corresponding to the development stage.

The industrial demonstration is where the firearm requires operating in a real environment. This level includes the requirement to reach the TRL6, MRL4, and IRL3 of the development stage and the requirements to reach the TRL7, MRL5. In addition, IRL4, corresponds to the deployment stage.

The deployment stage includes 11 sections. The first section was described above. The second section includes the MRL6 level, which in turn corresponds

| Stage | REQUIREMENTS TO REACH TRL/MRL/IRL | TRL/MRL/IRL | LEVEL |
|---|---|---|---|
| RESEARCH | User requirements, References of principles from basic research, Benchmarking, and Potential market segments<br>External constraints and Ideation | TRL1 | |
| RESEARCH | Internal constraints, Formulation of solution, Design concept, Value offer, and Capacity from design team | TRL2 | |
| RESEARCH | Operating principle, Experimental physical model and Feasibility evaluation<br>Initial materials and Manufacturing implications | TRL3<br>MRL1 | LABORATORY |
| DEVELOPMENT | Experimental physical model pilot and its effectiveness evaluation. Usability testing and Risk management plan<br>Manufacturing strategies to users (military-policies) needs<br>Business model | TRL4<br>MRL2<br>IRL1 | |
| DEVELOPMENT | Prototype with components of high reliability and Experimental lot<br>New manufacturing process developed with limited functionality<br>Market analysis, Market size and Competitive analysis | TRL5<br>MRL3<br>IRL2 | DEMONSTRATION Conditions that emulate real environment |
| DEVELOPMENT | Prototype by complete firearm testing in indoor and outdoor shooting<br>Manufacturing technologies, Producibility assessments, Facilities and Skills required<br>Problem-solution validation | TRL6<br>MRL4<br>IRL3 | INDUSTRIAL DEMONSTRATION Real environment |
| DEPLOYMENT | Prototype by complete system validation in outdoor shooting and movements objects, underwater, sand, snow and mud<br>Manufacturing refined and integrated with risk management plan; process are still in development<br>Materials and tooling defined to produce prototype components<br>VMP low fidelity and Testing with users and potential clients | TRL7<br>MRL5<br>IRL4 | |
| DEPLOYMENT | Capability to produce the prototype following process sheet using tooling and materials, Refining the process | MRL6 | PRODUCTION BY SYSTEM |
| DEPLOYMENT | Validate product and Market fit | IRL5 | |
| DEPLOYMENT | Validate revenue model | IRL6 | |
| DEPLOYMENT | Certified product with user manual, Technical support and maintenance organization<br>Pilot line and final materials, Production improvements, and Risk assessments are underway<br>VMP high fidelity | TRL8<br>MRL7<br>IRL7 | NATO AND COMMERCIAL CLEARANCES |
| DEPLOYMENT | Released product<br>Outlining new versions and scaling | TRL9 | |
| DEPLOYMENT | Initial production and minimal system changes<br>Optimization of manufacturing process | MRL8 | |
| DEPLOYMENT | Continuous production and Process statistical control | MRL9 | |
| DEPLOYMENT | Manufacturing process is the required based on quality and costs | MRL10 | FULL-RATE PRODUCTION |
| DEPLOYMENT | Validate value delivery | IRL8 | |
| DEPLOYMENT | Metrics to growth | IRL9 | MARKET EXPANSION |

**FIGURE 3.2** Requirements to reach TRL, MRL, and IRL by stage and level.

to the level of production by the system. The third and fourth sections include the IRL5 and IRL6 levels, respectively. The fifth section includes TRL8, MRL7, and IRL7, which in turn correspond to NATO and commercial clearances levels. The sixth, seventh, eighth, and tenth sections include TRL9, MRL8, MRL9, and IRL8, respectively. The ninth section includes MRL10, which corresponds to the level of full-rate production. Finally, the eleventh section includes IRL9, which corresponds to the level of market expansion [4].

## 3.4 REQUIREMENT IDENTIFICATION

The design gives solutions to opportunities, needs, or problems, depending on the design team approach. The necessity requires fast solutions; a problem is solved in a long time due to the planning of the research project; so, the identification of

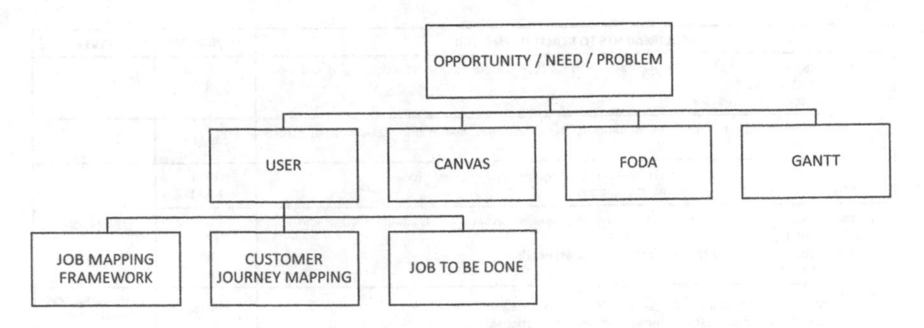


**FIGURE 3.3** Interaction of tools to get the user and project requirements.

market opportunity is the main motivation to define the requirements. Figure 3.3 shows the interaction of tools to get the user requirements as a job mapping framework, customer journey mapping, and job to be done, as well as the requirements to project as business model CANVAS, FODA, and GANTT.

The job mapping framework is a visual representation that describes the job of customers; for instance, in the vigilance job are depicted the specific steps, which allow the design team to capture the customer requirements throughout the process [5]. The customer journey mapping (CJM) is a group of several schemes describing the interaction of the client to purchase flow; in some stores, there are steps to achieve the client to purchase his first firearm, ammunition, fixing, and replacement for new models [6]. This CJM helps to identify the expectation of customers in the acquisition of a firearm and the team design considers these requirements to new designs [7]. The job to be done is a framework for customer requirements; the perspective of this methodology is to understand the customer motivations to use a product [8]. The client buys a product for the task it solves but not for the product itself. So, a client thinks in a firearm to perform better his job, sport, or training. Maybe the clients are looking for esteem, status, comfort, and values that a firearm represents. Clayton Christensen spread the concept of Jobs To Be Done (JTBD) in his book *The Innovator's Dilemma* [9]. The steps to implement JTBD are as follows: (1) define the market segments in the function of the job, (2) list of customer expectations, (3) quantify the attention of competency (analyze the actual competitors) to job, (4) list hidden opportunities, and (5) list requirements for the new firearm.

The functional architecture focuses specifically on operating firearms. For this, the interaction of different aspects is taken into account (Figure 3.4). The evaluation of needs includes the function of a firearm, which refers to the operating modes and security aspects. The external constraints refer to the conditions of use, anthropometric aspects, and accessories. The sequential analysis in the system functioning indicates the diagrams to moving parts in the motion phases and the times for every function phase. The reference product is distinguished based on the market segment, which defines the competence and identifies the attributes of our product. The design criteria point out the headspace, recoil, momentum, standard, and others described in Chapter 2.

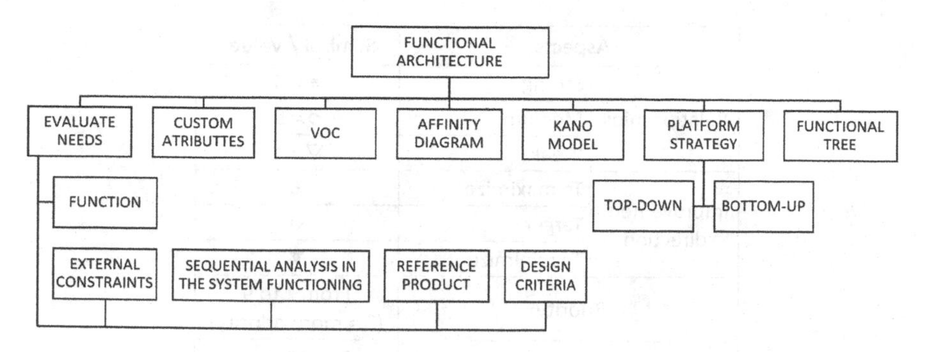


**FIGURE 3.4** Interaction of topics to planning the functional architecture.

The custom attributes are the dimensional parameters and tactical issues. The voice of the customer (VOC) corresponds to the responses by questionnaires realized to focal groups [9]. The affinity diagram is the list of custom preferences by market segment. The Kano model is an approach to prioritizing five types of requirements: performance, basic, excitement, indifferent, and reverse [10]. Platform strategy is the business strategy with multiple products based on a modular architecture; a platform is designed either top-down, from the platform to several products, or bottom-up, from a product by scaling to the platform [11].

A functional tree is a hierarchical structure; it considers functional requirement (FR) and design parameter (DP). FR is a function with a required level of performance. Usually, the user requirements are continuously increasing and then the technical evolution occurs. When the FR is satisfied and it does not change, the increasing technical property becomes slow; for instance, the magazine capacity keeps the same level because users only want a certain capacity. DP is referred to as design component (DC), which is the method to accomplish a given FR. Thus, DP is a certain parameter of a specific component. Functional tree includes the following nine steps: extract functions from existing engineering systems, categorize functions (function module), change functions to function requirements, arrange the FR, expand function tree, analyze function performance, select scenario for functional tree, build new functional tree, and evaluate the new design [12].

## 3.5 DETERMINATION OF FR AND ASSESSMENT OF ITS DIFFICULTY

The QFD usually includes four houses of quality (HOQ), which concerns the client, product design, process planning, and control [13]. In this chapter five HOQs are developed, namely HOQ1, HOQ2, HOQ3, HOQ4, and HOQ5.

The QFD is a diagram constituted by several sections and elements organized to visualize requirements and solutions compared to competitors. Figure 3.5 shows aspects, symbols, and/or values of HOQ. The aspects include relationships, improvement direction, priority, and difficulty. The relationships are divided into three categories: strong, medium, and weak. Improvement direction considers if

| Aspects | | Symbol / Values |
|---|---|---|
| Relationships | Strong | ● = 5 |
| | Medium | ○ = 3 |
| | Weak | ∇ = 1 |
| Improvement direction | To maximize | ▲ |
| | Target | ◇ |
| | To minimize | ▼ |
| Priority | | From 1 to 9 (9 = more priority) |
| Difficulty | | From 1 to 9 (9 = more difficulty) |

**FIGURE 3.5** Aspects, symbols, and values of HOQ.

it maximizes, targets, and minimizes. The values are multiplied to calculate the technical importance rating (TIR).

The HOQ1 is where the determination of FR is done from VOC, which is a list of requirements. Figure 3.6 shows an example of HOQ1 to design a firearm platform that includes two products. A list of eight items numbering the VOC includes its priority. List from 1 to 3 is a group to product 1, list from 6 to 8 is a group to product 2, and items 4 and 5 are requirements shared by both products.

Consequently, for instance, the lightweight requirement is solved by an FR of material property; then, five FRs solve eight requirements. Every FR is conditioned to improve the direction, being necessary to diminish any material property such as density to lighten the firearm and reach the improvement target by essential parts, modular architecture, and force regulator. The most useful life is indicated with the improvement direction. Once analyzed the relationship between VOC and FR, the TIR is obtained. According to the priority and relationship, the most TIR is a material property, and the less is useful life. The target of function in the FR must be numerically expressed. This HOQ1 considers the difficulty (D), which is analyzed by the design team according to its installed capacity; so, the new priority (NP) is obtained by dividing the TIR between D; NP is an index that includes the difficulty value and helps to outline the design strategy.

HOQ1 does not show the competitive customer and technical assessments because benchmarking graphics would not allow an in-depth appreciation of these aspects.

The HOQ2 assesses the FR difficulty. It could result in two ways: (1) where the greater difficulty level makes it necessary to feedback the HOQ1, and (2) where the difficulty is solved with an action planned by the design team. The latter is the case presented in Figure 3.7. The NP is multiplied per relationship values. Infrastructure, training, technology transfer, collaboration agreement, and test bench are the actions to solve the difficulties analyzed by the design team. The importance of each column is the result of the sum of multiplications between NP

| | | Improvement | ▼ | ◇ | ◇ | ◇ | ▲ | Customer competitive assessment |
|---|---|---|---|---|---|---|---|---|
| | Priority | FR / VOC | Material property | Essential parts | Modular architecture | Force regulator | Useful life | |
| 1 | 8 | Lightweight firearm | • | ○ | ▽ | ○ | ▽ | |
| 2 | 9 | 1 million of units by year | ▽ | • | ○ | ▽ | ▽ | |
| 3 | 5 | Model for women | ▽ | ▽ | • | ○ | ▽ | |
| 4 | 6 | Adjustable trigger pull-force | ▽ | ▽ | ▽ | • | ▽ | |
| 5 | 4 | Engraving name | ▽ | ▽ | • | ▽ | ▽ | |
| 6 | 7 | Cheap magazine | • | ▽ | ▽ | ▽ | ▽ | |
| 7 | 9 | 2 million of units by year | ▽ | • | ○ | ○ | ▽ | |
| 8 | 8 | Durable barrel | • | ▽ | ▽ | ▽ | • | |
| | | Target | 70% of the components are made of polymeric material. Placement of coating material on the barrel inside. | DFA > 60% | C > 60% | Force from 20 and 40 N | Acceptable operating conditions > 12,000 shots | |
| | | Technical Importance Rating (TIR) | 148 | 144 | 128 | 124 | 88 | |
| | | Difficulty (D) | 9 | 6 | 4 | 8 | 5 | |
| | | New priority (NP)= TIR/D | 16.4 | 24 | 32 | 15.5 | 16.4 | |
| Technical competitive assessment | | | | | | | | |

**FIGURE 3.6** HOQ1 to obtain FR from VOC.

| | | Improvement | ▲ | ▲ | ◇ | ◇ | ◇ |
|---|---|---|---|---|---|---|---|
| | NP | Action / Difficulty | Infrastructure | Training | Technology transfer | Collaboration agreement | Test bench |
| 1 | 16.4 | 70% of the components are made of polymeric material. Placement of coating material on the barrel inside. | • | ○ | ▽ | ▽ | ▽ |
| 2 | 24 | DFA > 60% | ▽ | • | ▽ | ▽ | ▽ |
| 3 | 32 | C > 60% | ▽ | • | ▽ | ○ | ▽ |
| 4 | 15.5 | Force from 20 to 40 N | ○ | ○ | • | • | • |
| 5 | 16.4 | Acceptable operational conditions > 12,000 shots | ▽ | ▽ | • | ▽ | • |
| | | Target | To instrument the ballistic laboratory | Specialists in sintering process | Agreement with patent owner | Agreement with factories and universities | Testing cookoff and lifecycle of barrel |
| | | Importance | 200.9 | 392.1 | 231.9 | 230.3 | 231.9 |

**FIGURE 3.7** HOQ2 to the assessment of FR difficulty.

and every action. In the target section, it is possible to observe in detail the action to do by the design team. The action is related to improvement direction, so the action is reading as more infrastructure, more training, technology transfer, collaboration agreement, and test bench.

## 3.6 DETERMINATION OF DPs USING AXIOMATIC DESIGN

Once the FRs and difficulty analysis are obtained, the DPs are attained from FRs; the axiomatic design is used. An axiom is a fundamental truth for which there are no counterexamples or exceptions. It cannot be derived from other laws of nature or principles [14]. It is important to mention that the implementation of axiomatic design is done before the HOQ3 because it is here where the DPs regarding FRs are proposed and refined.

The axiomatic design includes two axioms: the independence axiom and the information axiom. The independence axiom analyzes the effects of the relationships FR among DP, as presented in Equation 2.1 of Chapter 2. During the relationships, collateral effects appear, which result in three types of the design matrix. The first matrix is referred to as coupled design when the design matrix is completely filled with collateral effects. The second matrix is referred to as decoupled design when the design matrix is semi-filled with collateral effects. The third matrix is the ideal matrix, which is referred to as uncoupled design, meaning that the collateral effects are null or almost null and each DP accomplishes every FR. It should be noted that although the uncoupled design is ideal, most designs in a real environment reach the decoupled design after refining the design. Usually, the first DP results in a coupled design and, once the design parameters are refined or replanned, the design matrix represents a decoupled or uncoupled design. Overall, the design is disguised because the coupled design is a poor design, decoupled design is a good design, and uncoupled design is the better design [15]. Figure 3.8 shows the kinds of design matrices according to collateral effects.

The FRs are taken to propose the corresponding DPs, which have been refined to reach an uncoupled design. Figure 3.9 shows the independence axiom equation and the relationships between FR and DP. As can be seen, insert molding is related to components made of polymers and 10% more essential parts. DFA>60% is related to insert molding, 10% more essential parts, and 15% more common parts. C>60% is only related to the spring-elastomer mechanism.

Coupled design Decoupled design Uncoupled design

$$\begin{bmatrix} X & X & X & X & X \\ X & X & X & X & X \\ X & X & X & X & X \\ X & X & X & X & X \\ X & X & X & X & X \end{bmatrix} \quad \begin{bmatrix} X & X & X & X & X \\ 0 & X & X & X & X \\ 0 & 0 & X & X & X \\ 0 & 0 & 0 & X & X \\ 0 & 0 & 0 & 0 & X \end{bmatrix} \quad \begin{bmatrix} X & 0 & 0 & 0 & 0 \\ 0 & X & 0 & 0 & 0 \\ 0 & 0 & X & 0 & 0 \\ 0 & 0 & 0 & X & 0 \\ 0 & 0 & 0 & 0 & X \end{bmatrix}$$

**FIGURE 3.8** Types of design from the design matrix.

$$\begin{array}{ccc} \text{FR} & \text{MATRIX} & \text{DP} \\ \left\{\begin{array}{c} \text{Components of polymers} \\ \text{DFA} > 60\% \\ \text{C} > 60\% \\ \text{Force from 20 to 40 N} \\ \text{Acceptable operating} > 12{,}000 \text{ shots} \end{array}\right\} = & \begin{bmatrix} X & X & 0 & 0 & 0 \\ X & X & X & 0 & 0 \\ 0 & 0 & X & 0 & 0 \\ 0 & 0 & 0 & X & 0 \\ 0 & 0 & 0 & X & X \end{bmatrix} & \left\{\begin{array}{c} \textit{Insert molding} \\ \textit{10\% more of essential parts} \\ \textit{15\% more of common parts} \\ \textit{Spring – elastomer mechanism} \\ \textit{Internal barrel covering} \end{array}\right\} \\ & \text{Uncoupled design} & \end{array}$$

**FIGURE 3.9** Design matrix of the independence axiom.

A force of 20–40 N is only related to the spring elastomer mechanism. Acceptable operating >12,000 shots is related to the spring-elastomer mechanism and internal barrel covering.

The aim of the information axiom is to minimize the information content of the design; this is recommended to simplify the design. Information content, I, is defined as the probability of satisfying a given FR, which is expressed in Equation 3.1 [16].

$$I = \log_2 \frac{1}{P} = \frac{\text{System range}}{\text{Common range}} \tag{3.1}$$

where $I$ is the information expressed in bits and $P$ is the probability of success. The system range is the range of values of FR, DP, or PV according to the application of the information axiomatic equation; the common range is the common area (intersection) between system range and common range. Design range is the range of DP values that will satisfy the FR.

Among DPs that satisfy FR, the DP with the minimum information content has the highest probability of success; this axiom provides a quantitative way to reach the optimum from DP. Figure 3.10 shows the graph realized with the data obtained from FR for firearm design. The elements shown in the graph are the design range from 20 N to 40 N and the system range from 25 N to 60 N, which are the measures in competitive firearms. Bias is a measure of the "exactitude" of the measurement and representation system to the systematic error of the system. The contribution is the total error due to the combined effects of all sources of variation, known or unknown [17]; bias, in this case, is the value between the target of FR (30 N) and the mean value of the system range (42.5 N). Using Equation 3.1, the information content is calculated as:

$$I = \log_2 \frac{(60 - 25)}{(40 - 25)}$$

$$I = \log_2 \frac{35}{15}$$

$$I = 1.222 \text{ bits}$$

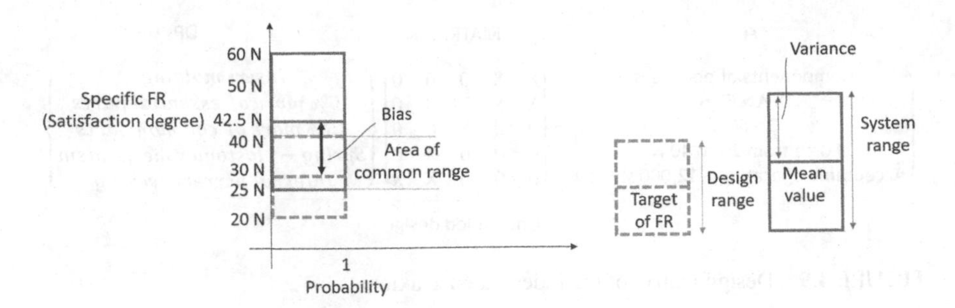


FIGURE 3.10 Graph of the information axiom.

## 3.7 RELATIONSHIP DP TO FR

DPs were obtained in Section 3.5, which are analyzed in the HOQ3 (Figure 3.11). The HOQ3 is where the FRs are related to DPs, adjusting the priority after having analyzed the difficulty and refining the DPs using the axiomatic design.

The result of most importance is the mechanism and that of the minor importance is the insert molding. The insert molding is related to the use of polyamide in components of moving parts and encasement and the use of Makrolon® in the magazine. In addition, the target design in the physical domain indicates 10% more essential parts, 15% more common parts, spring-elastomer mechanism, and hard chrome plating of gun barrels. The target DPs are the components necessary to solve the FRs.

The DPs are related to improvement direction. The DP is the solution to reach insert molding to at least 70% of firearm components, 10% more of essential parts, 15% more of common parts, to reach the mechanism with regulated force from 20 N to 40 N, and to reach an internal barrel covering using hard chrome for reaching an acceptable firearm performance. Figure 3.11 shows that the priority is multiplied per relationships values: insert molding, essential parts, common parts, mechanism, and cover inside the barrel.

## 3.8 RELATIONSHIP CRITICAL DESIGN PARAMETER (CDP) TO TEST BENCH FEATURE (TBF)

A critical design parameter (CDP) is a requirement that must be driven with high accuracy and evaluated on its behavior repeatability to verify the permanence of performance in an operating range. Test bench features (TBFs) are conditions to satisfy each CDP, which are implemented in the instrumentation of the test bench, experimental protocol, and certification of the event. Figure 3.12 shows the analysis of relationships among CDPs and TBFs.

The HOQ4 is related to each CDP to every TBF, considering the priority of CDP defined by the design team. The priority is multiplied per relationships values. Devices to avoid deformation of plastic parts guarantee the dimensional parameters to correct firearm operating, allowing interchangeability among

| | Priority | Improvement / DP / FR | Insert molding | Essential parts | Common parts | Mechanism | Internal barrel covering |
|---|---|---|---|---|---|---|---|
| | | Improvement | ◇ | ▲ | ▲ | ◇ | ◇ |
| 1 | 7 | 70% of components made of polymer | ● | ○ | ○ | ○ | ▽ |
| 2 | 6 | DFA > 60% | ▽ | ● | ○ | ○ | ▽ |
| 3 | 6 | C > 60% | ▽ | ○ | ● | ○ | ▽ |
| 4 | 8 | Force from 20 to 40 N | ▽ | ▽ | ▽ | ● | ▽ |
| 5 | 9 | Acceptable operating conditions > 12,000 shots | ▽ | ▽ | ▽ | ● | ● |
| | | Target | Insert molding using polyamide and Makrolon® | 10% more of essential parts | 15% more of common parts | Spring-elastomer mechanism | Hard chrome plating of gun barrels |
| | | Importance | 64 | 86 | 86 | 142 | 72 |

**FIGURE 3.11** HOQ3 of design parameters from functional requirements.

| | Priority | Improvement / TBF / CDP | Devices to avoid deformation of plastic parts | Shooting test | Ultrahigh- speed sensors | Pull-force verification | Data acquisition |
|---|---|---|---|---|---|---|---|
| | | Improvement | ▲ | ◇ | ◇ | ◇ | ◇ |
| 1 | 5 | 70% of components made of polymer material | ● | ○ | ▽ | ▽ | ▽ |
| 2 | 8 | Cook-off | ▽ | ● | ○ | ▽ | ○ |
| 3 | 9 | Operating firearm | ▽ | ○ | ● | ○ | ○ |
| 4 | 7 | Force from 20 to 40 N | ▽ | ○ | ▽ | ● | ○ |
| 5 | 6 | Acceptable operational conditions > 12,000 shots | ▽ | ○ | ○ | ○ | ● |
| | | Target | Interchangeability | Conditions in breech | Repeatability of operating sequence | Reliability range | Durability range |
| | | Importance | 55 | 121 | 99 | 93 | 107 |

**FIGURE 3.12** HOQ4 of test bench features from critical design parameters.

| | | Improvement | ◇ | ◇ | ◇ | ◇ | ◇ |
|---|---|---|---|---|---|---|---|
| | Priority | PV / DP | Moving parts and covers. Magazine | Algorithm to essential parts | Market segments | Material to spring and elastomer | Manufacturing process of hard chrome |
| 1 | 9 | Insert molding using polyamide and Makrolon® | • | ○ | ▽ | ▽ | • |
| 2 | 6 | 10% more of essential parts | ▽ | • | ○ | ▽ | ○ |
| 3 | 4 | 15% more of common parts | ▽ | ○ | • | ○ | ○ |
| 4 | 8 | Spring-elastomer mechanism | • | ▽ | ▽ | • | • |
| 5 | 5 | Hard chrome plating of gun barrels | ▽ | ▽ | ▽ | ▽ | • |
| | | **Target** | (PA 66) Vestamid® and Astamid®. Makrolon® (3107) | Commercial parts and manufacturing parts | 2 Product families, 3 products | Material to spring and elastomer | Process sheet to barrel durability |
| | | Importance | 100 | 82 | 60 | 72 | 140 |

**FIGURE 3.13** HOQ5 of process variables from design parameters.

components and groups. The cook-off is evaluated through exhaustive shot testing to reach the repeatability to define the safe operating range and assure the permissible conditions in the breech. Ultrahigh-speed sensors and transparent covers allow verifying the operating of the firearm, reaching the repeatability of the operating sequence. Pull-force verification satisfies a force from 20 N to 40 N, reaching the reliability range in the pressure of the trigger. Data acquisition satisfies the acceptable operating conditions >12,000 shots, reaching the expected durability range.

The multiplication result is the importance of setting up the elements in the test bench or several test benches and planning a schedule to converge the times to testing with a test bench supplier or the same design team.

## 3.9 RELATIONSHIP PROCESS VARIABLE (PV) TO DP

Process variable (PV) is a process value or parameter measured in a particular part of the manufacturing or testing process in the development stage. This PV is monitored and controlled; for instance, the temperature in the breech is monitored and controlled to reach the desired temperature known as set point [18].

In the HOQ5, the PVs are established in a general way, which includes specific PVs according to each DP requirement. Figure 3.13 shows the analysis of relationships among DPs and PVs. The HOQ5 is related to each DP to every PV, considering the priority of DP defined by the design and manufacturing teams.

The priority is multiplied per relationship values. The injection parameters to make the moving parts and covers of polyamide and magazine of Makrolon® must be defined, specifying in the target the materials to satisfy the insert molding. Using the DFA algorithm to define the essential parts, the required 10% more is satisfied, reaching the classification in the bill of materials (BOM) of the commercial and manufacturing parts. Market segments satisfy the 15% more required to modular architecture according to the product platform. Material for spring and elastomer satisfies the mechanism to regulate the pull force. The manufacturing process of the barrel satisfies the hard chrome plating of the gun barrel, reaching the durability required.

The multiplication result is the importance of setting up the PVs in the manufacturing process and planning a schedule to converge the times to manufacture the parts and acquire the commercial parts. The improvement of this HOQ5 has no direction because the PVs are aiming at a specific measure.

## 3.10 DETERMINATION OF INSTRUMENTS TO TECHNOLOGY TRANSFER

Technology transfer is the transmission of rights implemented by the intellectual property and author rights or copyrights. As mentioned, the requirements for designing a firearm are distinct depending on the approach of the design team. For this, some designers focus on the design, starting from intellectual property. The intellectual property registers documents to certify the idea owner in invention, design, and utility patents; names change according to the country. These patents are the instruments to carry out the technology transfer based on a strategy to reach expected revenues.

Figure 3.14 shows the analysis of relationships between technology transfer and patent valuation. In the HOQ6, the patent valuation is established in a general way, which includes specific patents according to each requirement of technology transfer. HOQ6 is related to each patent valuation to every technology transfer, considering the priority of technology transfer defined by design, manufacturing, and commercial teams. The priority is multiplied per relationships values. Trademark for model and product family satisfies the transferable product portfolio [19]. Invention patent by each design in the firearm, utility, and invention allows transferring the operating principle and the requirements of styling and customization [20]. Distributor and manufacturing licensing allow planning business models and guiding design requirements for current and future product generations [21].

## 3.11 CLOSING REMARKS AND PERSPECTIVES

The requirements are a group of conditions to guide the project, which depends on the approach of the design team and all teams involved, such as manufacturing and commercial teams. The research, development, and deployment stages have

| | | Improvement | ◇ | ◇ | ◇ | ▲ | ◇ |
|---|---|---|---|---|---|---|---|
| | Priority | Patent valuation<br>Technology transfer | Trademark | Invention patent | Design patent | Distributor | Manufacturing license |
| 1 | 6 | Product portfolio | ● | ○ | ○ | ○ | ○ |
| 2 | 3 | Operating principle | ▽ | ● | ▽ | ▽ | ○ |
| 3 | 7 | Styling and customization | ▽ | ▽ | ● | ○ | ▽ |
| 4 | 8 | Sell | ○ | ○ | ○ | ● | ▽ |
| 5 | 9 | Manufacturing | ○ | ○ | ○ | ▽ | ● |
| | | **Target** | Platform and product family | Radical innovation | Incremental innovation | Federal firearms license | Federal firearms license |
| | | Importance | 91 | 91 | 107 | 91 | 87 |

**FIGURE 3.14** HOQ6 to technology transfer of the firearm.

requirements according to technology, manufacturing, and investment readiness. The design methodologies such as the axiomatic design, QFD, TRL, MRL, and IRL help establish the requirements to reach the objectives in every phase and stage. The evolution of technology in the firearm design impulses the creation of new ways to define the user requirements every time with more feedback, experimental physical model, prototypes, and increase the proposals in the market.

Current markets demand the design of firearms that include state-of-the-art technological levels to ensure efficient operation in the environment. Speed in establishing design requirements is critical and affects firearm development times. Methodologies such as lean design, lean manufacturing, design thinking, and agile design focus on the speed of design while maintaining quality and safety in the performance of each system.

## REFERENCES


1. Lehtola, L., M. Kauppinen, and S. Kujala. Linking the business view to requirements engineering: Long-term product planning by roadmapping. *In 13th IEEE International Conference on Requirements Engineering (RE'05)*, Paris, France, 2005, IEEE.
2. Fernandez, J.A., Contextual role of TRLs and MRLs in technology management. Sandia National Lab, California, SAND2010-7595, 2010.
3. Sauser, B., et al. From TRL to SRL: The concept of systems readiness levels. In *Conference on Systems Engineering Research*, Los Angeles, CA, 2006, Citeseer.

4. Kim, H.W., S. Woo, and B.K. Jang, A study on readiness assessment for the acquisition of high quality weapon system. *Journal of the Korean Society for Quality Management*, 2013. **41**(3): pp. 395–404.
5. Bettencourt, L.A. and A.W. Ulwick, The customer-centered innovation map. *Harvard Business Review*, 2008. **86**(5): p. 109.
6. Crosier, A. and A. Handford, Customer journey mapping as an advocacy tool for disabled people: A case study. *Social Marketing Quarterly*, 2012. **18**(1): pp. 67–76.
7. Marquez, J.J., A. Downey, and R. Clement, Walking a mile in the user's shoes: Customer journey mapping as a method to understanding the user experience. *Internet Reference Services Quarterly*, 2015. **20**(3–4): pp. 135–150.
8. Christensen, C.M., et al., Finding the right job for your product. *MIT Sloan Management Review*, 2007. **48**(3): p. 38.
9. Christensen, C.M., *The Innovator's Dilemma: The Revolutionary Book That Will Change the Way You Do Business*. 2003, Harper Business Essentials: New York.
10. Cooper, R.G. and A. Dreher, Voice-of-customer methods. *Marketing Management*, 2010. **19**(4): pp. 38–43.
11. Xu, Q., et al., An analytical Kano model for customer need analysis. *Design Studies*, 2009. **30**(1): pp. 87–110.
12. Simpson, T.W., J.R. Maier, and F. Mistree, Product platform design: Method and application. *Research in Engineering Design*, 2001. **13**(1): pp. 2–22.
13. Romli, F.I., A.S.M. Rafie, and S. Wiriadidjaja. Conceptual product design methodology through functional analysis. *Advanced Materials Research*, 2014. **834–836**: pp. 1728–1731.
14. Moubachir, Y. and D. Bouami, A new approach for the transition between QFD phases. *Procedia Cirp*, 2015. **26**: pp. 82–86.
15. Kulak, O., S. Cebi, and C. Kahraman, Applications of axiomatic design principles: A literature review. *Expert Systems with Applications*, 2010. **37**(9): pp. 6705–6717.
16. Suh, N.P., Axiomatic design theory for systems. *Research in Engineering Design*, 1998. **10**(4): pp. 189–209.
17. Helander, M.G. and L. Lin, Axiomatic design in ergonomics and an extension of the information axiom. *Journal of Engineering Design*, 2002. **13**(4): pp. 321–339.
18. Critchley, L.A. and J.A. Critchley, A meta-analysis of studies using bias and precision statistics to compare cardiac output measurement techniques. *Journal of Clinical Monitoring and Computing*, 1999. **15**(2): pp. 85–91.
19. Hanumaiah, N., B. Ravi, and N. Mukherjee, Rapid hard tooling process selection using QFD-AHP methodology. *Journal of Manufacturing Technology Management*, 2006. **17**(3): pp. 332–350.
20. Cooper, R.G., S.J. Edgett, and E.J. Kleinschmidt, New product portfolio management: Practices and performance. *Journal of Product Innovation Management: An International Publication of the Product Development & Management Association*, 1999. **16**(4): pp. 333–351.
21. Liu, Y., et al. Integrating requirements analysis and design around strategy for designing around patents. In *2011 IEEE 2nd International Conference on Computing, Control and Industrial Engineering*, Wuhan, China, 2011, IEEE.

# 4 CAD Modeling and CAE Simulation

## 4.1 INTRODUCTION

The speed in the development of a weapon largely depends on the synergy between the design experience and the interaction of technological tools. The use of these tools represents the scientific rigor with which the design is executed. For this, it is essential to have a team trained to continuously adapt to technological changes in the product life cycle management. This design team must overcome obstacles such as having experience dealing with outdated software. A competitor acquires the modeling and simulation tool and modifies it; the software is no longer compatible with some type of hardware or requires add-ons and updates, besides that its mode of operation and user interface (UI) are different.

The challenges posed by the use of technologies for modeling and simulation have to do with the constant training of the design team and the collaboration with engineering consulting companies and universities as part of the concurrent development. Additionally, firearm developers sometimes prefer to use traditional tools rather than invest in technological tools that could represent a hiatus in product development due to training or hiring skilled personnel. This chapter addresses the implementation of CAD and CAE tools in firearm design, considering the evaluation environment and its performance validation. Several case studies are presented in this chapter to exemplify the potential of different simulation software.

## 4.2 MODELING AND SIMULATION

Computational modeling and simulation are one of the fastest-growing areas in engineering. This is used in a wide range of areas to predict or explain the behavior of a product or a system. It is also used for gaming entertainment, which is probably by far the largest market. The fundamental aspects of using these tools are predicting product performance through simulation, already at the design stage with a higher degree of precision and fidelity.

So now, it has the potential to reduce product development time drastically and is a fundamental part of what is called the base model design. In recent years, there has been a rapid development in system simulation, primarily due to the development of hardware, so no other area of technology has had such

DOI: 10.1201/9781003196808-4

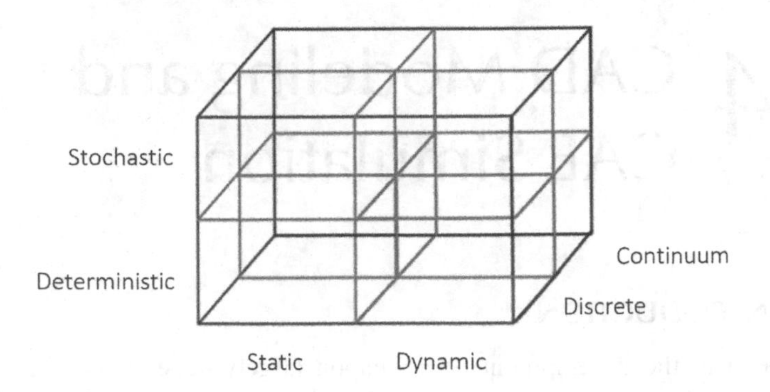


**FIGURE 4.1** System classification to apply modeling and simulation.

a drastic sustained increase for decades. Likewise, advances in software have also been significant, such as in the case of compilers and algorithm development [1].

One consequence of this is that the behavior of much larger and more complex systems can be analyzed than before, and this advantage has prompted the development of a new generation of simulation software with graphical interfaces much easier to use, which makes it more effective to work with heavier systems. Currently, larger systems that integrate several physical domains are simulated in multi-domain software [2].

Modeling and simulation are a tool that provides support for the planning, design, and evaluation of systems. Its importance will continue to grow at a remarkable rate; this growth is also a consequence of the increasing availability of important computing resources and the human capacity that allows taking advantage of this computational power. However, any practical use of a tool, especially a multifaceted tool such as modeling and simulation, implies an increasingly rapid learning curve.

In the modeling phase, the system to be analyzed is classified, which determines the characteristics of the required simulation. Depending on the type of model, it can be deterministic when its behavior is known or stochastic when its behavior can be measured and approximated. According to the variability of time, a system can be static when time is not a variable in the model; further, it can be a continuum dynamical system when it evolves over time or a discrete dynamical system when it occurs in time steps [3]. According to the variables involved in the mathematical model, the system classification is visualized based on the diagram shown in Figure 4.1.

The software for modeling and simulation is included in the 3D computer graphics software, and due to its evolution and the new requirements of immersion, interaction, and imagination, today it is related to the virtual reality (VR) software. This software is classified as shown in Figure 4.2, where the usability testing and training focus on the user model and the firearm and test bench refer to the environment model.

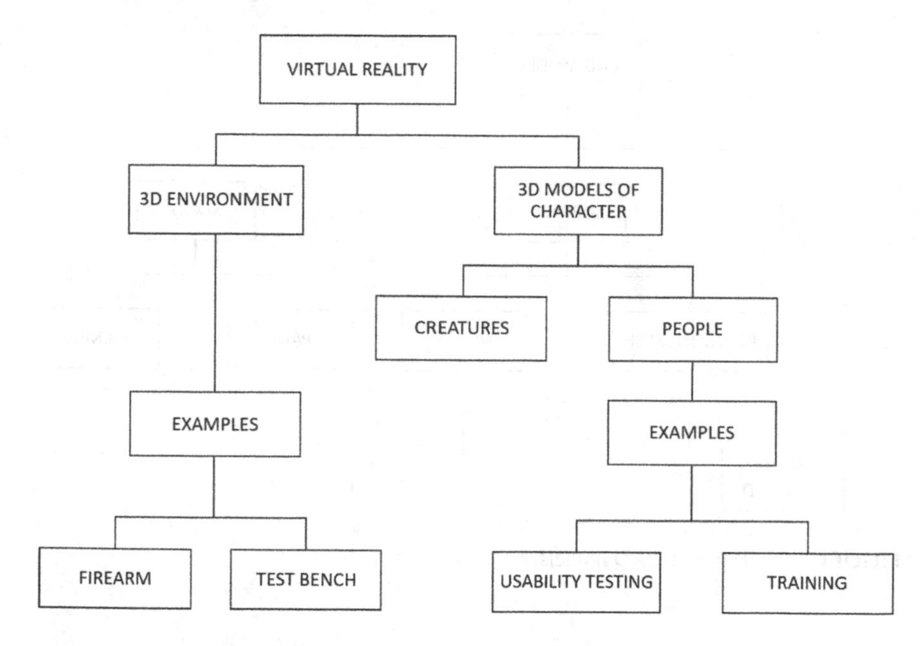


**FIGURE 4.2** Virtual reality software to firearm design.

## 4.3 CAD MODELING

A model constitutes a representation of a certain aspect of reality. On the one hand, its structure involves the elements that characterize the modeled reality and, on the other hand, the relationships between them. A CAD model is a physical non-tangible representation carried out in CAD software. There are three types of models (Figure 4.3): Wireframe, Surface, and Solid. Surface is divided into polygon mesh and non-uniform rational B-spline (NURBS), and Solid is divided into parts and assembly. A surface model is created using NURBS, while 3D scans are exported as a polygon mesh.

Each type of model is created using a specific modeling technique (Figure 4.4). The wireframe is a skeletal description; it consists only of curves, lines, and points. The polygon mesh consists of small triangles. NURBS are mathematical representations that describe complex 3D organic free form consisting of points connected by curves. Surface modeling focuses on the external faces, and the solid model is a parametric graphic that includes the features of operations realized from the sketch. Solid modeling is widely used in sketching, engineering analysis, animation, simulation, prototyping, and rendering [4].

### 4.3.1 Digital Model Obtaining

Digital modeling is the process of getting non-tangible models using several modeling techniques such as 3D sculpting and photogrammetry, as well as box,

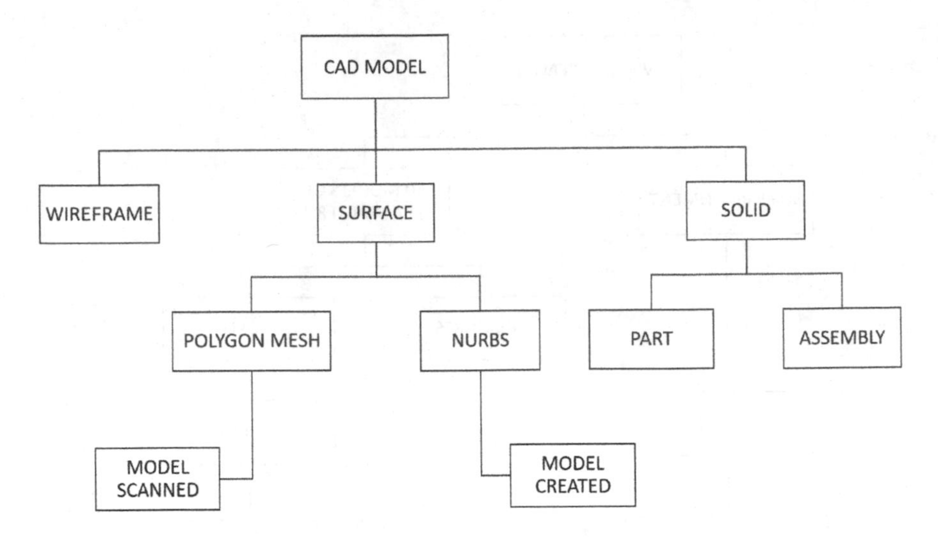


**FIGURE 4.3** Types of CAD models.

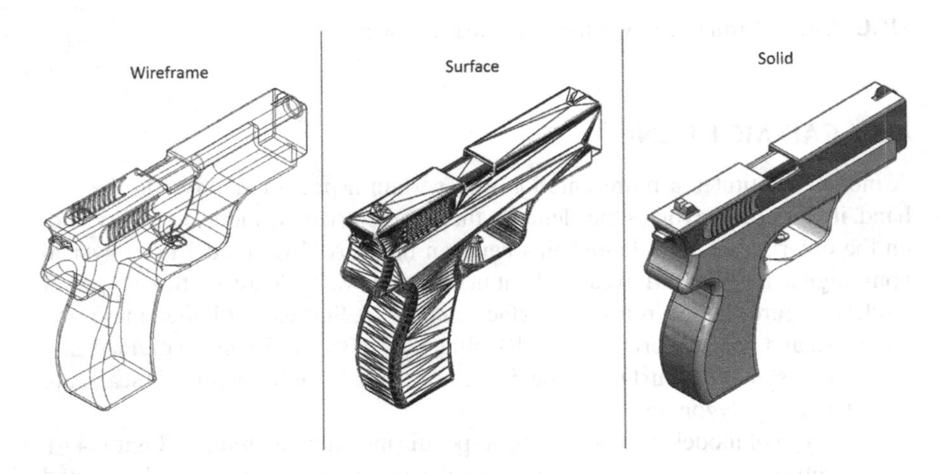


**FIGURE 4.4** Comparison of the kinds of CAD models for a pistol.

polygon, procedural, NURBS, and curve modeling. As shown in Figure 4.3, the CAD model can be obtained either by model creation in CAD software or by model scanning. The 3D scanning is suitable to obtain data and shapes of firearm components made of plastic or worn metal parts. The process begins with data acquisition by 3D scan, later importing objects to analyze data with software as GOM Inspector Pro®. The objects can then be imported into CAD software to handle the surfaces by means of an operation detection tool or remodeling using solid tools [5].

### 4.3.2 Materials Database Storage

In the design process in CAD-CAE software, it is required to include physical properties, which depend on the materials. CAD and CAE software have a material database, called material library, where the properties of materials are stored; this database is necessary to design and evaluate the behavior of the system. A new firearm design requires considering the real parameters of the materials; therefore, they are added to the material library as custom materials once they are characterized. The accuracy of the evaluation is highly dependent on the parameters of the materials. For this reason, the parameters of materials are obtained by testing laboratories; the subject of materials will be addressed in Chapter 7.

### 4.3.3 Bill of Material Property Manager

CAD software includes a tool to manage the materials used in the firearm design. This bill of material property manager considers each material of every component that belongs to the firearm assembly. These material data are used to prepare the experimental lots, detailing the preliminary manufacturing process, the logistics to organize the assembly zones, suppliers, rules to purchase, and standards for firearm operating.

### 4.3.4 CAD Animation

The CAD software has tools to animate the firearms giving movement to their components and recording events, being the main characteristic of this tool to show the virtual functioning and rendering appearance.

As reviewed in Chapter 2, the animation is suitable for proposing a design expectation, which requires compatibility to translate natives between CAD and animation software. The animation software has a better rendering capability than CAD and incorporates better textures, illumination, and materials. At present, some software such as SolidWorks®, Unreal Engine®, Maya®, and 3D Max® include VR tools to improve imagination, immersion, and interaction. A SolidWorks® tool called extended reality allows, for instance, to generate a pistol in CAD, saving the file in extended reality (*.gltf) or extended reality binary (*.glb). These extended reality files can be linked in PowerPoint® to animate slides; in smartphone App eDrawings® using augmented reality (AR) icon; and, in eDrawings® PC with virtual environments using VR icon. Figure 4.5 shows an animation example for a pistol.

### 4.3.5 Interference Detection

During the modeling process, it is necessary to ensure proper movement of the parts to achieve the required interaction among the components, verifying the principle of operation. A task to evaluate the design is the detection of interference in assemblies. This detection highlights the overlap and measures the overlapped surface.

**FIGURE 4.5** CAD animation using extended reality tool.

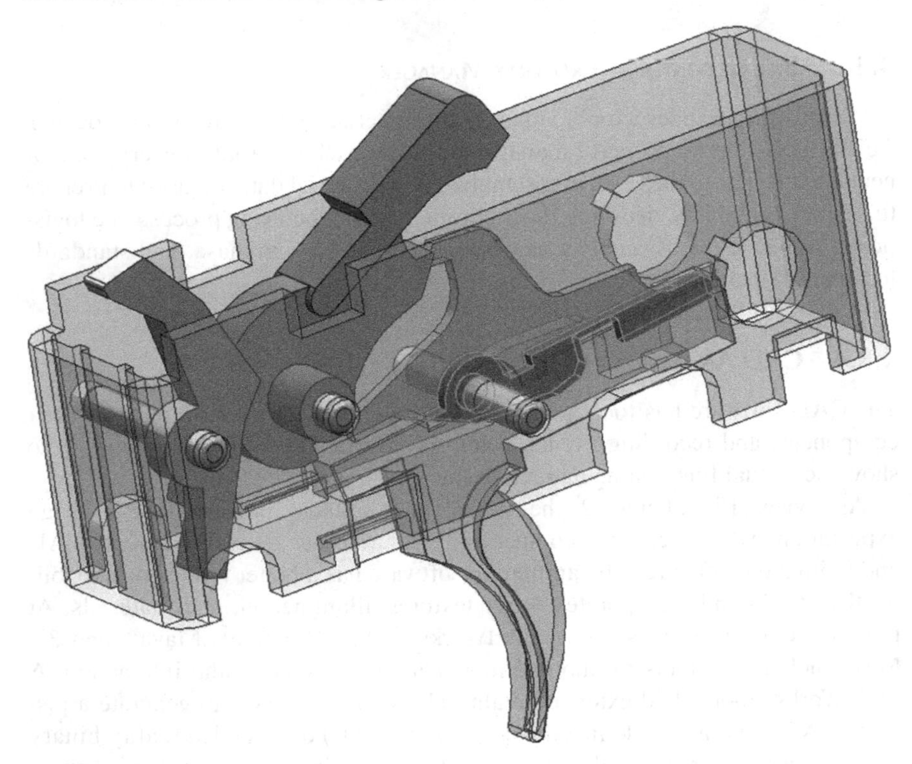

**FIGURE 4.6** Interference detection on a trigger mechanism.

Figure 4.6 presents a trigger mechanism as an example, where the trigger is in interference with the disconnector.

Another task to evaluate the design is the clearance verification in assemblies. This verification marks the clearance in a defined position, showing a measured gap. Such interference detection as clearance verification aims to estimate the permissible gap. Figure 4.7 shows the clearance verification between hammer and disconnector.

Considering the tolerances of manufacture, calibration, controllable and critical dimensions, interference detection, and clearance verification are analyzed to decide permissible measures. Figure 4.8 shows the clearance verification between hammer and trigger, hiding the rest of the parts.

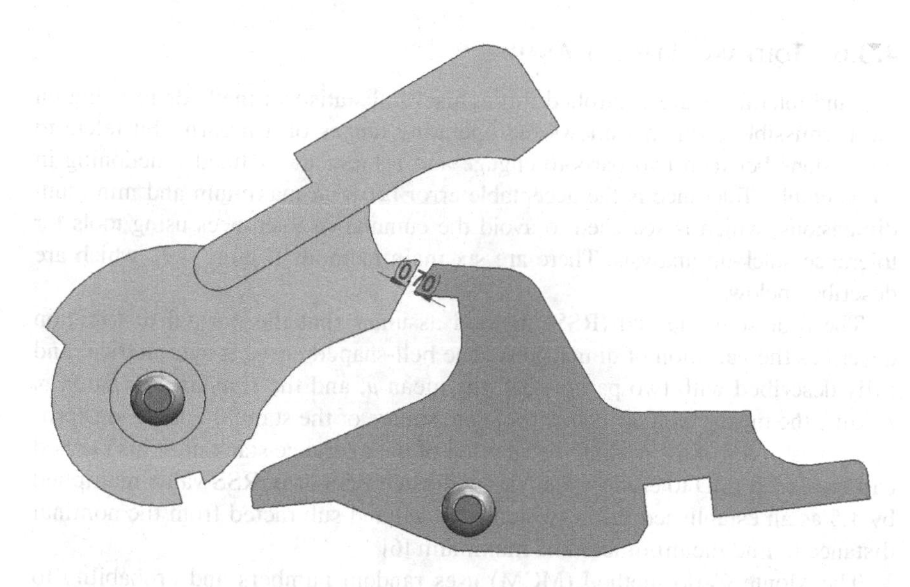


**FIGURE 4.7** Clearance verification between hammer and disconnector.

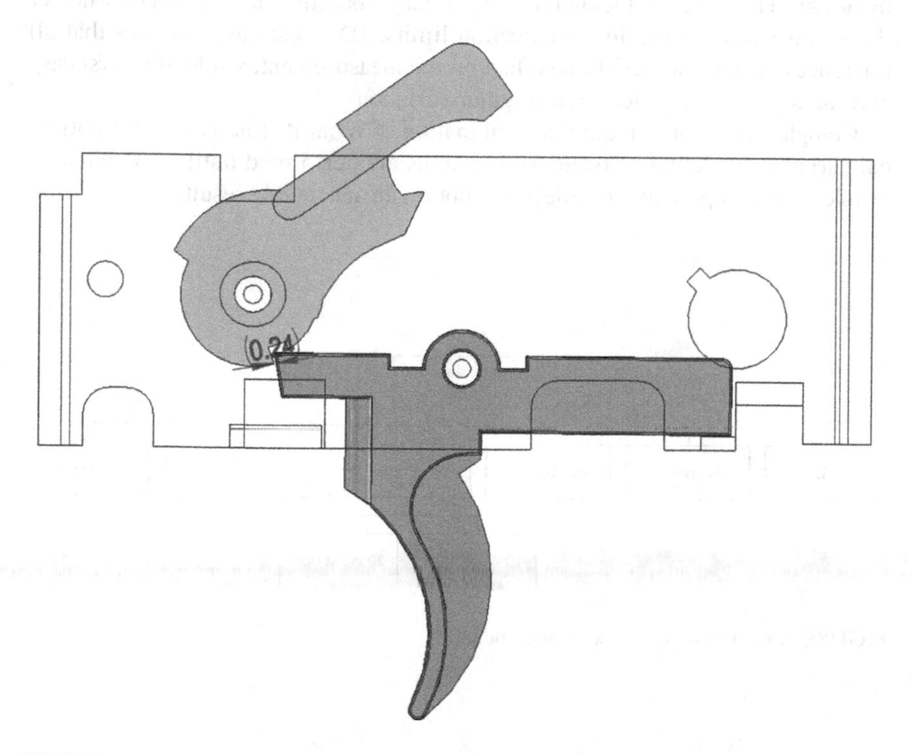


**FIGURE 4.8** Clearance verification between hammer and trigger.

### 4.3.6 Tolerance Stack-up Analysis

Fits and tolerances are controlled through several statistical methods to maintain the permissible error that allows the operating ranges of a firearm. Fit refers to dimensions between two parts to engage and achieve the defined functioning in an assembly. Tolerance is the acceptable error between maximum and minimum dimensions, which is searched to avoid the cumulative tolerances using tools for tolerance stack-up analysis. There are six main methods (Figure 4.9), which are described below.

The root sum squared (RSS) method assumes that the normal distribution describes the variation of dimensions. The bell-shaped curve is symmetrical and fully described with two parameters, the mean $\mu$, and the standard deviation $\sigma$. Adding the means and taking the root sum square of the standard deviations provide an estimate of the normal distribution of the tolerance stack-up. This method can be used for 3D tolerances analysis. Adjusted RSS is an RSS value multiplied by 1.5 as an established factor, which is added and subtracted from the nominal distance to find the minimum and maximum [6].

The Monte Carlo method (MCM) uses random numbers and probability to solve problems. The MCM includes a diagram showing the components and operations sequence for evaluating measurement uncertainty [6].

Worst-Case Tolerance Analysis is a method that derives from setting all the tolerances at their limits to make a measurement the largest or smallest possible by design. This process does not use statistical probability and focuses on whether the product falls within its specification limits. 1D worst case assumes that all tolerances have a one out of one effect on the measurement, while 3D worst case considers geometrics effects and angular variation.

Complex assemblies require a combination of What-ifs tolerances and statistical variation modeling software, so iterations are performed until a combination of tolerance components is achieved to obtain an acceptable result.

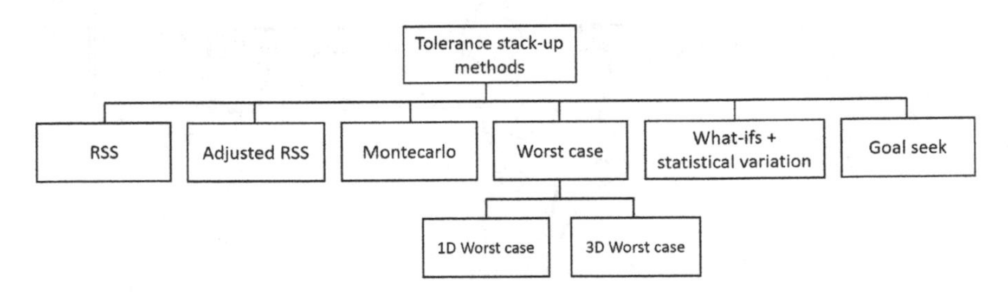


**FIGURE 4.9** Tolerance stack-up methods.

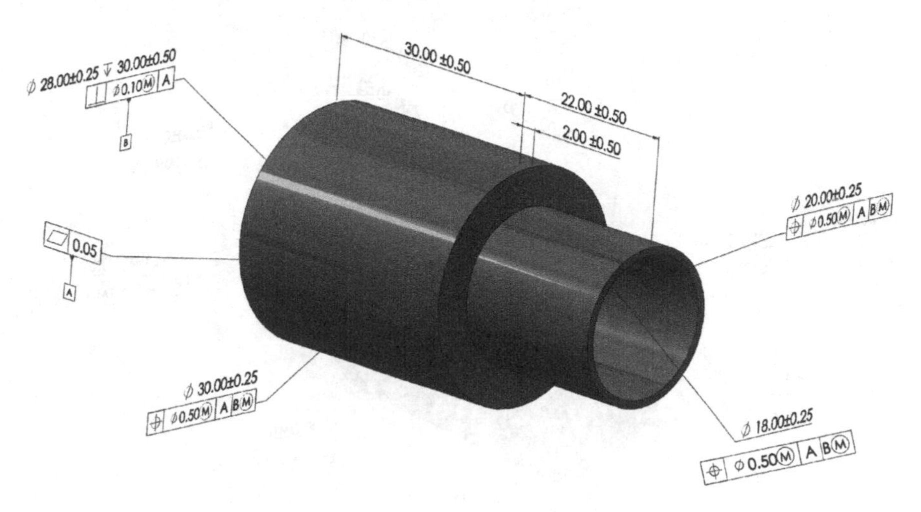


**FIGURE 4.10** Tolerance assigned to part 1 using DimXpert®.

Goal Seek of Excel® is a function that allows determining the tolerance value. The analyst can set the desired mounting tolerance value and have the program iterate to find the exact solution [6].

Opening tolerances is a technique for changing mates, such as swapping butt joints for lap joints, or changing the geometry of the surface to make misalignment less obvious using shims to reduce the number of parts than the cumulative total. Tolerances opening and Tolerances stack-up tool work together to reach the suitable tolerances for functioning and manufacturability.

TolAnalyst® is a tolerance analysis tool used to study tolerance effects and assembly methods that have on dimensional stack-up between two features of an assembly [7]. The result of each study is a minimum and maximum tolerance stack, a minimum and maximum RSS tolerance stack, and a list of contributing features and tolerances reported in Excel® sheet. A tolerance study uses four steps: (1) Define the measurements as linear distances between two DimXpert® features; (2) Assembly Sequence to establish a tolerance chain between the measurement features; (3) Assembly Constraints to define how the part is placed or constrained; and (4) Analysis Results where the minimum and maximum worst-case tolerance stacks are evaluated and reviewed. This tool uses type plus and minus or geometric tolerances (Gtols) concerning three datums named A, B, and C. Figures 4.10 and 4.11 detail an example using DimXpert® and TolAnalyst®. The tolerances analyses can be performed in a component that highlights the critical zone to make an assembly, as shown in the results to part 1 in Figure 4.11.

For the assembly analysis, the tolerances of part 1 are deleted and the tolerance analysis is performed to part 2 using DimXpert®, as shown in Figure 4.12.

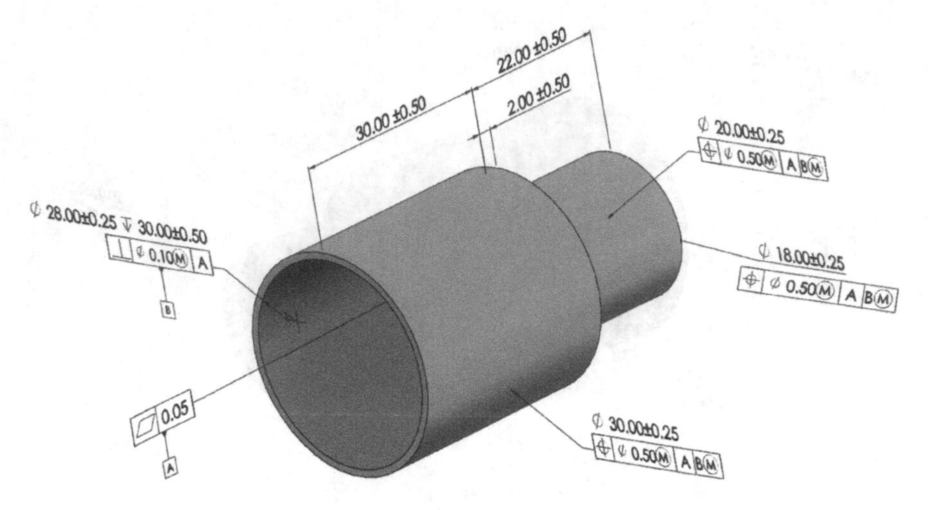


**FIGURE 4.11** Result of tolerance analysis applied to part 1 highlighting the critical assembly.

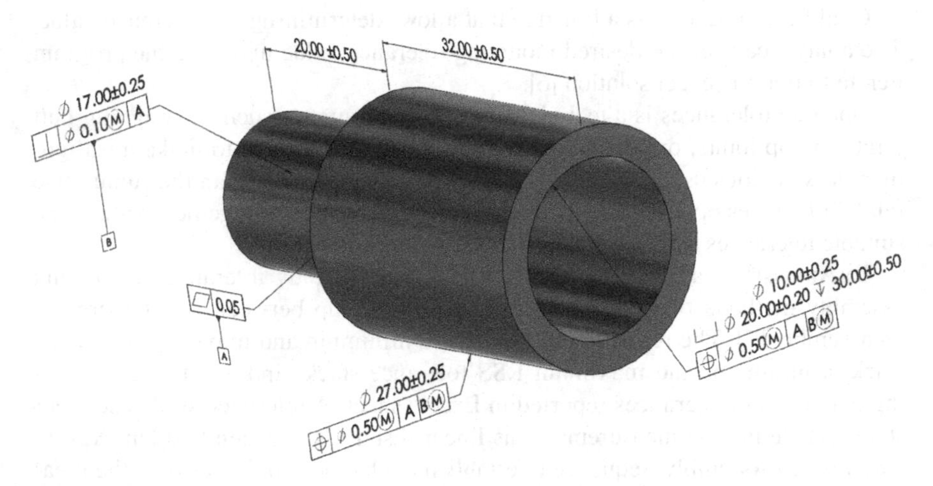


**FIGURE 4.12** Tolerances of part 2 using DimXpert® with Gtols.

Part 2 assembles into part 1, which is selected as the source part that contains a DimXpert® scheme. The target part is part 1 that does not have a DimXpert® scheme; both parts are shown in Figure 4.13.

When a tolerance requirement is known in the final assembly, and the tolerances will be determined as the final requirement allows, then the CAD model can be refined using oversized slots or holes to adjust the assembly. Figure 4.14 shows the tolerances recommended for part 1 being the target part.

**FIGURE 4.13** Definition of source and target parts.

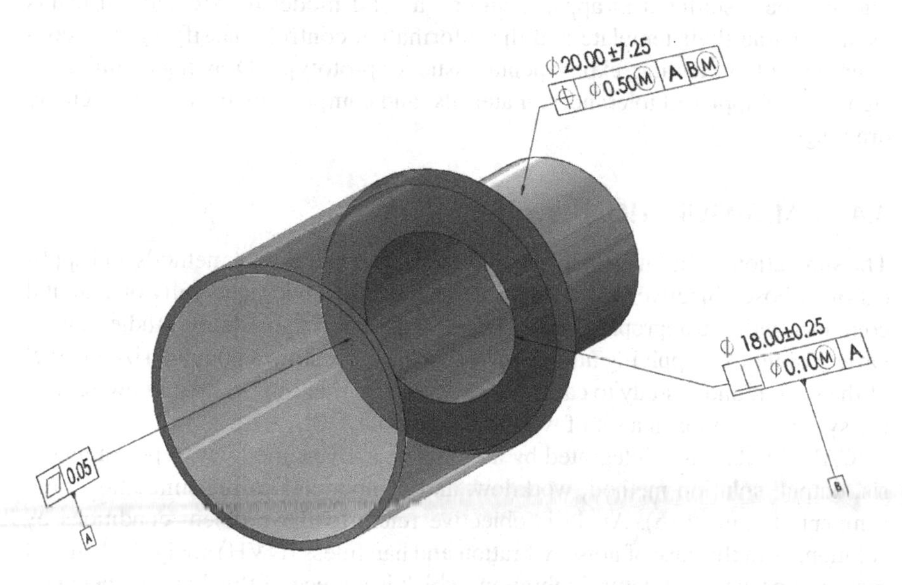


**FIGURE 4.14** Result of tolerance analysis applied to part 1.

**TABLE 4.1**
**Consolidated Tolerance Analysis Results**

| Study | Trigger Tolerances |
|---|---|
| TolAnalyst® measurement | (mm) |
| Nominal value | 0.5 |
| Minimum value | –5.322 |
| Maximum value | 7.238 |
| RSS minimum | –4.47 |
| RSS maximum | 6.341 |

Table 4.1 shows consolidated tolerance analysis results extracted from Excel® report in CAD software using TolAnalyst® tool.

It is essential to mention that setting so close tolerances increases the manufacturing cost; for this, the tolerances analysis must consider the tolerances that a defined manufacturing process can reach for each component.

### 4.3.7 CAD Drawings

Communication with other areas, such as manufacturing, experimental testing, quality control, intellectual property, technology transfer, and resources material, is done through engineering plans, drawings, or drafting. Currently, drawings can be visualized in apps, including a solid model in AR. Design teams usually define their template and the information control, classifying it as confidential, only to testing, experimental issue, or prototype. Drawings require the signature of approval tolerances, materials, and components in case of assembly drawings.

## 4.4 CAE SIMULATION

The simulation technique comprises an extensive collection of methods and applications whose objective is to reproduce the actual behavior, generally on a digital computer with the appropriate software [8,9]. Computer simulation studies a wide variety of models applying numerical techniques, creating a computerized model of the system under study to carry out experiments that improve the knowledge of the system behavior in a set of working conditions [10].

CAE simulation is integrated by analysis objectives, modules, types of analysis, output, solution method, workflow, and components in the simulation environment (Figure 4.15). Analysis objective refers to the problem conditions by solution, as in the case of noise, vibration and harshness (NVH) analysis. A modal mass is a property of natural vibration, which is a piece of the dynamic behavior of vibration systems (modal analysis). A modal mass can be understood as a mass that is activated in specific modes of vibration. A vibration mode is a property

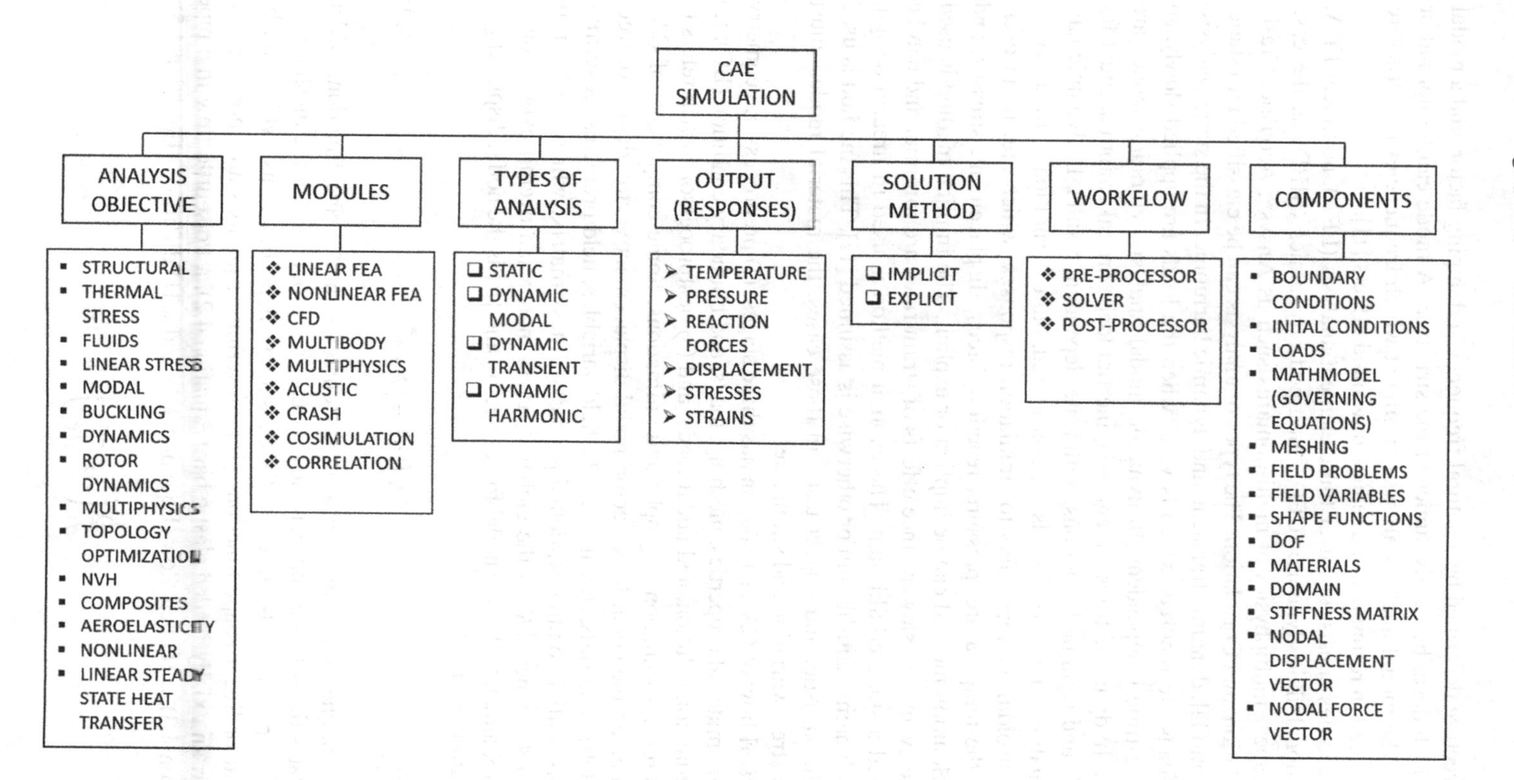


**FIGURE 4.15** Framework of CAE simulation.

of the structure described by a natural frequency, a damping factor, and a modal shape, which describe the dynamics of the structure. A mode can be normal or complex. The modal analysis allows defining at which frequencies the structure can be excited in resonance and reduce unwanted effects [11].

The modules consist of linear finite element analysis (FEA), nonlinear FEA, computational fluid dynamic (CFD), multibody dynamics (MBD) as the case of Simscape®, multiphysics of multi-domains such as Ansys®, Acoustic, Crash, Co-simulation, and Correlation. The types of analysis can be classified into static, dynamic modal, dynamic transient, and dynamic harmonic. In the static analysis, inertial loads are ignored, which is valid when the loads are applied slowly, so results are time-independent. In dynamic modal analysis, vibration modes are analyzed. In dynamic transient analysis, inertial loads are taken into account for fast loads and rigid body motions, with time-dependent results. In dynamic harmonic analysis, the inertial loads are considered, but external loads are assumed to be sinusoidal. Output refers to simulation responses, which are, in an overall way, the temperature, pressure, reaction forces, displacement, stresses, and strains. Solution method can be implicit or explicit. The implicit method is used when the events are slower, since effects of strain rates are minimal and can be considered a static equilibrium. The explicit method is used in firearm design, where a dynamic equilibrium or otherwise is assumed [12]. This method is used for crash, ballistic, and impact tests; in these cases, the material models must consider stress variation and strain rate.

The workflow of CAE simulation has three stages: (1) preprocessor, geometry treatment, material properties, meshing, loads, and boundary condition (BC); (2) solver, equation solution, and nodal loads; and (3) postprocessor, which refers to analysis result evaluation in graphs and animations. Some components of CAE simulation are the nodal force vector, nodal displacement vector, degree of freedom (DOF), materials, meshing, loads, field variables, field problems, governing equations, initial conditions, domains, and stiffness matrix, which is shown in Equation 4.1, where $[K]$ is the global stiffness matrix (an overall expression to linear problems), $\{F\}$ is the nodal forces vector, and $\{u\}$ is the nodal displacement vector (unknown) [13].

$$[K]\{u\}=\{F\} \tag{4.1}$$

Other components are BCs, governing equations, and shape function, which interpolates the solution between the discrete values obtained at the mesh nodes. Low-order polynomials are typically chosen as shape functions, which can be nonlinear or linear shape functions [14]. Governing equations are shown as follows for: an axially loaded elastic bar (Equation 4.2), a Poiseuille flow in a pipe (Equation 4.3), a 1D heat flow (Equation 4.4), and a 1D diffusion (Equation 4.5).

$$\frac{dy}{dx}\left(AE\frac{du}{dx}\right)+b=0 \tag{4.2}$$

where $u$ = displacement, $A$ = area, $E$ = Young's modulus, $b$ = axial loading

$$\frac{dy}{dx}\left(A\frac{D^2}{32\mu}\frac{dp}{dx}\right)+Q=0 \tag{4.3}$$

where $p$ = pressure, $A$ = area, $D$ = diameter, $\mu$ = viscosity, $Q$ = fluid supply

$$\frac{dy}{dx}\left(Ak\frac{dT}{dx}\right)+Q=0 \tag{4.4}$$

where $k$ = Thermal conductivity, $A$ = area, $T$ = temperature, $Q$ = heat supply

$$\frac{dy}{dx}\left(AD\frac{dC}{dx}\right)+Q=0 \tag{4.5}$$

where $C$ = ion concentration, $A$ = area, $D$ = difussion coefficient, $Q$ = ion supply

Figure 4.15 shows a framework of CAE simulation. Some simulations are named according to the elements used in the framework. For instance, explicit dynamic transient simulation comes from the fact that it contains an explicit solution method and belongs to the type of dynamic transient.

Implicit and explicit problems are expressed through partial differential equations (PDEs) in matrix equations; they can be linear (Equation 4.1) or nonlinear, which is expressed as Equation 4.6.

$$[K(u)]\{u\}=\{F\} \tag{4.6}$$

The nonlinear equations can be of different nature: material nonlinearity, where deformations and strains are large, as in the case of polymeric materials; geometric nonlinearity, where strains are small, but rotations are large as thin structures; and boundary nonlinearity as, for instance, in contact problems [15]. For dynamic problems as that of a bullet within a barrel, Equation 4.7 is used, where $M$ denotes the mass matrix; $U$ is the displacement; $\dot{U}$ is the velocity; $\ddot{U}$ is the acceleration; $C$ is the damping matrix; $K$ is the stiffness matrix; and $F$ is the loading vector.

$$M\ddot{U}+C\dot{U}+KU=F \tag{4.7}$$

Figure 4.16 shows how loads, mesh, materials, and BCs interact. Preprocessing defines the loads and can be concentrated, distributed, or shear; the mesh can be classified into adaptive (refining method of a simulation mesh based on the solution) and grid (types of mesh: structured, unstructured, and hybrid); the materials can be classified into form and properties. BCs are also defined in the preprocessing stage. Meshing is part of the discretization necessary to provide a solution through the simulation process. This discretization generates a set of points and cells [16].

BCs define the interactions of systems (firearms) with the environment and include the BC types (thermodynamics, solid mechanics, and fluid dynamics), BC assignments (assigned to the domain boundary), and BC values (pressure, temperature, force, velocity). Thermodynamics includes fixed value temperature, convective heat flux, surface heat flux, and volume heat flux. Solid mechanics includes pressure, force, nodal load, fixed value, surface load, volume load, centrifugal force, remote force, symmetry plane, remote displacement, fixed support, rotating motion, elastic support, and bolt preload. Fluid dynamics includes velocity, pressure, natural convection, wall, periodic, symmetry, and wedge [17].

BCs depend on constraints applied to the governing equations of a firearm or ballistic phenomenon. These are the following (Figure 4.16): Dirichlet BC, which specifies the value that the unknown function needs to take along the boundary of the domain (for instance, a no-slip condition in fluid mechanics where the value of the velocity is set to zero, in solid mechanics prescribing a particular load or displacement, and in heat transfer setting the temperature at a surface); Neumann BC, which specifies the values that the derivative of a solution is going to take on the boundary of the domain (for instance, fluid mechanics is the fully developed condition at an outlet where the gradient of flow variables is set to zero, traction conditions in solid mechanics, and insulated surfaces in heat transfer); Robin BC, when in a differential equation $(mu + n\left(\frac{\partial u}{\partial x}\right) = y$; $m$ and $n \neq 0)$, a linear combination of the values of a function ($y$) and the values of its derivative $(\frac{\partial u}{\partial x}$; $u$ is the unknown solution defined on $\Psi$ domain) on the boundary ($\partial\Psi$) are specified; Cauchy BC, which is a condition on both the unknown field and its derivatives (implies the imposition of two constraints, Dirichlet+Neumann); and mixed BC, which consists of applying different types of BCs in different parts of the domain [18].

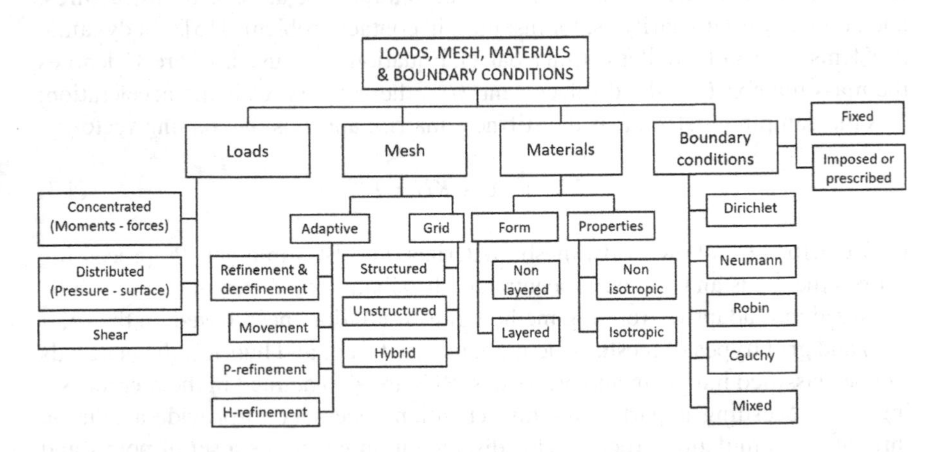


**FIGURE 4.16** Interaction scheme of loads, mesh, materials, and BCs.

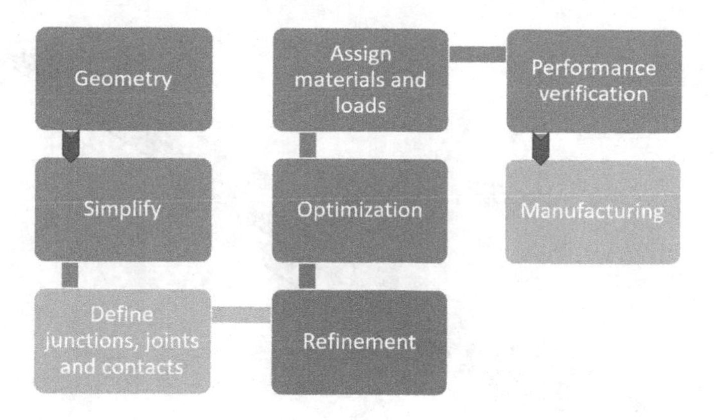


**FIGURE 4.17** Workflow of optimization study.

Topology optimization is a technique encompassed within the field of structural analysis. It is based on the mechanical analysis of a component or structure. Its main objective is structural lightening while maintaining the mechanical functionalities of the target component [19]. Figure 4.17 shows a workflow to make the topology optimization.

### 4.4.1 CAD Model Treatment

CAD model is prepared according to the experience of the design team and capabilities of CAE software, considering simplification, model layers, small chamfer, and rounding. Sometimes, this treatment requires the experimental protocol using case studies and tasks, as well as suppressing sharp edges and softening curves. Figure 4.18 shows the assessment and simplification of an assembly using a section cut to show the steel core of a bullet.

Geometric forms facilitate meshing and avoid errors during the simulation process. Figure 4.19 shows the meshing in the assembly components.

### 4.4.2 CAE Analysis by FEA

FEA is the simulation of any given physical phenomenon using the numerical technique called finite element method (FEM). Engineers use FEA software to complement the product design, experiments, and optimize components, improving the solution certainty. FEA focuses on individual parts (microscopic behavior), the structural performance of a part, small motion, stress, strain, and deformation of individual parts, as well as static or dynamic transient analysis for a very short duration [20].

An analysis of a barrel caliber 5.56 mm is presented here. This barrel is manufactured by cold radial forging; this process is described in Chapter 9. A hard

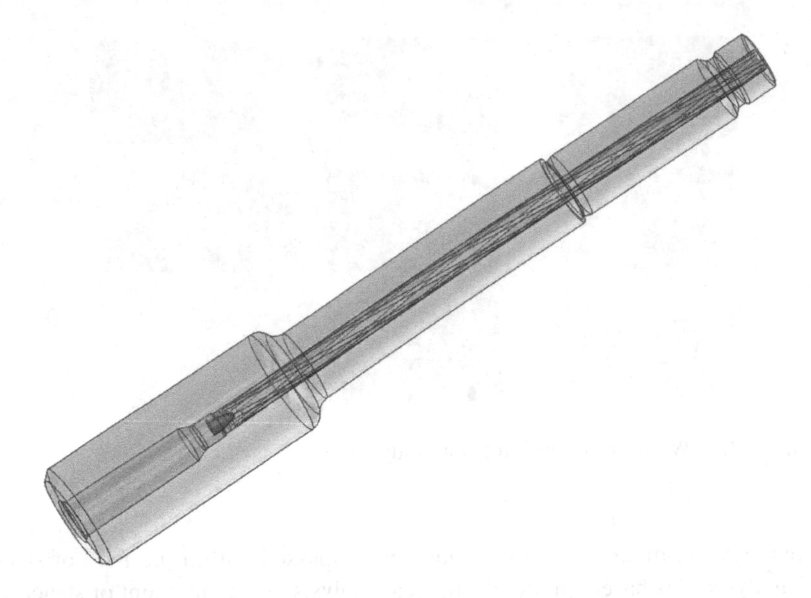

**FIGURE 4.18** Assessment and simplification of an assembly.

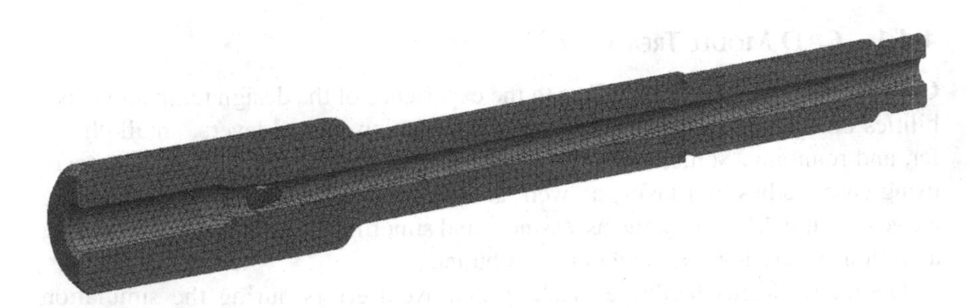

**FIGURE 4.19** Configuration of meshing in the assembly.

chrome coating is applied to the inside of the barrel. After having performed a metallographic analysis, the barrel profile shows a material hardened by heat treatment, in addition to the cold deformation process. Chemical analysis revealed that the material corresponds to a DIN 32CrMoV1210 steel, which has the following mechanical properties: yield strength of 923 MPa, tensile strength of 980 MPa, modulus of elasticity of 196 GPa, and hardness of 260 HV; it withstands a breech pressure of 380 bar.

**FIGURE 4.20** Structural analysis of a barrel.

The simulation results shown in Figure 4.20 reveal that the barrel performance does not exceed its elastic limit, even considering a safety factor of 20%. The maximum radial displacement of the breech walls is 3.0 hundredths (mm). From the data generated by the simulations, the increase in the outer diameter of the barrel due to deformation would be 0.06 mm (3 radial tenths).

### 4.4.3 CAE Motion

Firearms are composed of mechanisms such as trigger, bolt breech, and recoil mechanisms. Motion® provides complete, quantitative information about the kinematics, including position, velocity, and acceleration, and the dynamics, including joint reactions, inertial forces, and power requirements, of all the components of a moving mechanism. Motion® verifies interferences in real time and provides the exact spatial and time positions of all mechanism components, as well as the exact interfering volumes [21]. Figure 4.21 shows the interference detection of a trigger mechanism.

Sometimes, the motion simulation is configured by applying torque to moving parts in contact until the simulation time is reached, measured in the angular displacement. Figures 4.22 and 4.23 present the angular displacement and the acceleration magnitude, respectively, for a hammer.

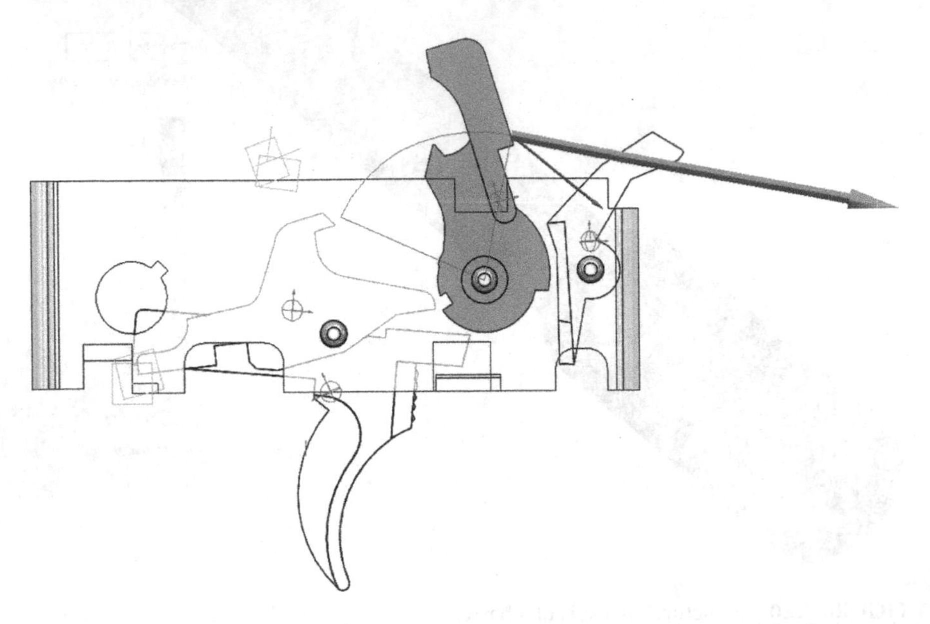

**FIGURE 4.21** Motion analysis and interference detection of a trigger mechanism.

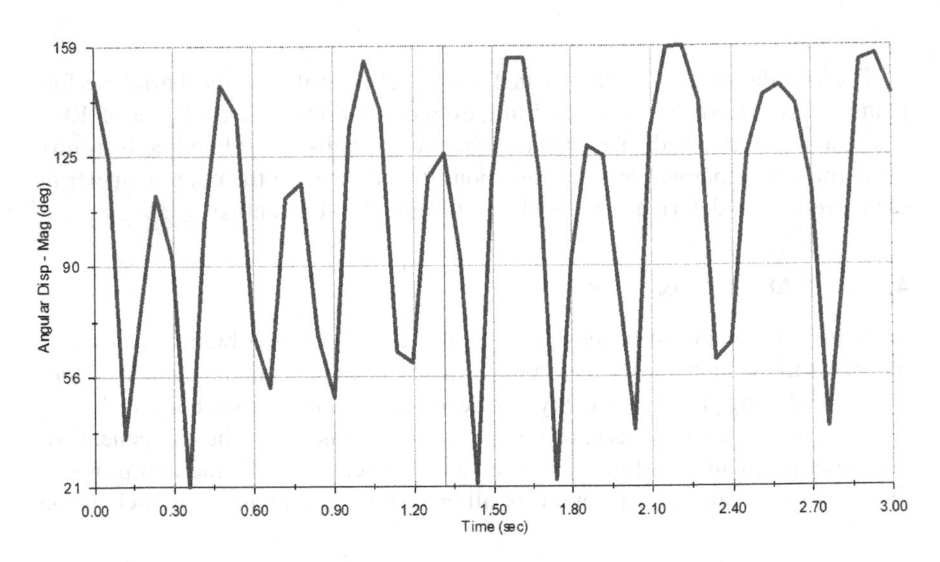


**FIGURE 4.22** Magnitude of hammer angular displacement.

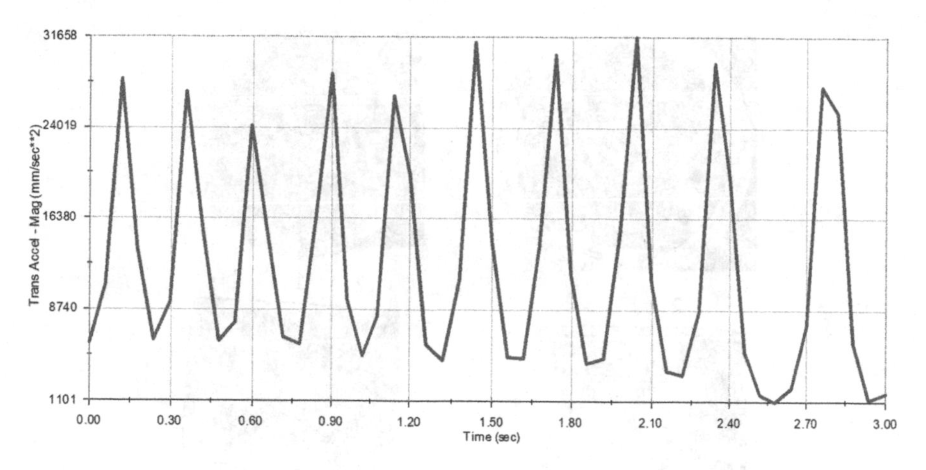


**FIGURE 4.23** Magnitude of hammer acceleration.

### 4.4.4 Flow CFD

Several packages are able to perform CFD, including CAE software. Ansys® has various CFD packages. For example, Ansys Fluent® is a CFD software that includes a solver of high-performance computing (HPC) technologies; Ansys CFX® contains multi-stage CFD modeling, transient blade row methods, and harmonic balance methods; and Ansys Chemkin-Pro® is for modeling complex, chemically reacting systems [22]. Figure 4.24 shows a case study concerning the simulation performed in Ansys Fluent® for a polygonal barrel and a groove barrel. The FEM with equal boundary and loading conditions was used in both types of barrels, specifying the actual materials of the projectile and the barrels. Subsequently, experimental tests were carried out on various weapons with 9 mm ammunition of 115, 122, and 124 gr. The results show that the 9 mm bullet fired in a polygonal barrel undergoes a maximum deformation towards its exterior of 0.178 mm and interior of 0.158 mm, with a stress up to 295.85 MPa. These results are compared with a maximum external deformation of 0.025 mm and an internal deformation of 0.112 mm for 9 mm projectiles fired in a grooved barrel, with a stress up to 269.79 MPa. The deformation in the polygonal barrel extends to a greater area, but the rifling impression left is less deep.

Another case study consists of simulating a bullet into a fluid, where the fluid is configured as a 3D deformable object. Figure 4.25 shows meshing of fluid and bullet with a refinement of the bullet meshing. Figure 4.26 presents

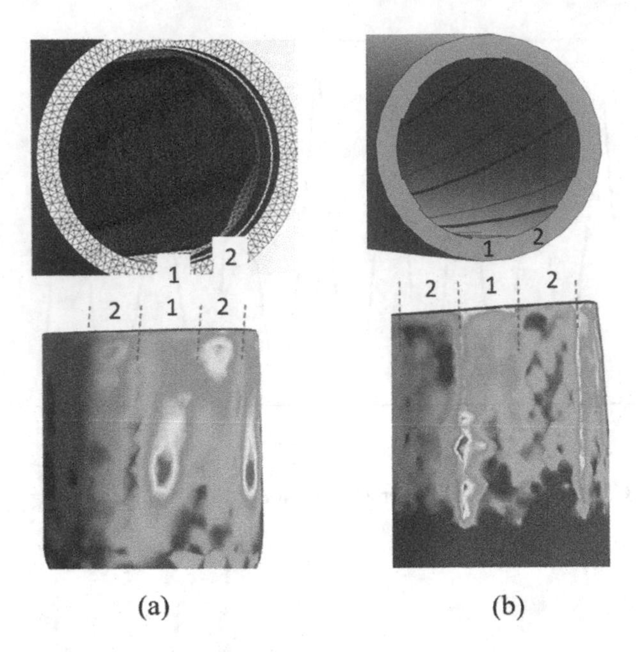


**FIGURE 4.24** Comparison of the internal ballistic in (a) a polygonal rifled barrel and (b) a grooved rifled barrel. (Courtesy of Prof. Usiel Sandino Silva Rivera, Military School for Engineers, Mexico.)

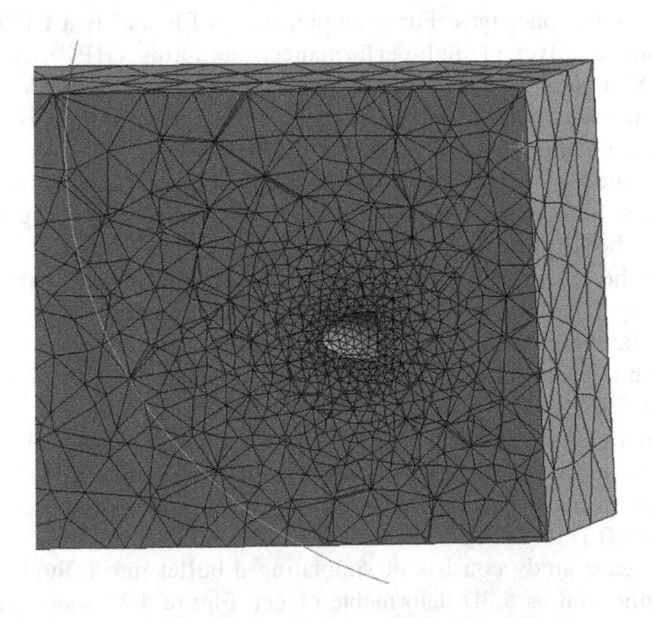

**FIGURE 4.25** Meshing of fluid and bullet refining the bullet meshing.

**FIGURE 4.26** Final meshing of bullet and fluid.

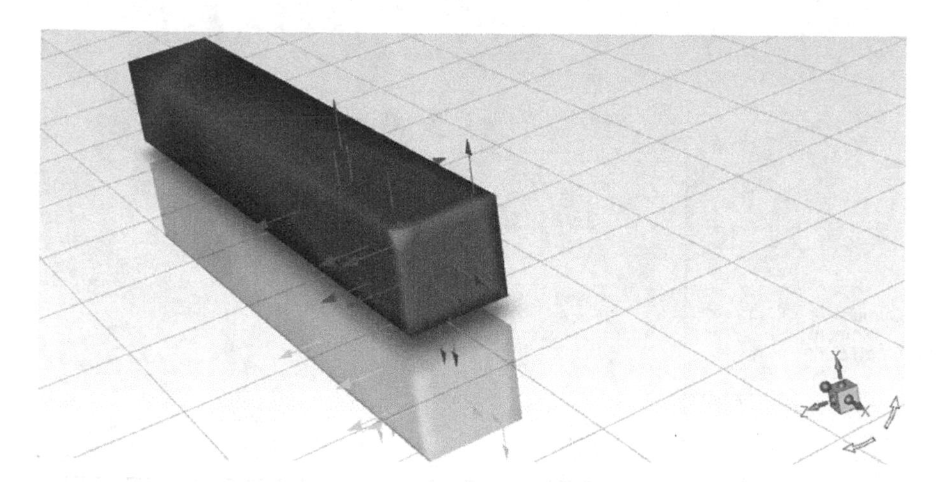

**FIGURE 4.27** Configuration of meshing in Fluent® module.

the final meshing of bullet and fluid in the mesh module of Ansys®. Figure 4.27 shows the configuration of meshing in Fluent® module considering six DOFs, the bullet being a rigid body, and the fluid a deforming body. Figure 4.28 shows the simulation results as velocity in x, y, and z components in Fluent® module. Finally, Figure 4.29 shows the cumulative moment coefficient in Fluent® module.

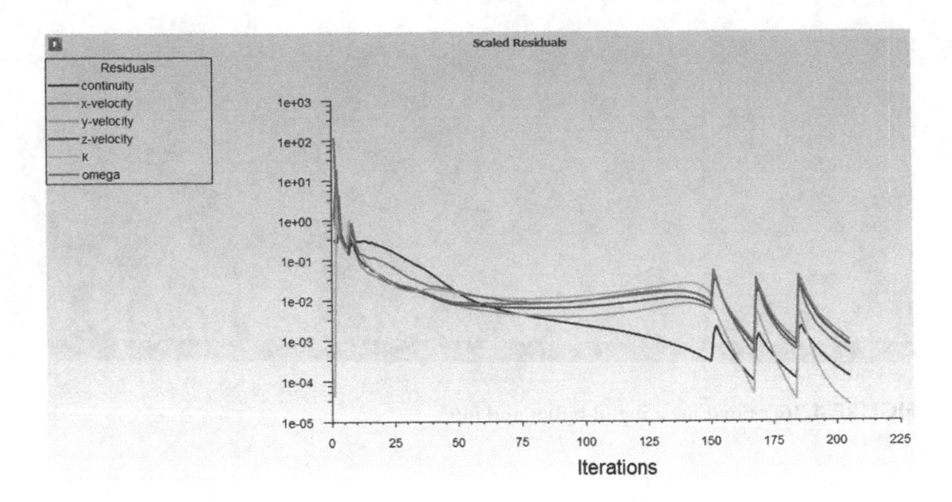


**FIGURE 4.28** Simulation results of velocity in Fluent® module.

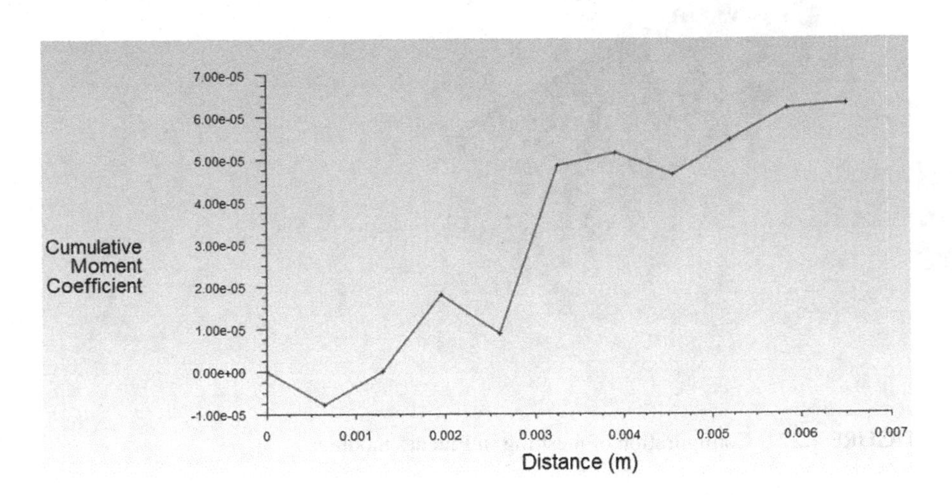


**FIGURE 4.29** Cumulative moment coefficient versus distance in Fluent® module.

One more case study consists of simulating a bullet into a fluid, where the fluid is configured as a 2D deformable object. The following figures show different parameters obtained from the flow simulation of a bullet across a fluid performed in Fluent® module: meshing (Figure 4.30), velocity magnitude (contour 1) (Figure 4.31), velocity (Figure 4.32), particle pathlines (Figure 4.33), velocity magnitude (Figure 4.34), sound speed (Figure 4.35), strain rate (Figure 4.36),

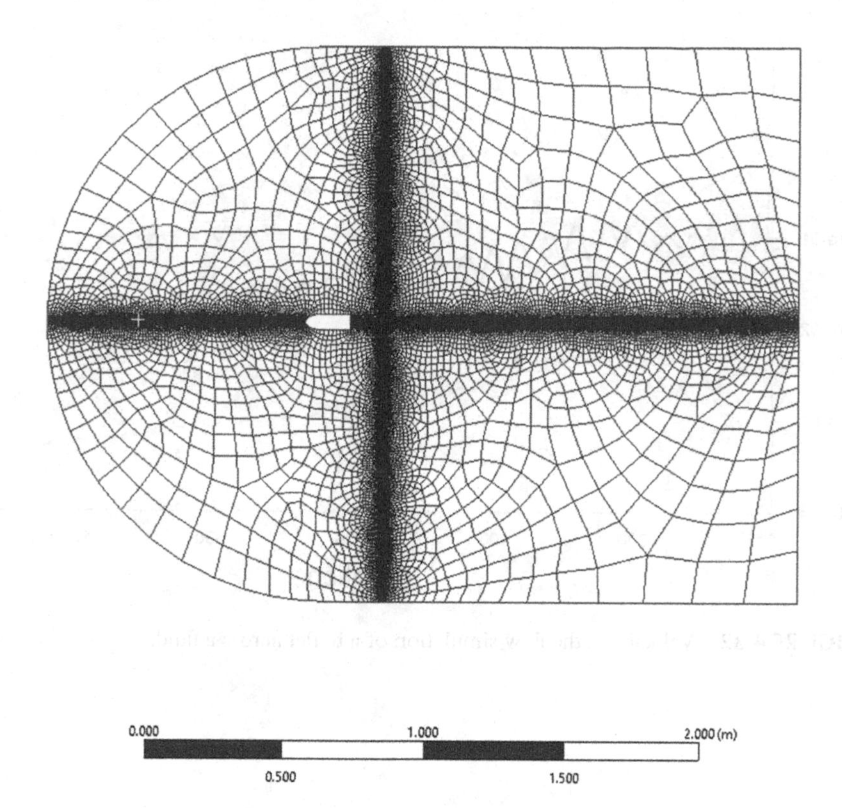


**FIGURE 4.30** Meshing to simulate the flow of a bullet through a fluid.

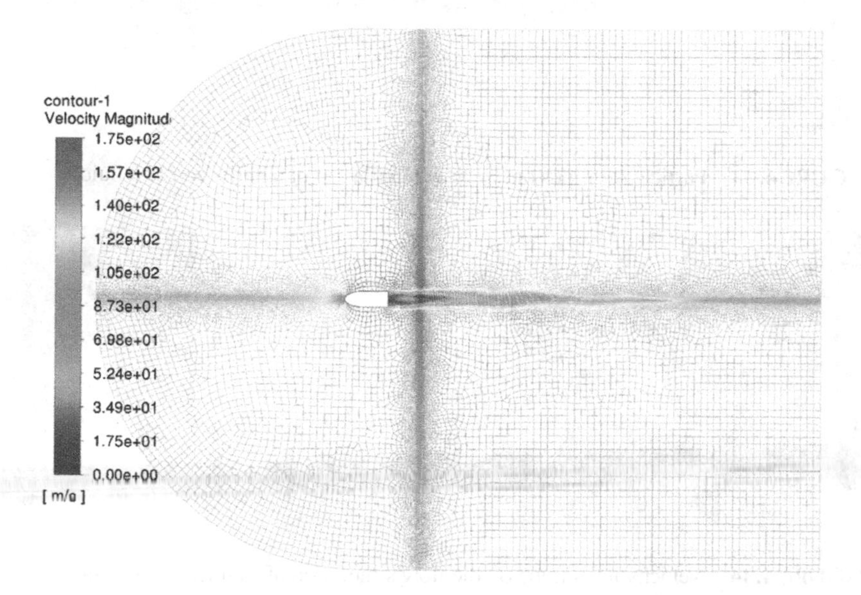


**FIGURE 4.31** Velocity magnitude (contour 1) on the flow simulation of a bullet across a fluid.

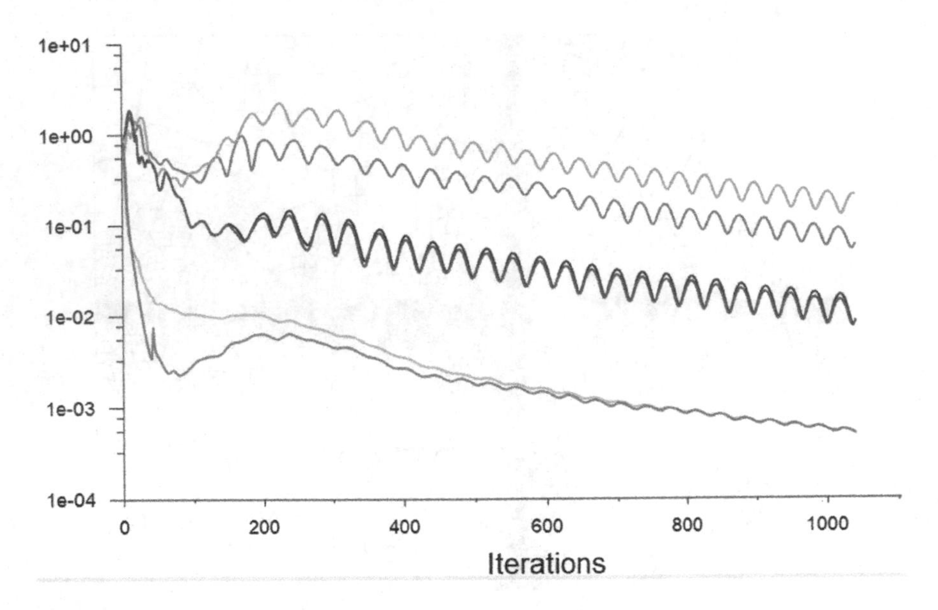


FIGURE 4.32 Velocity on the flow simulation of a bullet across a fluid.

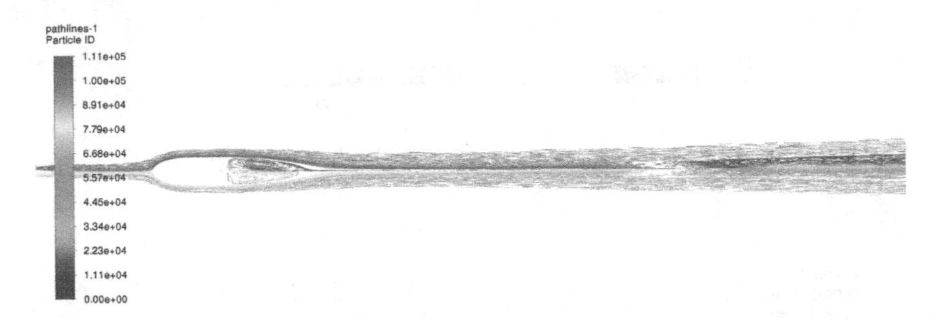


FIGURE 4.33 Particle pathlines on the flow simulation of a bullet across a fluid.

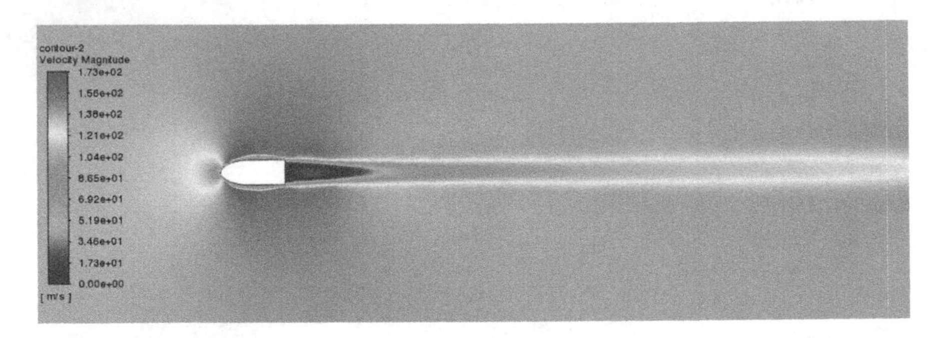


FIGURE 4.34 Velocity magnitude on the flow simulation of a bullet across a fluid.

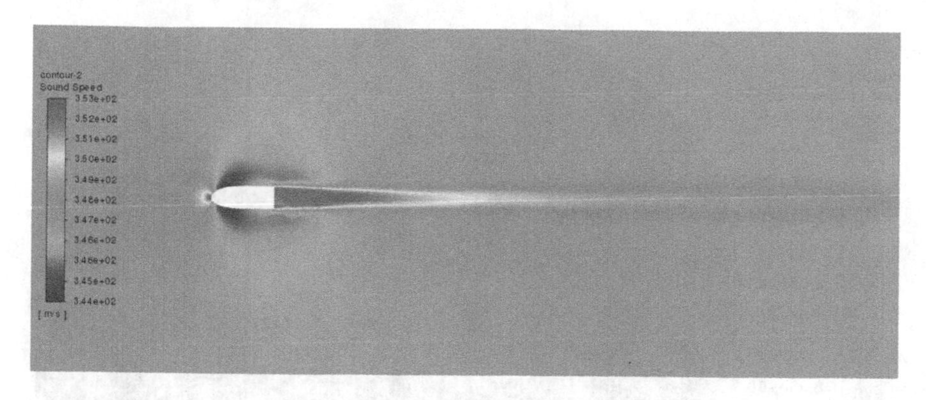


**FIGURE 4.35** Sound speed on the flow simulation of a bullet across a fluid.

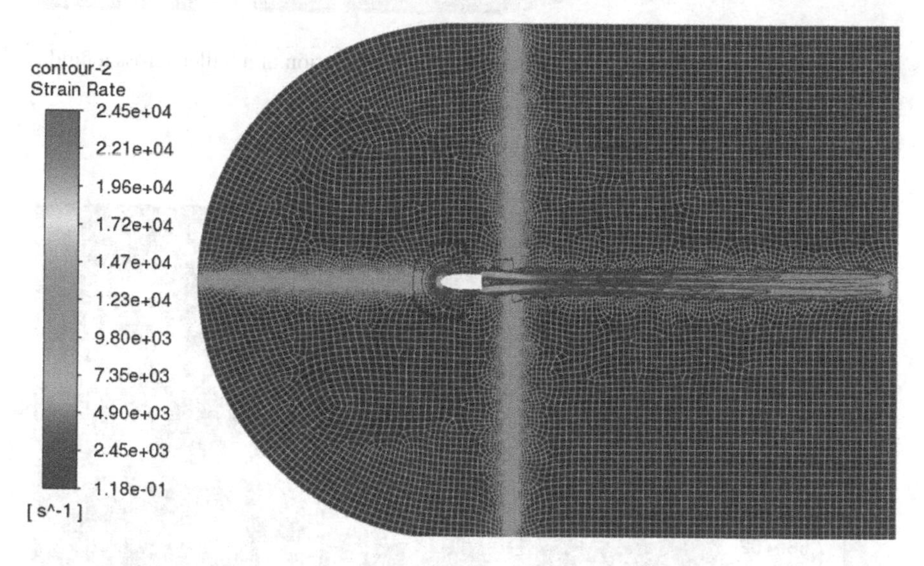


**FIGURE 4.36** Strain rate on the flow simulation of a bullet across a fluid.

skin friction coefficient (Figure 4.37), turbulent intensity (Figure 4.38), turbulent dissipation (Figure 4.39), turbulent kinetic energy (Figure 4.40), static temperature (Figure 4.41), total temperature (Figure 4.42), internal energy (Figure 4.43), total energy (Figure 4.44), match number (Figure 4.45), vorticity magnitude (Figure 4.46), cell Reynolds number (Figure 4.47), tangential velocity (Figure 4.48), and radial velocity (Figure 4.49).

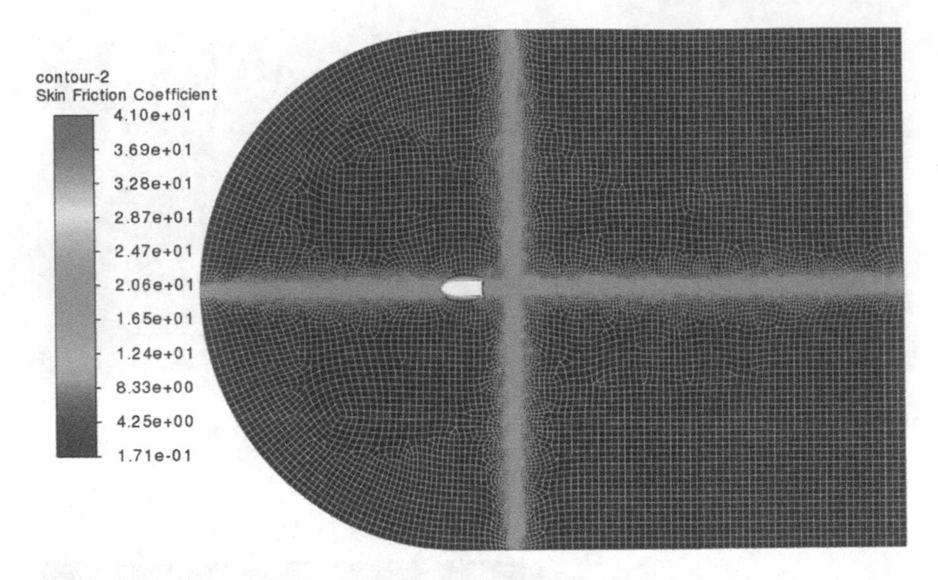


**FIGURE 4.37** Skin friction coefficient on the flow simulation of a bullet across a fluid.

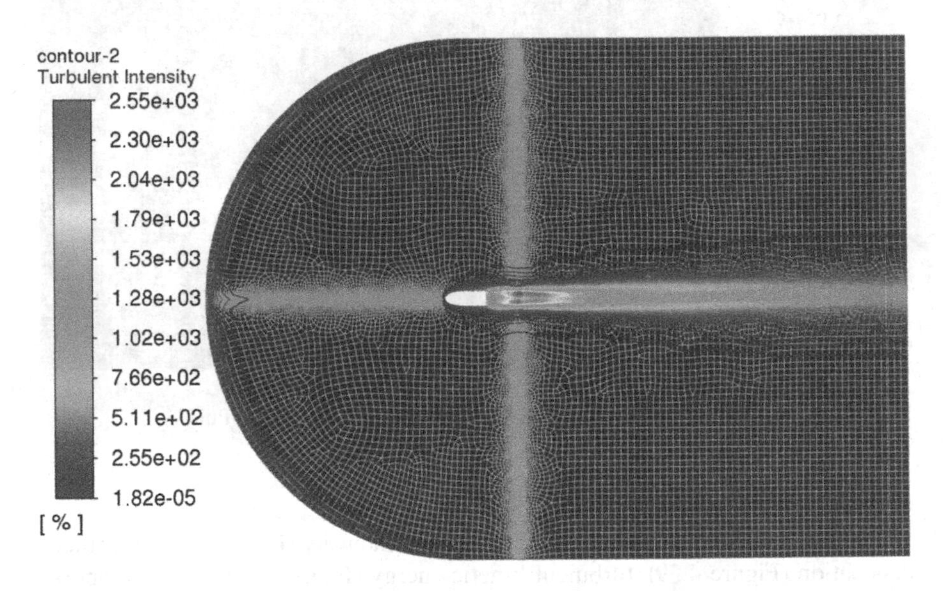


**FIGURE 4.38** Turbulent intensity on the flow simulation of a bullet across a fluid.

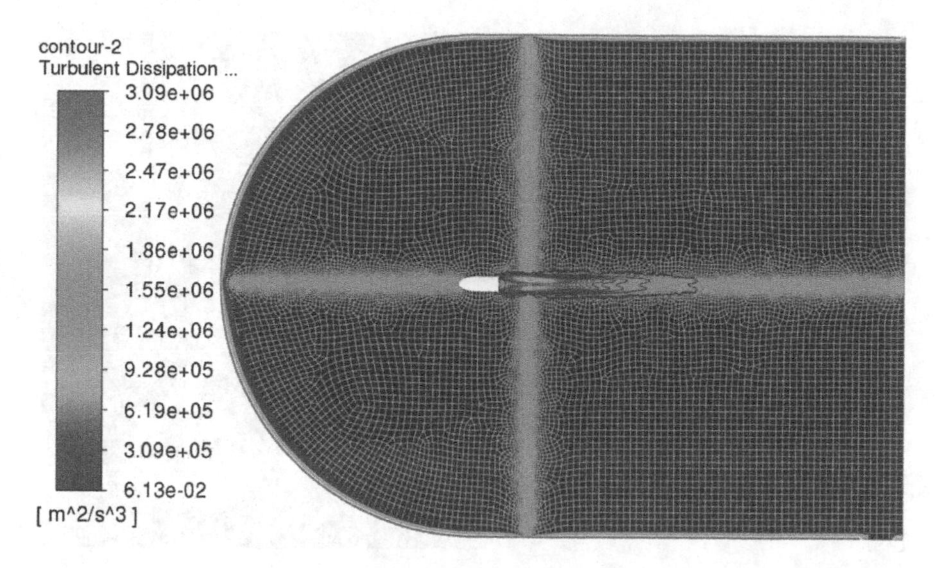


**FIGURE 4.39** Turbulent dissipation on the flow simulation of a bullet across a fluid.

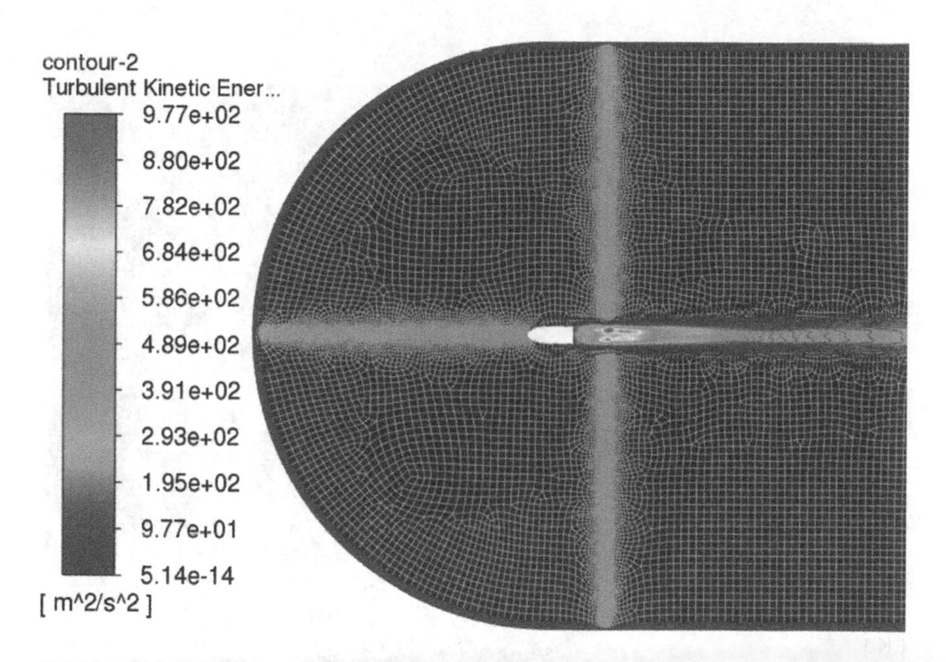


**FIGURE 4.40** Turbulent kinetic energy on the flow simulation of a bullet across a fluid.

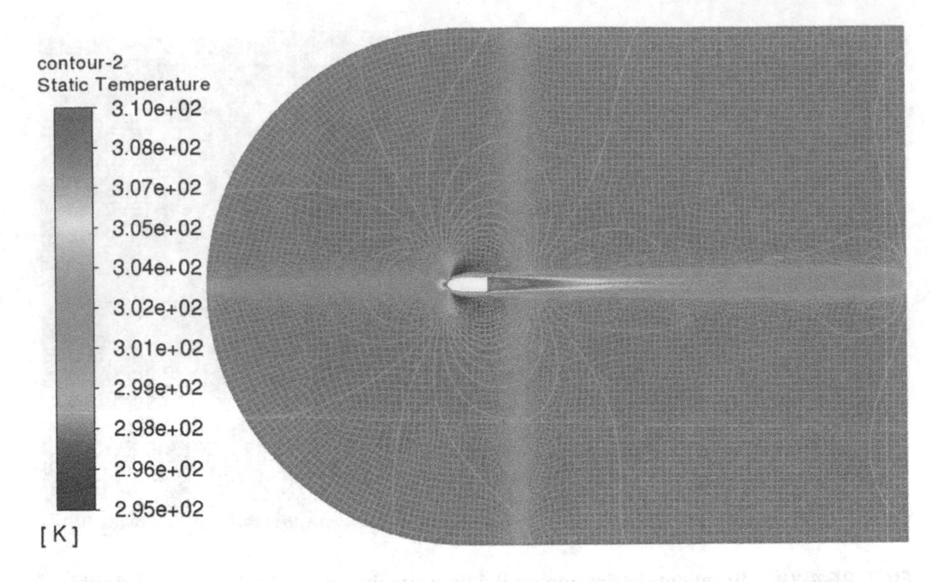


**FIGURE 4.41** Static temperature on the flow simulation of a bullet across a fluid.

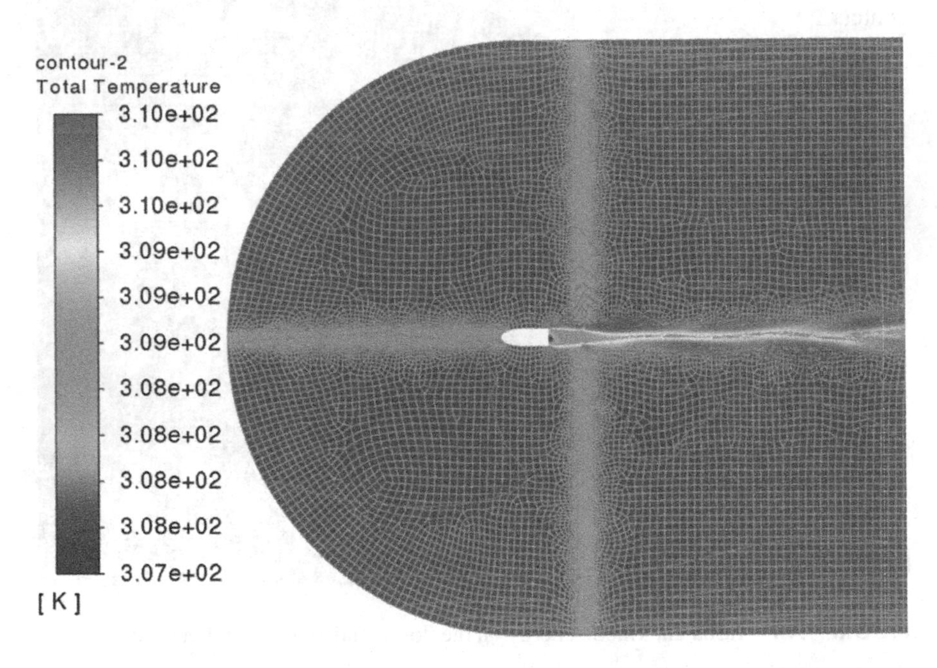


**FIGURE 4.42** Total temperature on the flow simulation of a bullet across a fluid.

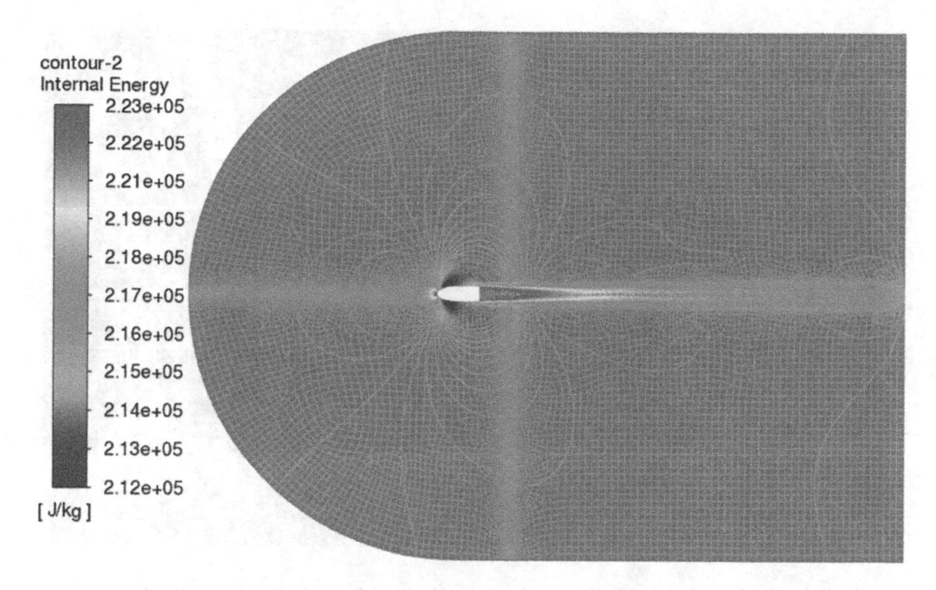


**FIGURE 4.43** Internal energy on the flow simulation of a bullet across a fluid.

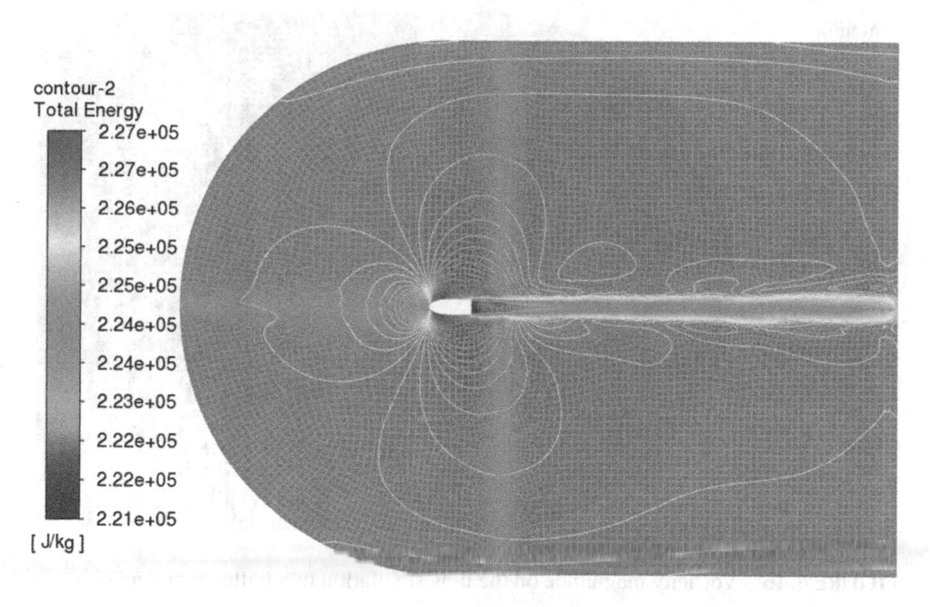


**FIGURE 4.44** Total energy on the flow simulation of a bullet across a fluid.

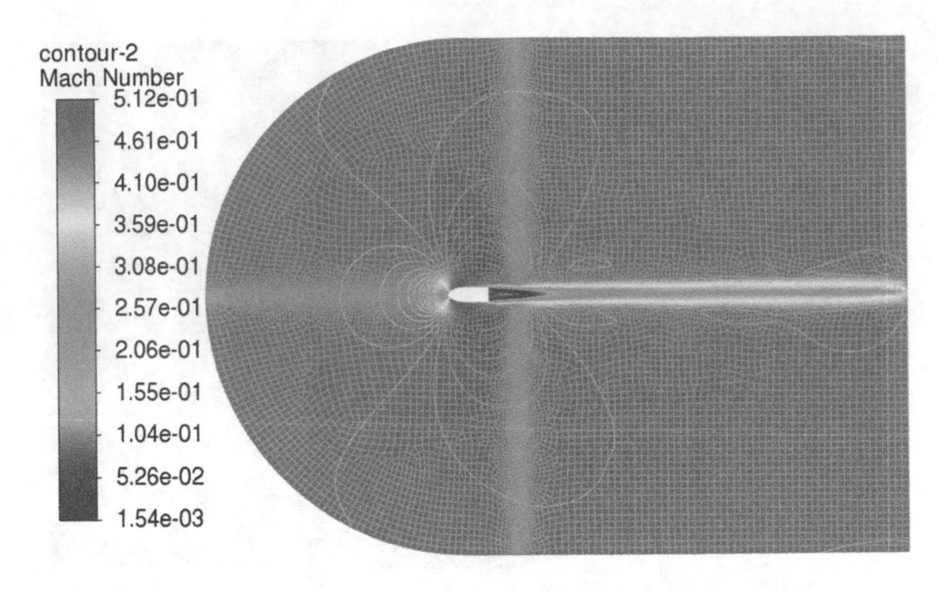


**FIGURE 4.45** Match number on the flow simulation of a bullet across a fluid.

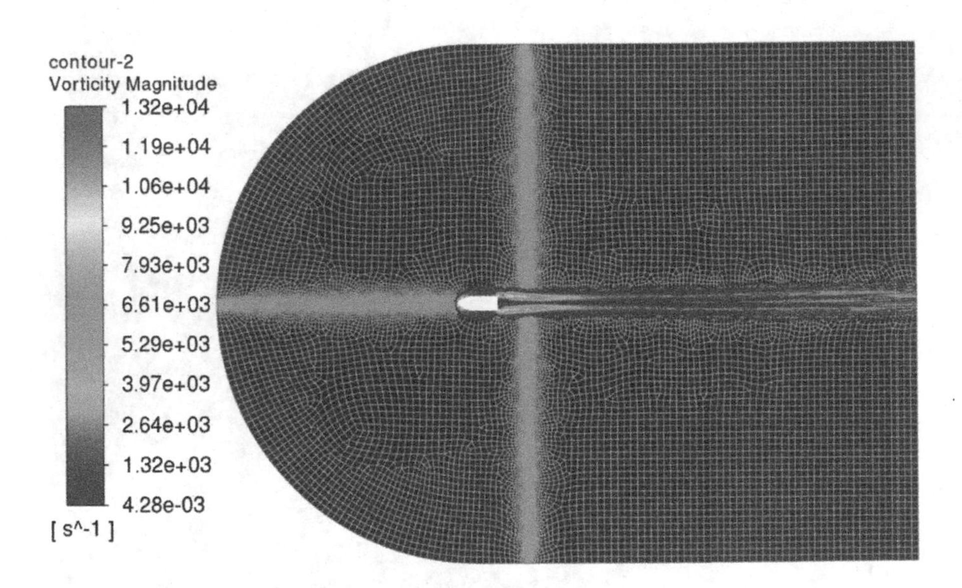


**FIGURE 4.46** Vorticity magnitude on the flow simulation of a bullet across a fluid.

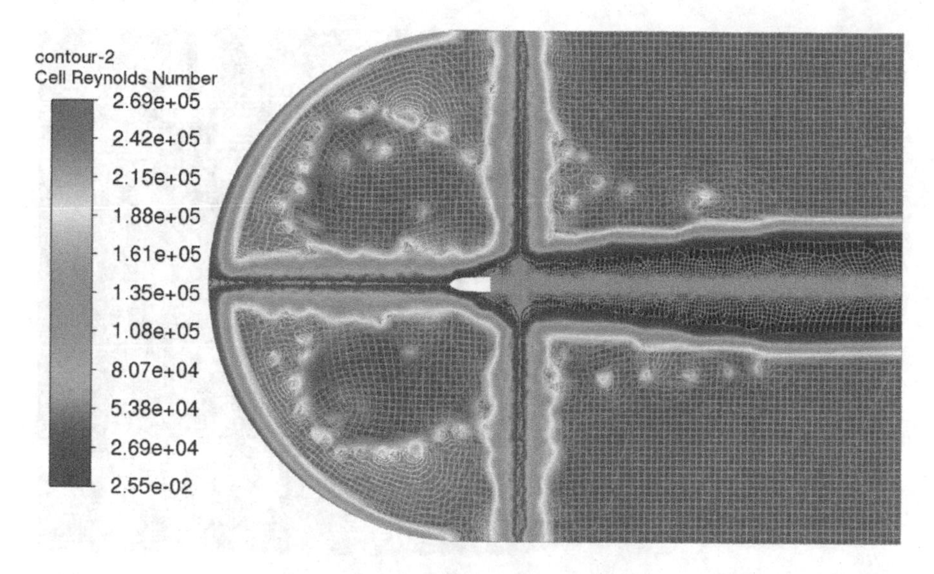


**FIGURE 4.47** Cell Reynolds number on the flow simulation of a bullet across a fluid.

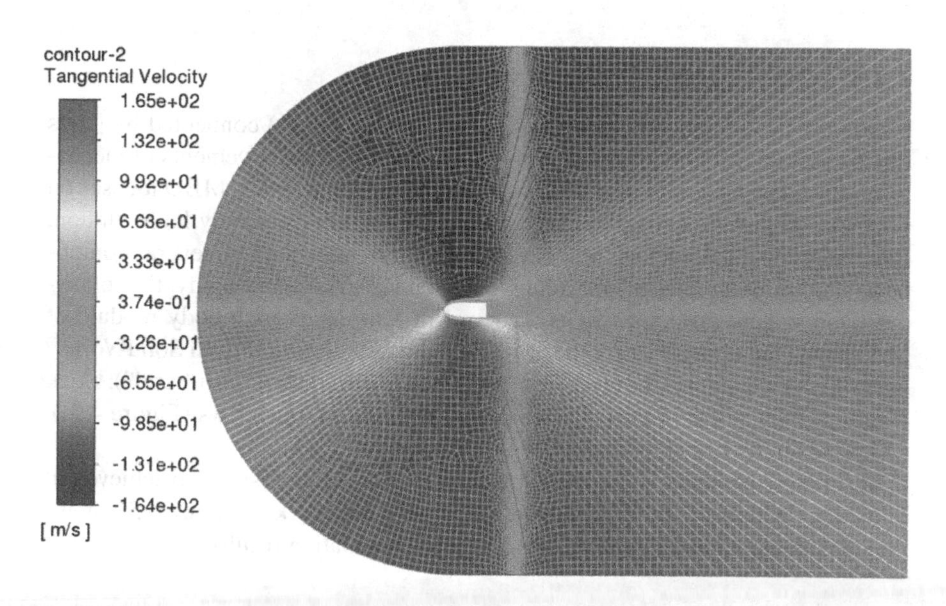


**FIGURE 4.48** Tangential velocity on the flow simulation of a bullet across a fluid.

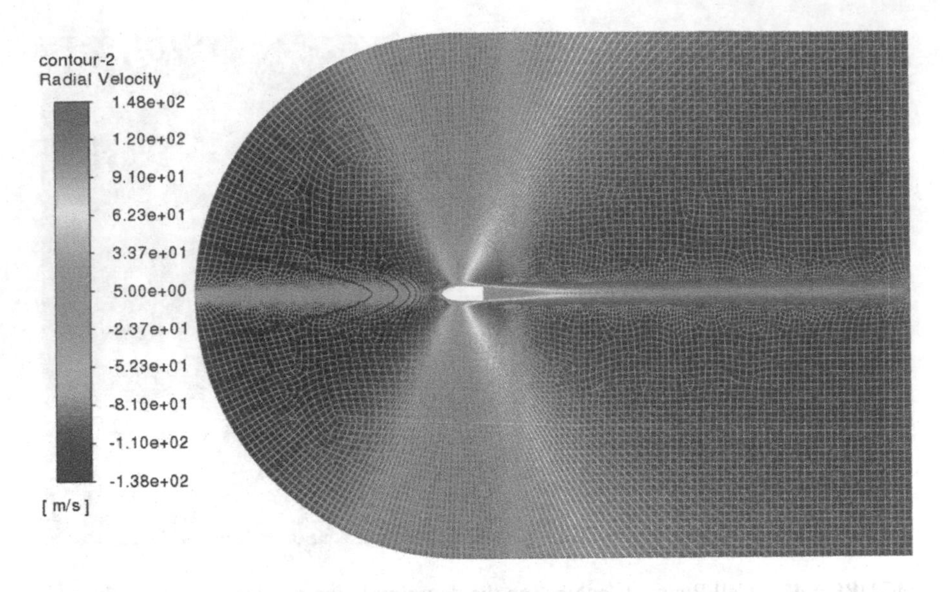


**FIGURE 4.49** Radial velocity on the flow simulation of a bullet across a fluid.

### 4.4.5 CAE Multibody Dynamics

Multibody dynamics (MBD) consists of solid bodies (links) connected by joints that restrict their relative motion. This study describes the movements of mechanism caused by forces relative to forward and reverse dynamics. MBD focuses on the entire firearm (macroscopic behavior), performance, relatively large motion, calculation of displacement, velocity, acceleration, and loads of several components and transient analysis for a long duration [23]. As a case study, the following figures show different results achieved in Simscape® multibody module of MATLAB® for a pistol, where the CAD models were imported from SolidWorks® using the code to import the model into MATLAB® (smimport, “pistol.xml”): block diagram of the pistol (Figure 4.50), pistol simulation in four views (Figure 4.51), and pistol simulation in one view (Figure 4.52).

A further case study is the simulation of a trigger mechanism, also achieved in Simscape® with the CAD models imported from SolidWorks®. Figure 4.53 shows the block diagram and Figure 4.54 presents the simulation results.

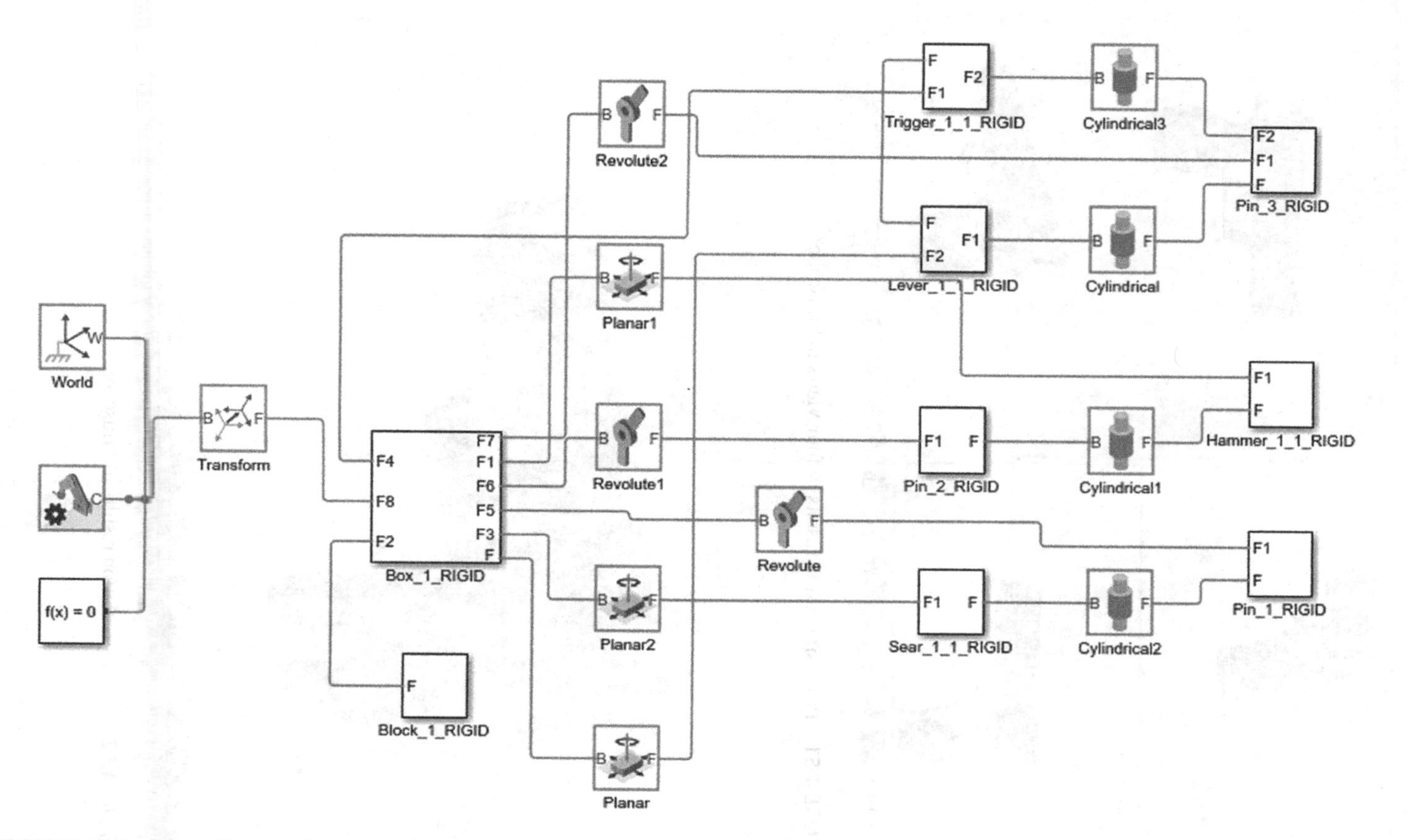


**FIGURE 4.50** Block diagram for a pistol in Simscape®.

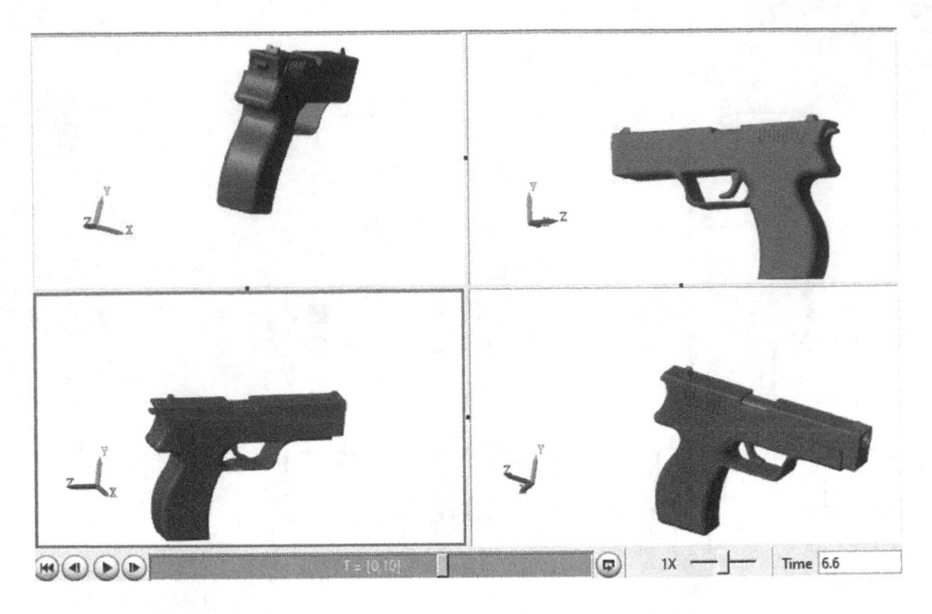


FIGURE 4.51 Pistol simulation results in four views in Simscape®.

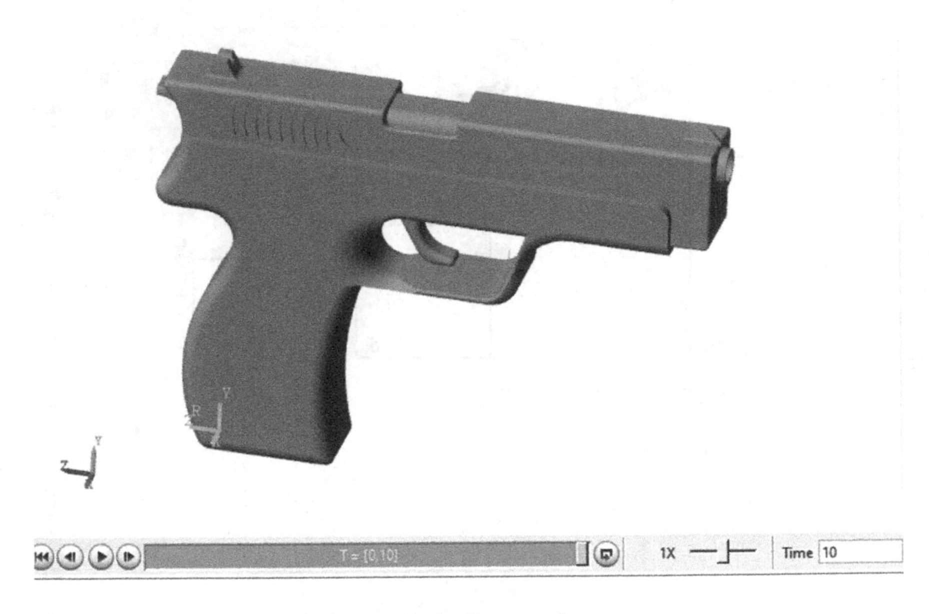


FIGURE 4.52 Pistol simulation results in Simscape®.

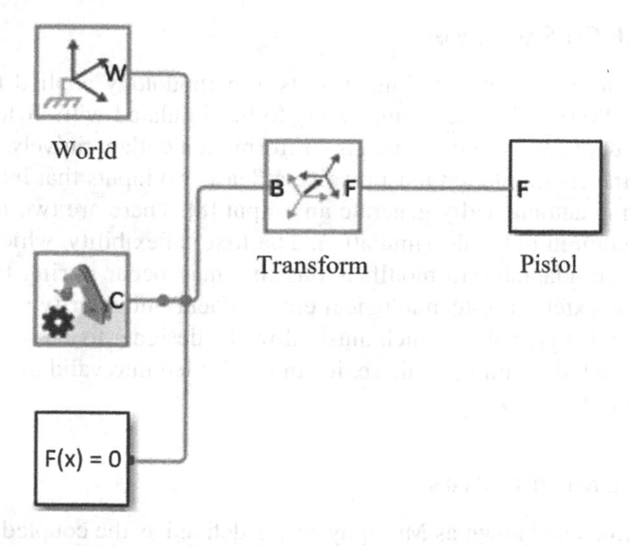


**FIGURE 4.53** Trigger mechanism block diagram in Simscape®.

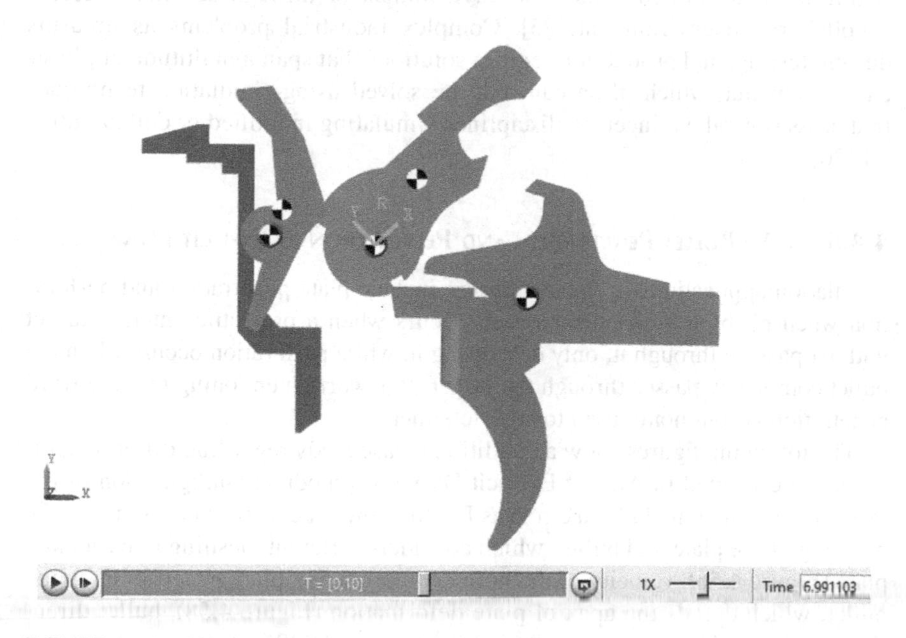


**FIGURE 4.54** Trigger mechanism simulation results in Simscape®.

### 4.4.6 CAE Co-Simulation

Co-simulation or cooperative simulation is a methodology applied to simulation, which allows individual components to be simulated with different tools running simultaneously and exchanging information collaboratively. The environment in the co-simulation must receive at least two inputs that interact with each other and automatically generate an output [8]. There are two main characteristics required in the co-simulation. The first is flexibility, which is a tool available to be adaptable to modifications that may occur during the design, such as in the external or technological environment: modularity and scalability. The second is precision, which must allow the designer to choose the levels of accuracy and, depending on these, it can be divided into validation time and functional validation [9].

### 4.4.7 CAE Multi-Domain

Multi-domain, also known as Multiphysics, is defined as the coupled processes or devices involving more than one simultaneously occurring physical field [24]. There are several Multiphysics packages such as COMSOL®, MATLAB®, and Ansys Multiphysics®. The latter allows simulating the interaction among structural mechanics, heat transfer, fluid flow, acoustics, and electromagnetic within a single unified simulation environment or through add-ins to create a collaborative environment [25]. Complex industrial problems as firearms design, testing, and production require solutions that span a multitude of physical phenomena, which often can only be solved using simulation techniques that cross several engineering disciplines simulating in unified or collaborative UI [26].

### 4.4.8 CAE Bullet Penetration and Perforation by Explicit Dynamics

Simulation applications in firearm design include plate penetration and perforation when hit by a bullet. Penetration occurs when a projectile enters a target without passing through it, only deforming it, while perforation occurs when the bullet completely passes through the target. It is worth mentioning that the word penetration is commonly used to refer to either.

The following figures show an additional case study regarding different simulations performed in Ansys® Explicit Dynamics module: configuration of the target plate, which includes a circle as BC to allow set constraints (Figure 4.55), meshing of the plate and bullet, which considers different meshing to bullet and plate (Figure 4.56), velocity of the bullet (Figure 4.57), plate penetration by the bullet, which details the apex of plate deformation (Figure 4.58), bullet directional deformation in plate penetration set up to 0.0005 s (Figure 4.59), total deformation in plate penetration by the bullet set up to 0.0005 s (Figure 4.60), and plate perforation by the bullet on the $z$ component (Figure 4.61).

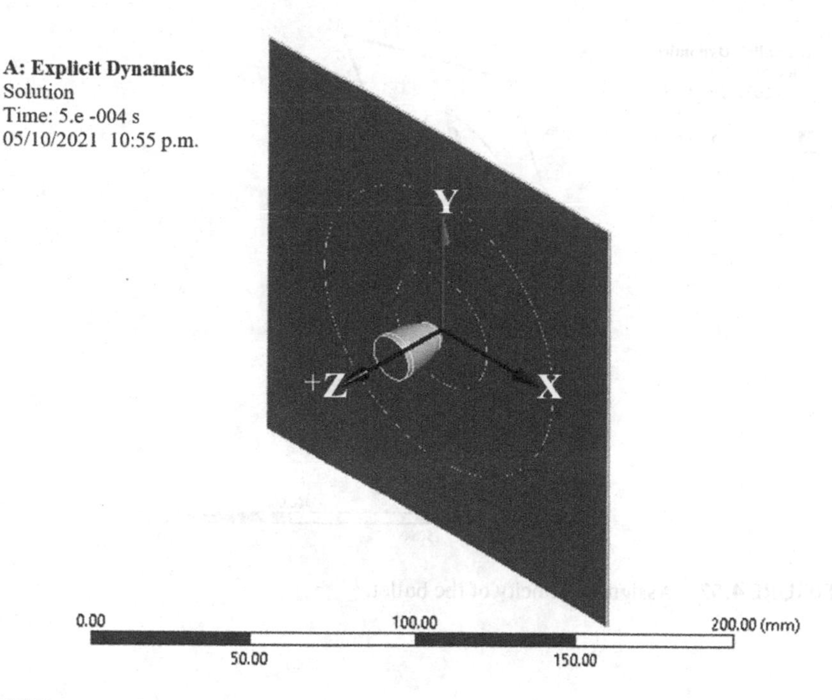


**FIGURE 4.55** Configuration of the target plate.

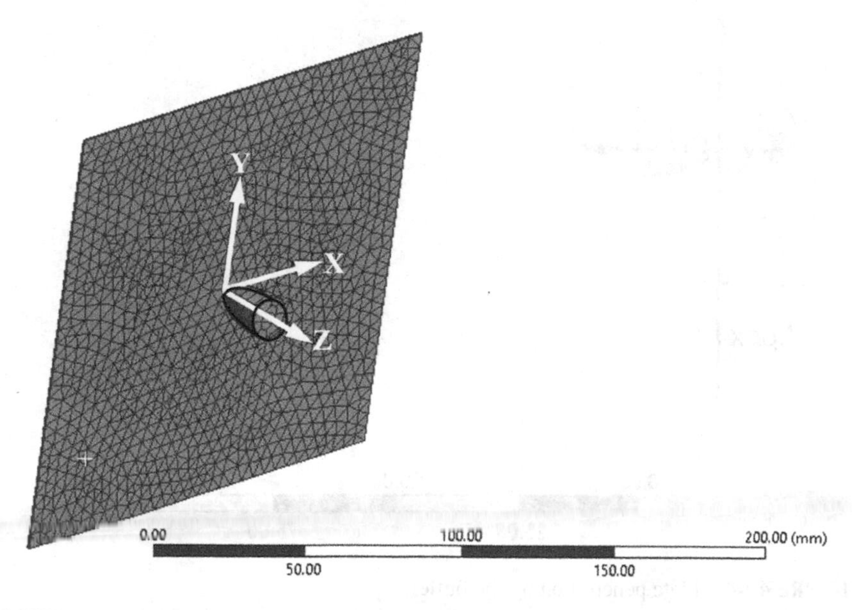


**FIGURE 4.56** Meshing of the plate and bullet.

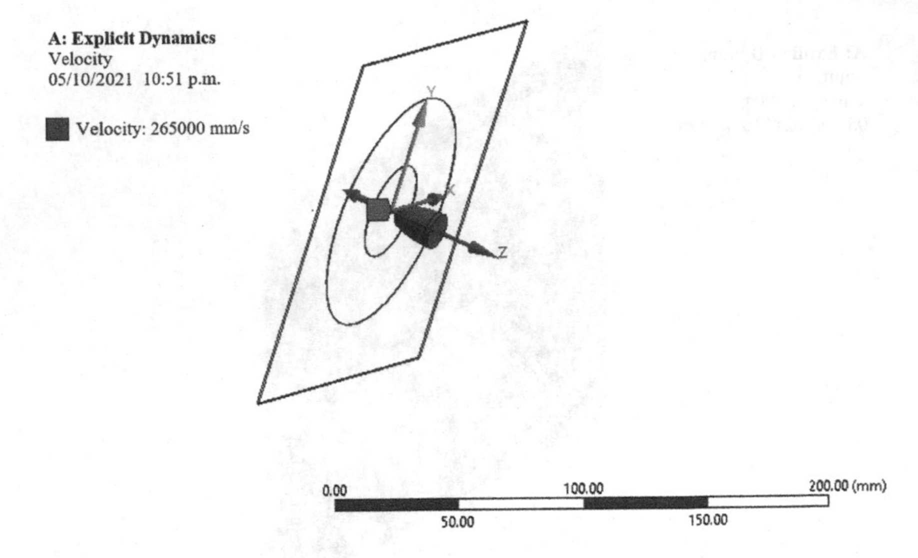


FIGURE 4.57 Assigning velocity of the bullet.

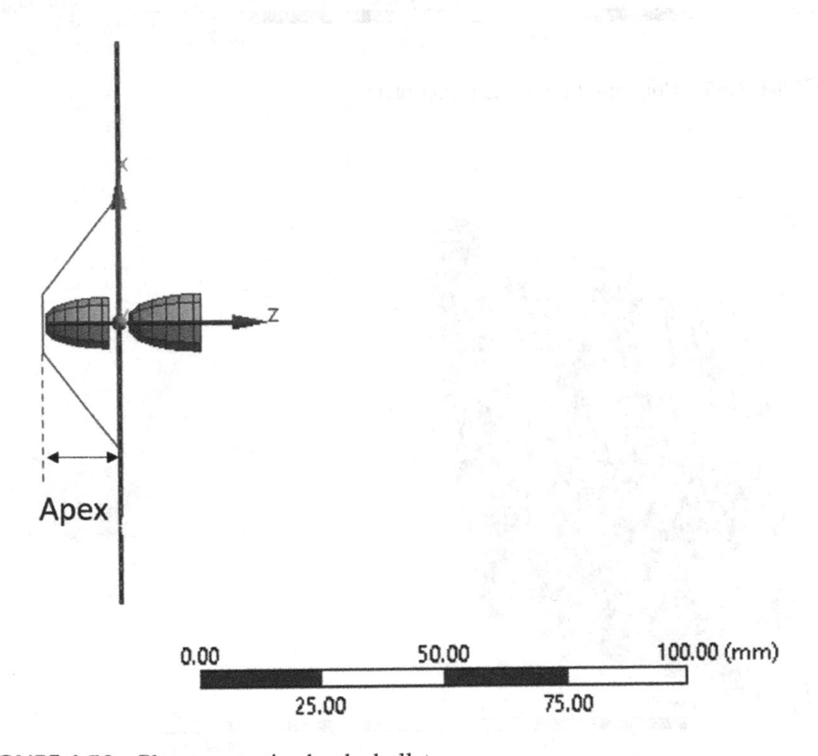


FIGURE 4.58 Plate penetration by the bullet.

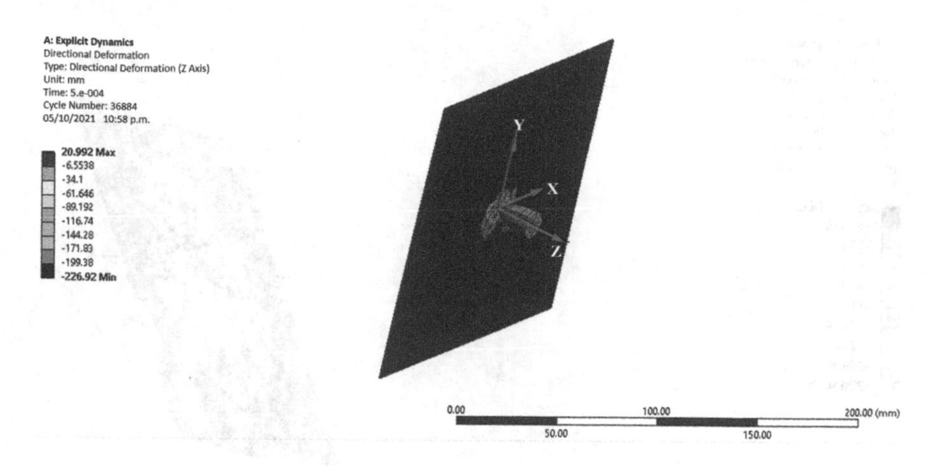


**FIGURE 4.59** Directional deformation in plate penetration by the bullet.

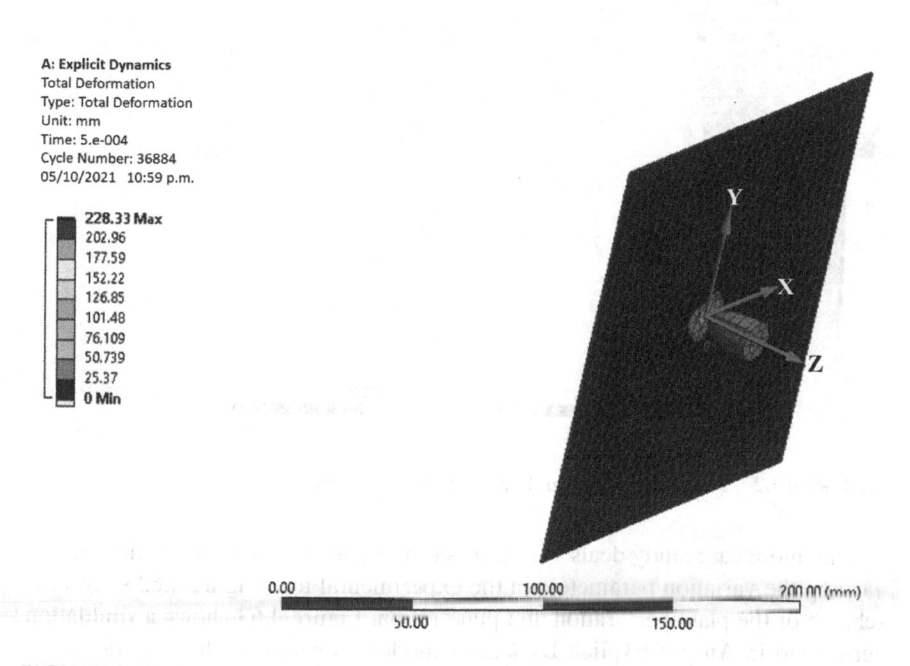


**FIGURE 4.60** Total deformation in plate penetration by the bullet.

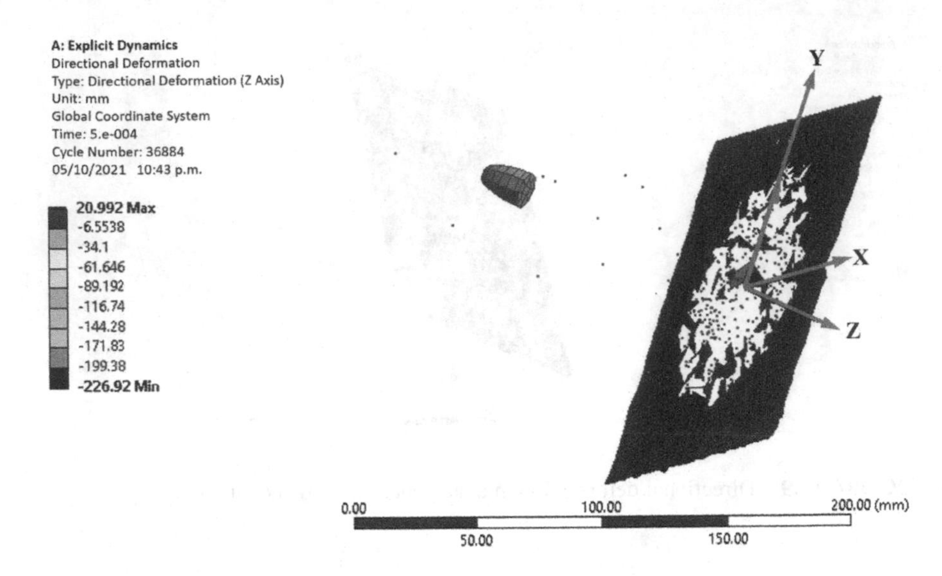


FIGURE 4.61 Plate perforation by the bullet.

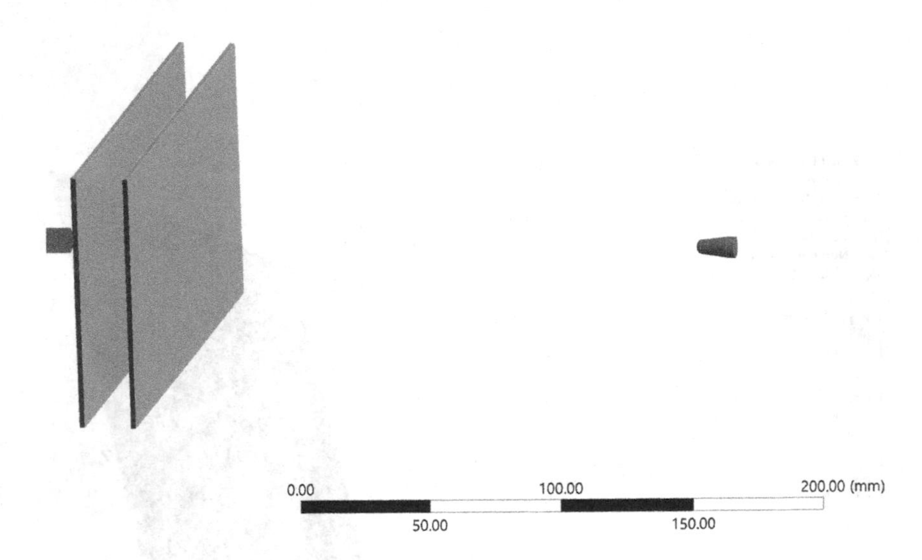


FIGURE 4.62 Plate perforation and penetration by a bullet.

One more case study deals with a shocking bullet to test the ballistic limit by varying the variation parameters in the experimental test. Figure 4.62 shows the scheme of the plate perforation and penetration. Figure 4.63 shows a simulation performed in Ansys® Explicit Dynamics module concerning the total deformation and equivalent elastic strain of the plate, which is penetrated and perforated by the bullet, after 0.00035 s.

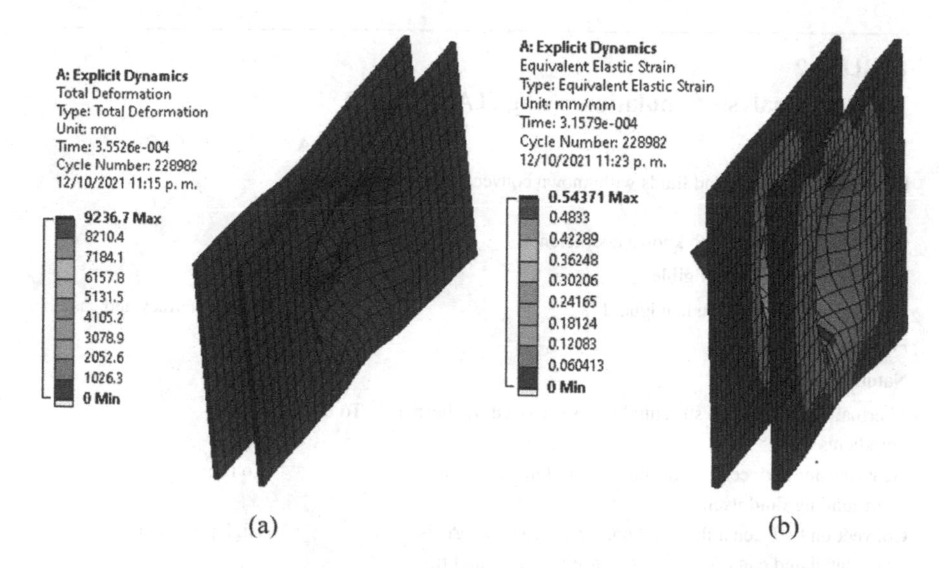


**FIGURE 4.63** Plate perforation and penetration by the bullet (a) total deformation and (b) equivalent elastic strain.

### 4.4.9 Thermal Simulation by FEA, CFD, and FSI

In several aspects of a firearm, design is complex to get an analytic calculation, which is solved by FEA and CFD or some of them. The performance of components and the operating process of firearms depend on dissipation or management of temperature [27]. The thermal analysis includes the fluid flows and heat transfer analysis, which can be simulated using only FEA, only CFD, or coupled FEA-CFD, as shown in Table 4.2.

As an example of thermal analysis simulation, Figure 4.64 presents the effective thermal conductivity caused by the energy generated by a bullet during its trajectory across a fluid; the simulation was achieved in CFD software.

### 4.4.10 CAE Emulation

The dataset in CAE comprises the input and output (results) data. The output data are analyzed in data technologies that focus on using bulk and complex data, employing data mining (existing dataset to find patterns) or machine learning (making sense of data and predictions about new datasets). The improvement of these datasets makes a difference between simulation and emulation.

**TABLE 4.2**
**Thermal Analysis Simulation Using FEA and CFD**

| Thermal Analysis | FEA | CFD |
|---|---|---|
| Heat transfer in parts and fluids with known convection coefficients | Faster solution | |
| Conduction problem with known coefficients | To use | |
| Parts with radiation negligible | To use | |
| Dissipation region of the handguard | | More accuracy and ease |
| Forced convection | To use | |
| Natural convection | | To use |
| Thermal expansion and structural stresses caused by thermal gradients | To use | |
| Heat transfer between the gas intake and fluid, including the surrounding fluid itself | | To use |
| Convection between a fluid and one or more solid parts is calculated and can take away the uncertainty around the variation of the heat transfer coefficient | | High precision |
| Convection impact by local velocities along the part surface and the turbulence in combination with temperature-dependent properties of the fluid | | High precision |
| Structural deformations and stresses due to temperature variations can be calculated with FEA. The thermal values can come from FEA or CFD analysis. CFD and FEA solvers are coupled, and results are shared | This coupled analysis is called a fluid-structure interaction (FSI) analysis | |

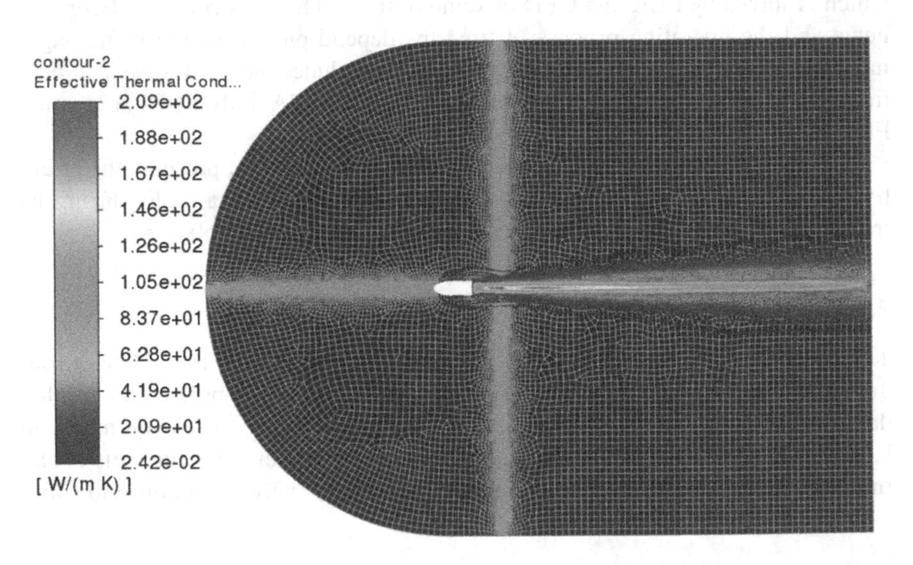


**FIGURE 4.64** Effective thermal conductivity caused by the energy generated by a bullet.

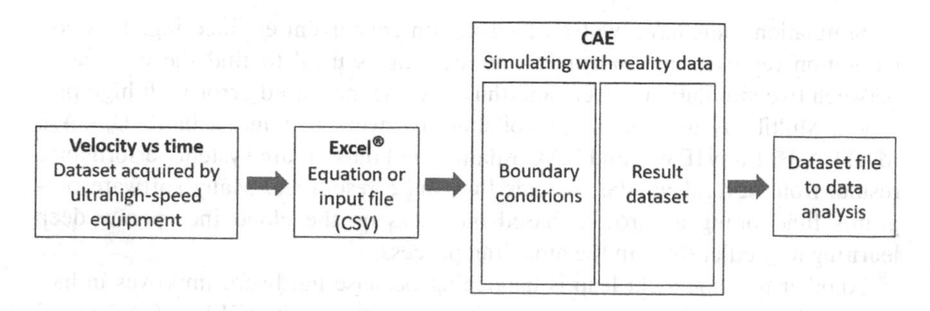


**FIGURE 4.65** Dataset sequence in CAE emulation.

The simulation tries to accurately reproduce (or predict) the behavior of the real system but only approximates it. Emulation, unlike simulation, does not approximate real system behavior but rather copy the real system behavior. For this, MSC Nastran® has spread the slogan "Simulating reality, delivering certainty" [28], which emphasizes the accuracy of the dataset. Furthermore, several companies of CAE software as NX Nastran® and Ansys® can take datasets as a datasheet in CAE software, input file, CSV from Excel® and equation of displacements, velocities, accelerations, temperatures, and pressure tabulated in Excel®.

For instance, in a CAE analysis of dynamic behavior of trigger mechanism, the dataset is acquired from ultrahigh-speed equipment on a test bench, tabulated in Excel®, and used like an input equation in the CAE software, as shown in Figure 4.65.

## 4.5 CAD-CAE DOCUMENTATION AND REPORT

Formerly, there was a documentation section in the software CAD, which allowed to generate a Word® document to explain important aspects of design. Today, CAD software as SolidWorks® includes tools to insert notes and markups on a CAD model. A report is a document to detail results and diagnosis. A CAD report shows the results of interference, sustainability, or geometric comparison reports, while a CAE report describes the simulation results. The location of this file depends on each software, but, in general, the reports are saved in the report manager. In some cases, the report manager is called the report generator, as is the name in Ansys®. NX Nastran® and Femap® call it the user-defined report generation. The customization of the report is fundamental to formalize, clearly display the results, and integrate the company logos.

## 4.6 CLOSING REMARKS AND PERSPECTIVES

The new technological trends in modeling tools are focused on the exchange of files, with which it is possible to compare files, refine the design, and apply software based on deep learning. The application of deep learning introduces predictive behavior in design software. Today, CAD software includes tools for generative design and topology optimization, so it is possible to identify the best shapes according to the design requirements.

Simulation tools have evolved focusing on concurrent engineering. This collaboration representing concurrent engineering is used to find the correlation between two simulation solvers and thus find the measured error with high precision. Multibody tools facilitate collaborative work to interact in CAD, CAE, MATLAB®, LabVIEW®, and MSC Adams® and thus obtain system performance results from several validation tools that share resources. Many software programs functioning as browser-based modelers on the cloud incorporate deep learning to predict steps in the modeling process.

Another technological leap is happening because hardware improves in harnessing the processing power of graphics processing units (GPUs); CAD users' expectations also change. The latest generation of GPUs allows working in real-time ray tracing thanks to the reduction in computing cost, in such a way that today the rendering result is so fast.

The goal of adding artificial intelligence (AI) to CAD and CAE is to ease the workflow based on deep learning to do generative design, design of experiments, CAD-CAE automation, transfer learning, visualization, and analysis. Using AI-CAD makes it possible to evaluate many 3D CAD models, estimate results, and find conceptual design candidates for the simulation stage. Also, it is convenient to evaluate shapes to optimize the design using AI-CAE.

## REFERENCES


1. Carson, J.S., Introduction to modeling and simulation, *Proceedings of the 2005 Winter Simulation Conference*, Orlando, FL, USA, 2005, IEEE.
2. Faruque, M., et al., Interfacing issues in multi-domain simulation tools. *IEEE Transactions on Power Delivery*, 2011. **27**(1): pp. 439–448.
3. Zuñiga Aviles, L.A., Methodology for modeling and simulation of mobile manipulators applied in the mechatronic design of an EOD robot (in Spanish), PhD thesis, Center for Engineering and Industrial Development. 2011, CIDESI: Mexico, p. 116.
4. Stroud, I. and H. Nagy, *Solid Modelling and CAD Systems: How to Survive a CAD System*. 2011, Springer Science & Business Media: Berlin/Heidelberg.
5. Eiríksson, E.R., et al., Precision and accuracy parameters in structured light 3-D scanning. *International Archives of the Photogrammetry, Remote Sensing and Spatial Information Sciences*, 2016. **5**: p. 10.
6. Fischer, B., Mechanical tolerance stackup and analysis. In *Mechanical Engineering*, 2nd ed, L.L. Faulkner, Editor. 2011, CRC Press Taylor & Francis Group: New York, p. 486.
7. Dawei, T. and K. Zakrisson, *Non-Rigid FE-Based Variation Simulation for Furniture*. 2016, Chalmers University of Technology: Gothenburg, Sweden.
8. Gomes, C., et al., Co-simulation: A survey. *ACM Computing Surveys (CSUR)*, 2018. **51**(3): pp. 1–33.
9. Gomes, C., et al., Co-simulation: State of the art. arXiv preprint arXiv:1702.00686, 2017.
10. Lee, H.-H., *Finite Element Simulations with ANSYS Workbench 2021: Theory, Applications, Case Studies*. 2021, SDC Publications: Mission, KS.
11. Dos Santos, F.L., et al., Multiphysics NVH modeling: Simulation of a switched reluctance motor for an electric vehicle. *IEEE Transactions on Industrial Electronics*, 2013. **61**(1): pp. 469–476.

12. Buhl, J., R. Israr, and M. Bambach, Modeling and convergence analysis of directed energy deposition simulations with hybrid implicit/explicit and implicit solutions. *Journal of Machine Engineering*, 2019. **19**: 94–107.
13. García, M., et al., Computational steering of CFD simulations using a grid computing environment. *International Journal on Interactive Design and Manufacturing (IJIDeM)*, 2015. **9**(3): pp. 235–245.
14. Stolzenburg, M.R. and P.H. McMurry, Equations governing single and tandem DMA configurations and a new lognormal approximation to the transfer function. *Aerosol Science and Technology*, 2008. **42**(6): pp. 421–432.
15. Peng, J.-S., et al., Nonlinear electro-dynamic analysis of micro-actuators: Effect of material nonlinearity. *Applied Mathematical Modelling*, 2014. **38**(11–12): pp. 2781–2790.
16. Hivet, G. and P. Boisse, Consistent 3D geometrical model of fabric elementary cell. Application to a meshing preprocessor for 3D finite element analysis. *Finite Elements in Analysis and Design*, 2005. **42**(1): pp. 25–49.
17. Boundary conditions. [August 31, 2021]; Available from: https://www.simscale.com/docs/simulation-setup/boundary-conditions/.
18. Singh, R., H. Garg, and V. Guleria, Haar wavelet collocation method for Lane–Emden equations with Dirichlet, Neumann and Neumann–Robin boundary conditions. *Journal of Computational and Applied Mathematics*, 2019. **346**: pp. 150–161.
19. Cuillière, J.-C. and V. Francois, Integration of CAD, FEA and topology optimization through a unified topological model. *Computer-Aided Design and Applications*, 2014. **11**(5): pp. 493–508.
20. Marcé Nogué, J., et al., Coupling finite element analysis and multibody system dynamics for biological research. *Palaeontologia Electronica*, 2015(**18**.2. 5T): pp. 1–14.
21. Nedelcu, D., et al. *The Kinematic and Kinetostatic Study of the Shaker Mechanism with SolidWorks Motion*. In Journal of Physics: Conference Series. 2020, IOP Publishing: Bristol.
22. Lange, C., et al., Impact of HPC and automated CFD simulation processes on virtual product development: A case study. *Applied Sciences*, 2021. **11**(14): p. 6552.
23. Sherman, M.A., A. Seth, and S.L. Delp, Simbody: Multibody dynamics for biomedical research. *Procedia Iutam*, 2011. **2**: pp. 241–261.
24. Liu, Z.L., *Multiphysics in Porous Materials*. 2018, Springer: Berlin/Heidelberg, pp. 29–34.
25. Multiphysics solutions features. [October 21, 2021]; Available from: https://www.ozeninc.com/ansys-multiphysics/multiphysics-solutions-features/.
26. Multiphysics simulation. [August 31, 2021]; Available from: https://www.plm.automation.siemens.com/global/en/products/simulation-test/multiphysics-simulation.html.
27. Dhande, D., G. Lanjewar, and D. Pande, Implementation of CFD–FSI technique coupled with response surface optimization method for analysis of three-lobe hydrodynamic journal bearing. *Journal of the Institution of Engineers (India): Series C*, 2019. **100**(6): pp. 955–966.
28. Bernhardt, R., H. Schafstall, and I. Hwang, Simulate reality-deliver certainty through the virtual weld. *Journal of Welding and Joining*, 2016. **34**(5): pp. 41–46.

# 5 CAM Assessment and Rapid Prototyping

## 5.1 INTRODUCTION

Approval of part design of a firearm is achieved once the interaction between components and assemblies is evaluated with satisfactory results. This evaluation is a repetitive process where feedback leads to design refinement, first in the appearance aspect, then executing the modes of operation to ensure blockage and part sequential movement, and finally in ballistic laboratories and field tests.

The information provided in this chapter is linked to Chapter 4 when comparing the CAD model with the rapid prototype (RP), to Chapter 6 when doing the experimentation using the experimental physical model, and to Chapter 9 when selecting the suitable manufacturing process.

The manufacture of tangible models requires a CAM assessment to evaluate the solution proposal and verify machining times and costs. Rapid prototyping is focused on processes such as CNC machining and 3D printing to obtain necessary parts and tools without an industrial approach or at a high price. These processes require metal inserts to evaluate the firearm design with or without ammunition in a ballistic laboratory using a test bench.

This chapter addresses the considerations to evaluate the manufacture of firearm components from their part features, which are validated by simulation and redesigned conforming to the defined manufacturing process.

## 5.2 TECHNICAL PRELIMINARIES

In the literature, there is information on rapid prototyping for the manufacture of firearms. This information requires certain clarifications, since it is information of disclosure of manufacturing processes that lacks scientific rigor and details of implementation. Some methodologies show the immediate manufacture of prototypes before their perfection, making the firearm development process more expensive [1].

Design and manufacturing are activities that must be balanced according to the infrastructure and technical capacity of the design team and manufacturing team. Suppose the designers do not have enough skill and knowledge. In that case, the manufacturing team is expected to compensate for this disadvantage with infrastructure and technical capacity to deduce design aspects and consider good manufacturing practices to achieve the design of the weapon components, reaching the required tolerances and the principle of

DOI: 10.1201/9781003196808-5

operation of the weapon when assembled. Now, suppose the manufacturing team does not have sufficient skills. In that case, the designers compensate for this disadvantage by performing more specialized computing work, from modeling, interference detection, tolerance analysis, Motion, FEA, CFD, and MBD simulations until an acceptable correlation is achieved comparing the solutions of each solver.

As described in Chapter 2, models or dummies are used during CAD modeling, which are tangible physical models that assist in formulating the solution. In addition, during CAE simulation, experimental physical models (EPMs) are used to validate the operating principle. For these activities, which correspond from technology readiness level 1 (TRL1) to TRL3 and later to TRL4 as described in Chapter 6, a CAM assessment is carried out to determine the materials and manufacturing process of RPs: models, dummies, experimental physical models, and prototypes controlled in a laboratory setting. Then, RPs refer to tangible physical models that formulate the solution and validate the operating principle. The solution formulation and the operating principle are developed from TRL1 to TRL 4. They do not refer to a prototype as the first type to be replicated in a TRL5, TRL6, or TRL7. The former is used for experimental lots and the last two are used in a real environment (Figure 2.5).

## 5.3 CAM ASSESSMENT

The manufacturing assessment is performed to obtain the RPs that are required in low-volume manufacturing and high-volume manufacturing [2]. The CAM assessment for low-volume manufacturing is performed based on the following considerations (Figure 5.1):

- Clamping devices, which include rapid holding systems and flexible positioning clamping [3].
- CAD drawings to be used as a template for laser cutting or dimensioning in optical comparators and approbation of RPs.
- Required parts to experimental process.

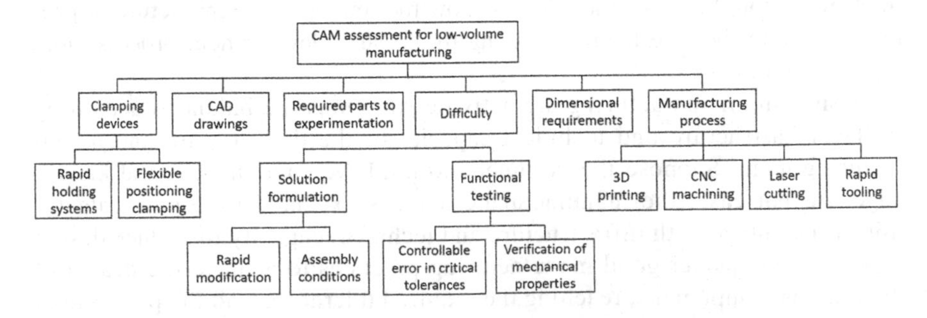


**FIGURE 5.1** Consideration to CAM assessment for low-volume manufacturing.

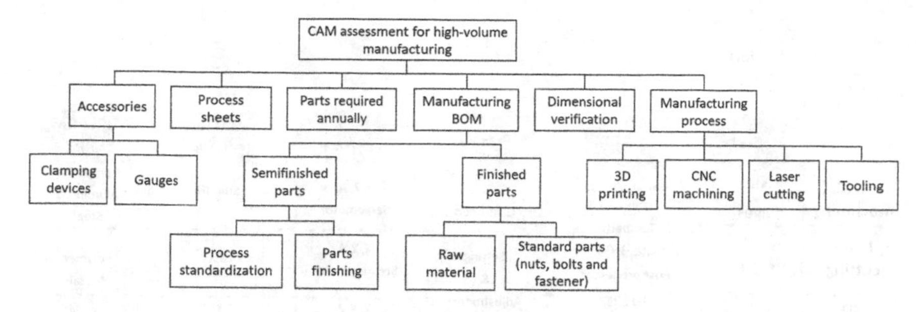


**FIGURE 5.2** Consideration to CAM assessment for high-volume manufacturing.

- Difficulty degree, considering the quick modification and assembly conditions for solution formulation, as well as controllable error in critical tolerances and verification of mechanical properties for functional testing.
- Dimensional requirements according to styling process, interaction, simplification, and assembly.
- Manufacturing processes, where polymer and metal 3D printing, computer numerical control (CNC) machining, laser cutting, and rapid tooling (RT) are taken into account.

The CAM assessment for high-volume manufacturing is accomplished based on the following aspects (Figure 5.2):

- Accessories, which include clamping devices and gauges.
- Process sheets to be used as a guide to fabrication and approbation of the product.
- Parts required annually at the beginning of the pilot lot and later for sale.
- Manufacturing bill of materials (BOM), considering the process standardization and part finishing for semifinished parts, as well as raw material and standard parts for finished parts.
- Dimensional verification in each component and permissible gaps in assembly using scanning sensors (vision systems) on coordinate measuring machine (CMM).
- Manufacturing process to high-volume manufacturing, considering polymer and metal 3D printing, CNC machining, laser cutting, and tooling.

Figure 5.3 shows the workflow to manufacture an RP of a firearm by CNC machining, laser cutting, or 3D printing, which considers the variation among each type of process used for this purpose. For the case of CNC machining, it is possible to obtain a model in an exclusive extension depending on CAD software: *.sldprt from SolidWorks® or generic extension as Parasolid *.x_t. The CAD file is converted to *.step or *.iges formats, which are valid input data as CAM files and are

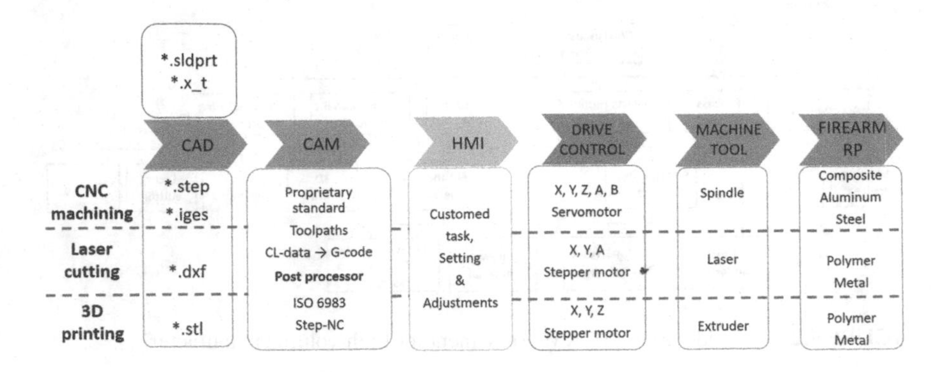


**FIGURE 5.3** Scheme of workflow to get RPs.

open in CAM software as Mastercam®, Edgecam®, SURFCAM®, or Cimatron®, where the manufacturing parameters as toolpaths are analyzed [4]. For the case of CAM software, the CAM file is converted to cutter location (CL) file, or CL-data and later from the postprocessing phase to G-code, which includes the G-code (axis movements) and M-code (auxiliar code) [5]. The G-code, also known as ISO 6983 or with the most modern version as STEP-NC, is used by CNC machines, laser cutters, and 3D printers to operate [6]. Today, the drive control can operate up to ten axes driving the spindle to work RPs in composite, aluminum, and steel. The CNC control panel is the human-machine interface (HMI) for movement programming, backup, and file upload and download. The drive control operates X, Y, Z, A, and B axes according to the number of CNC machine axes [7].

For the case of laser cutting, it is also possible to obtain a model in an exclusive extension depending on CAD software: *.sldprt from SolidWorks® or generic extension as Parasolid *.x_t. Depending on vectorizing requirements, CorelDRAW® with the *.cdr format can be used. The CAD file is converted to *.dxf format, which is a valid input data as a CAM file and is open in CAM software as SmartCarve®, DraftSight®, or SketchUp®, where the manufacturing parameters as toolpaths are analyzed. CAM software converts the CAM file to CL-data and later from the postprocessing phase to G-code, including the G and M codes [8]. The HMI assists to upload files, turn on the laser, and visualize the cutting progress. The drive control operates X and Y axes and the rotational movement A to move the laser during the cutting of an RP part in polymer or metal [7].

For the case of 3D printing, it is also possible to obtain a model in an exclusive extension depending on CAD software: *.sldprt from SolidWorks® or generic extension as Parasolid *.x_t. The CAD file is converted to *.stl format and transferred to a 3D printer, where a Slicer software is used. The Slicer file is opened in Slicer software as Slic3r®, Cura®, Simplify3D®, 3DXpert®, or 3D Sprint®, where the manufacturing parameters are analyzed, and the slice errors are verified and corrected. The Slicer file is converted to CL-data in the Slicer software and later from the postprocessing phase to G-code [9]. The drive control operates X, Y, and

Z axes in Cartesian or Delta types to position the extruder when printing an RP part in thermoplastic and metal materials.

## 5.4 CNC MACHINING

One of the ways to manufacture RPs is using CNC machinery, which includes milling, turning, grinding, drilling, router, and machining center (MC). A more extensive explanation of the manufacturing processes of firearm components is given in Chapter 9.

Numerical control (NC) is a system of movement programming on the HMI of the machine tool by an operator [10]. When this action is made through a CAM software, it is called CNC, and when this action is carried out by means of a network to allow simultaneous upload and download instructions of multiple CNC machines, it is called direct numerical control (DNC) [11]. In the manufacture of firearms, the evolution of the machinery began using the DNC; the creation of multi-operation machines (MMs) emerged later. An MM can perform up to 150 operations from component machining to pistol manufacturing. A disadvantage of MMs is that they are designed with an integral and custom design approach, making it difficult to repair, scale, and upgrade. Currently, this approach has changed using a modular design approach. Some of these modules work as MCs with multifunction programming, automatic tool changer, and tool storage. An MC comprises a control cabinet, tool magazine, head, HMI with programs backup, uninterruptible power supply (UPS), worktable, machine bed, servomotors to axis, drives, and automatic tool changer [12]. The DC servomotors were used formerly in the MCs; today, the features of the MCs have changed to supply AC 3×400 V using, for instance, Siemens® technology as SINUMERIK® 840D, conveyor, feeder, security network, and coolant system [13].

CNC machines are classified based on the following characteristics: motion control, point to point (PTP), continuous path; control loops, open and closed loop; power drives, hydraulic, electric, and pneumatic; positioning, absolute and incremental [14]. The assessment of the process allows considering some techniques to optimize the RPs, such as reducing machining time of the workpiece, dividing small linear toolpath to curve or spline, and using preform blocks. Some advantages of this process are greater accuracy, repeatability, and obtaining complex part geometries [15].

CNC programming is classified into three types (Figure 5.4): the G-code (includes M-code) RS-274, open architecture control (OAC), and STEP-NC, the most modern code [16]. G-code is classified into the following versions: binary cutter language (BCL); Deutsches Institut für Normung (German Institute for Standardization - DIN) 66025; PN-73M-55256 and PN-93M-55251; International Organization for Standardization (ISO) 6983; electronic industries alliance (EIA) RS274D; and the version EIA RS274X, which is used for the fabrication of printed circuit board (PCB); conversational programming, which is a wizard-like mode that hides G-code, for instance, the IPS® from Haas®, MAZATROL® from

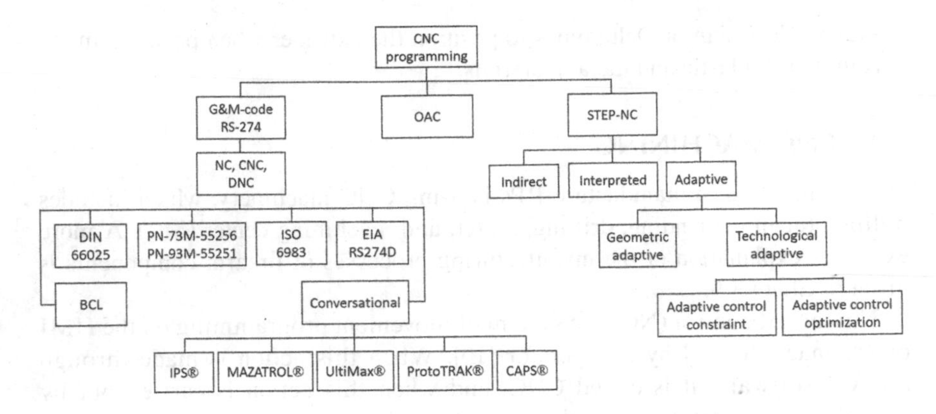


**FIGURE 5.4** Types of CNC programming.

Yamazaki Mazak Corporation®, UltiMax® from Hurco Companies®, ProtoTRAK® from Southwestern Industries®, and CAPS® from DMG Mori Seiki® [17].

OAC is growing due to the tendency to create open-source machine designs. OAC is not as suitable for high-volume manufacturing due to lack of compatibility, scalability, and technical support. However, OAC is an excellent option for RPs as it is possible to use open source by combining rapid manufacturing processes to low cost and with sufficient characteristics for the fabrication and testing of firearms.

STEP-NC is classified into the following types: indirect, interpreted, and adaptive; the latter is organized into geometric adaptive and technological adaptive [16]. Adaptive control constraint and adaptive control optimization are control modes of technological adaptive. STEP-NC-based exchange of product data in product lifecycle management (PLM) software helps in faster information processing and reduces errors owing to human intervention. A new format of STEP-NC, called STEP-NC AP238, was recently implemented and includes geometric dimension and tolerance, as well as product data management (PDM) [18]. STEP-NC AP238 allows adaptive manufacturing with bidirectional information. These data include the continuous feedback verifying the geometric dimensioning and tolerancing (GD&T), minimizing the errors, estimating process data, and converting to machine-specific code for the case of sending manufacturing data to CNC machines and 3D printers [18].

## 5.5 LASER CUTTING

Laser cutting is another beneficial manufacturing technology for firearm RPs, especially for the evaluation of design concepts in mechanisms such as the firing mechanism and recoil system. Its versatility makes its applications very convenient since the time to obtain components is faster than in a 3D printer; in addition, tolerances of ±0.03 mm for 10 mm thickness can be attained [19]. Also, considering the reduced dimensions of small weapon components, laser cutting is better to get RPs than water jet cutting and plasma cutting due to the average

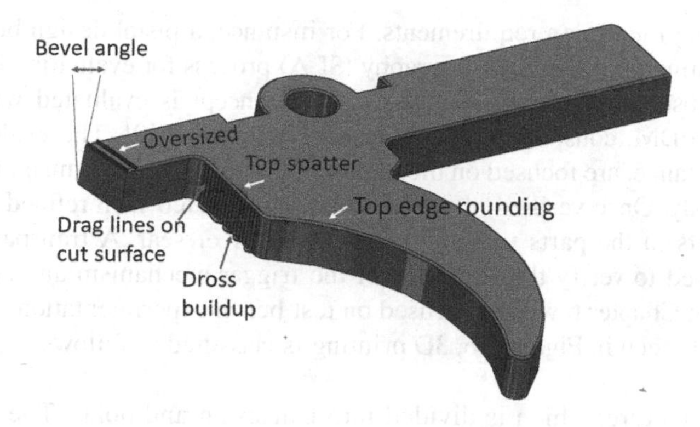


**FIGURE 5.5** Trigger features manufactured by laser cutting.

speed of 300 mm/s, cut finish, tolerance, and kerf thickness [20]. Figure 5.5 shows a schema of a trigger produced by laser cutting, where the features of the process are indicated. It is required to consider that the thicker the component to be cut, the larger the bevel angle, and oversized in the critical dimension to fit it later, thereby removing the dross buildup and the kerf thickness.

There are several kinds of laser cutting machines, but the most used for firearm RPs are $CO_2$ machines for cutting polymers and doped silica fiber machines for cutting metals [21]. For instance, depending on metal thickness, it can be used 750 W with the fiber laser generator Raycus® [22] or 10 kW considering the best quality of an IPG® fiber laser [23].

Advances in adaptive manufacturing and the advantages of STEP-NC have improved the versatility and performance of the laser cutting, thus compensating for errors and improving the characteristics of the obtained component. The stepper motors drive the movements of the laser tool in the X and Y axes, as well as in the A rotary axis within the workspace, for both $CO_2$ and fiber machines [24].

## 5.6 3D PRINTING

It is common to refer to additive manufacturing as a synonym for 3D printing. However, it is important to highlight its appropriate use according to the capabilities of this manufacturing process for RPs. 3D printing is suitable for evaluating the design concept, solution formulations, and operating principle, but it is not suitable for cook-off testing, shooting, or field testing. As mentioned before, important advances have been made in the implementation of adaptive manufacturing, which is why 3D printing has increased its uses and scope. The design of firearms depends to a large extent on the speed of proofs-of-concept validation and principles of operation of their mechanisms. In this sense, 3D printing is fundamental, starting from the design concept tests with the manufacture by powders to verify appearance, and with the fused deposition modeling (FDM) process to evaluate the interaction of mechanisms. These evaluations are compared with those carried out in the simulation, allowing the refinement of the design concept until meeting the design requirements. For instance, a pistol design begins with the 3D printing by the stereolithography (SLA) process for evaluating its appearance and cosmetic aspects; later, the design concept is evaluated with an RP printed by FDM, considering a tolerance of ± 0.1 mm [19]. The evaluations of RPs, for instance, are focused on the interaction of trigger mechanism, magazine, and bolt body. Once verified, the interaction is evaluated with refined FDM and metal inserts in the parts that operate as hammer or sear. A transparent pistol model is used to verify the operation of the trigger mechanism and acquire the testing data; Chapter 6 will be focused on test bench experimentation.

As can be seen in Figure 5.6, 3D printing is classified as follows:

- Architecture, which is divided into Cartesian and polar. The former includes Rectilinear that can be H-bot, CoreXY, or Belt; Delta; and Scara.
- Drive location, which can be of three types: Bowden if a Bowden extruder is used; direct drive if the extruder and nozzle are in the same assembly; and remote motor if the motor and extruder are separated from the nozzle [25].
- Firearm material, which refers to the application, properties, functioning, and features of materials used for the 3D printing. The application includes the concept models, tooling, and functional models. Properties are classified into the polymer that can be thermoplastic (TP), thermosetting (TS) or elastomer, metal, industrial use, filament characteristics, nozzle, powder, flexible, strength, high resolution, transparent, complex geometry, support layer, controlled heating in the chamber, heat bed, cooling, pausing, and impact resistant [26]. Functioning is divided into extrusion, photopolymerization, and powder bed fusion. Features are classified into rubber-like, polylactic acid (PLA), acrylonitrile butadiene styrene (ABS), Verowhite, Visiclear, ABS-like, Nylon, thermoplastic polyurethane elastomer (TPU), polycarbonate (PC), acrylic styrene acrylonitrile (ASA), and polyethylene terephthalate glycol (PETG) [27].

Looking at Figure 5.6, different considerations to evaluate a 3D printing can be taken into account. For instance, if the concept model is required with transparent appearance and high resolution, the extrusion functioning and the Visiclear material should be selected; if the functional model is required with strength, the extrusion functioning, and the Nylon material should be selected. According to the material used for 3D printing, there are features better performed with a specific material type; for instance, layer adhesion is better achieved with PLA, heat resistance with PC and ABS, impact resistance with TPU, visual quality with Nylon, maximum strength with PC, and ease of printing with PLA. Besides, multicolor is attractive; the use of black and white is preferable for RP manufacturing [28]. Concerning the types of processes, FDM and selective laser sintering (SLS) are the most used. Further, it is important to consider acquiring a 3D machine for spare parts, support, maintenance, and standardization to obtain RPs [29].

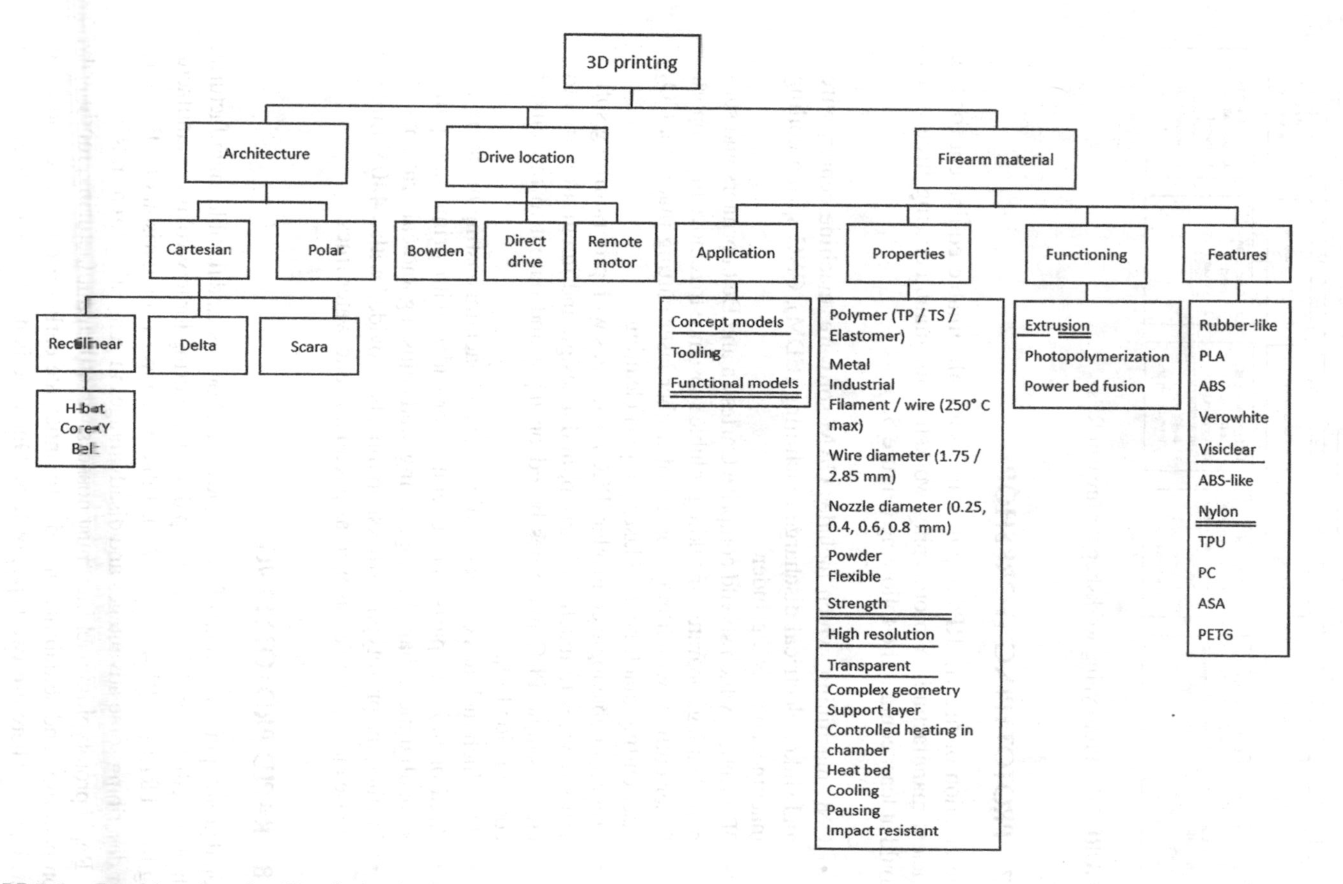


**FIGURE 5.6** Considerations for 3D printing assessment.

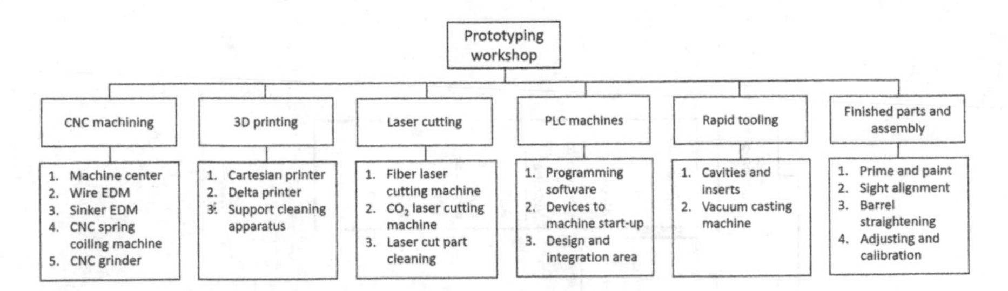


**FIGURE 5.7** Prototyping workshop for firearm RPs.

## 5.7 PROTOTYPING WORKSHOP

The creation of firearm RPs requires an area with suitable equipment for their efficient manufacture. A prototyping workshop for manufacturing firearm RPs should at least contain the following (Figure 5.7):

- CNC machining, which includes CAM software, machine center, wire and sinker electrical discharge machining (EDM), CNC spring coiling machine, and CNC grinder.
- 3D printing, which should comprise Cartesian and Delta printers and support cleaning apparatus, including soluble substance to remove the support.
- Laser cutting, which may consist of a fiber laser cutting machine, $CO_2$ laser cutting machine, and laser cut part cleaning.
- Programmable logic controller (PLC) machines with programming software, devices to machine start-up, and a design and integration area to create simple PLC machines based on input and output to movement coordination [30].
- RT, which includes cavities and inserts and vacuum casting machines.
- Area of finished parts and assembly, provided with prime and paint, sight alignment, barrel straightening, and adjusting and calibration.
- Further, the prototyping workshop must be provided with a 440 V three-phase electric power, as well as pneumatic and water lines.

## 5.8 RAPID PROTOTYPING

Rapid prototyping is a group of techniques that allow fast and flexible manufacturing which, together with RT, is part of rapid manufacturing for low-volume manufacturing from TRL1 to TRL4 (Figure 2.5). Table 5.1 shows a comparison of RP features for describing some advantages and disadvantages in generating firearm RPs.

Rapid prototyping machines must possess versatile manufacturing, modularity, compatibility, and scalability in order to anticipate early obsolescence and underutilization. There are two types of RPs. One type is used to get concept design, solution formulation, and test the operation principle using an experimental physical model tested under controlled laboratory parameters. The other type of RP is

**TABLE 5.1**
**Comparison of RP features**

| Item | Applications | Disadvantage |
|---|---|---|
| 1 | Quick solution formulation, design concepts, and EPMs. | Thermal shrinkage and warping of the model. |
| 2 | Multiple iterations for optimization. | Model size limitation due to printer or laser cutter workspace. |
| 3 | Systematic refinement. | Postprocessing time and resources are required. |
| 4 | Ideation traceability. | Limited mechanical properties of raw material. |
| 5 | Reduced time and cost. | Problems due to part surface by layering. |
| 6 | Fast detection of critical dimensions. | Limited control of material variability. |
| 7 | Customizing. | High manufacturer dependence for raw materials, upgrades, and technical support. |

used to define the high-volume manufacturing process. It is worth mentioning that there are companies that supply components for RPs, such as Ballistic Advantage, which manufactures rifle barrels and other firearm components for the AR-15 platform [31]. Other companies have developed adaptive machining processes using custom CNC machines that reduce operator intervention and machining errors.

## 5.9 RAPID TOOLING AND MANUFACTURING DEVICES

RT refers to mold cavities that are either directly or indirectly fabricated using rapid prototyping techniques. Figure 5.8 makes a classification of manufacturing devices and RT. When an RP is being validated, material properties, accuracy, cost,

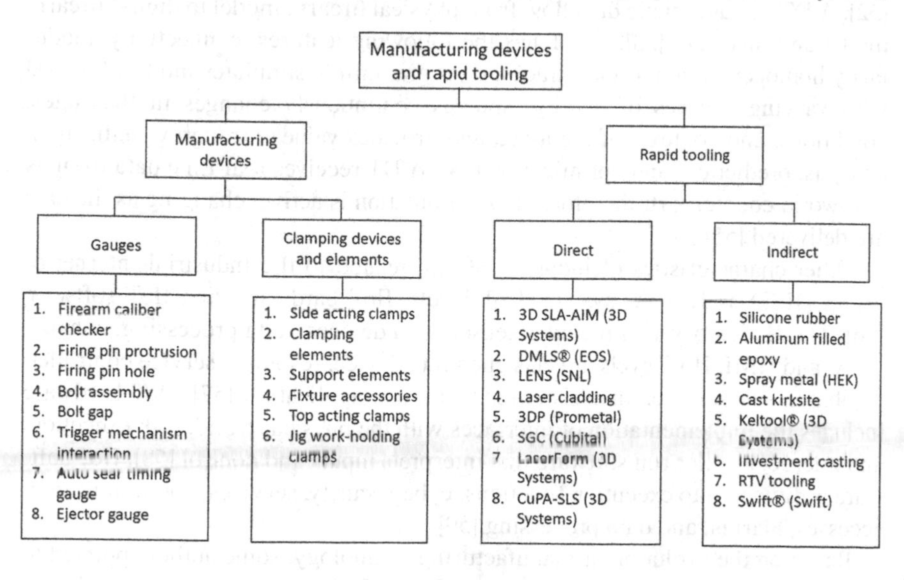


**FIGURE 5.8** Manufacturing devices and rapid tooling for firearm RPs.

and lead time are determined to design production tools. Manufacturing devices are classified into gauges and clamping devices and elements. Gauges include several measuring instruments for different firearm components [32]. The group of clamping devices and elements covers different clamps and accessories [3]. RT includes direct and indirect tools. The former is tooling obtained by a manufacturing process, while the latter use an RP pattern as a model for molding, casting, and die making [33]. Direct RTs are classified into 3D SLA-AIM (3D Systems) [34], DMLS® (EOS) [35], LENS (SNL) [36], laser cladding [37], 3DP (Prometal) [38], SGC (Cubital) [39], LaserForm (3D Systems) [40], and CuPASLS (3D Systems) [41]. Indirect RTs are divided into silicone rubber [42], aluminum-filled epoxy [43], spray metal (HEK) [44], cast kirksite [45], Keltool® (3D Systems) [46], investment casting [47], RTV tooling [48], and Swift® (Swift) [49].

## 5.10 INDUSTRY 4.0

As new technologies advance in the third internet wave (IW), RP manufacturing evolves, adapting to the technological pace of Industry 4.0, which represents the fourth industrial revolution. Figure 5.9 shows the characteristics of Industry 4.0, which is considered to have started in 2006 and became broadly known in 2011 [50]. Industry 4.0 is focused on manufacturing digital transformation, lean and agile manufacturing, smart reconfigurable manufacturing machines, and adaptive manufacturing. The impact of industry 4.0 is reflected in product fabrication from manufacturing technologies, supply chain monitoring, and PLM [51]. Likewise, digital twin (DT) technology is a significant characteristic to get an RP. A digital model has manual data flow. The digital shadow occurs when there is automatic data flow from the physical firearm model to the digital firearm model [52]. A DT has automatic data flow from physical firearm model to digital firearm model and inversely [53]. A DT has the following features: connectivity, modularity homogenization, digital traces, reprogramming, simulates models forward with varying degrees of fidelity, updates continuously changes in the states, conditions, and contexts of the asset, and provides values through visualization, analysis, prediction, and optimization [54]. A DT receives real-time data from its real-world counterpart; therefore, a DT simulation is active, changing as the data are delivered [55].

Other characteristics of industry 4.0 are related to the industrial internet of things (IIoT) as IIoT ecosystem, IIoT levels, IIoT hardware, and IIoT software [56]. The IIoT ecosystem includes sensors and devices, data processing, connectivity, and HMI. IIoT levels are classified into device, resource, service controller, database, web service, analysis component, and application [57]. IIoT hardware includes the implementation of interfaces with the physical world, task execution, and microcontroller run software that interprets inputs and control [58]. IIoT software is divided into execution functions, cybersecurity, services, communication, accesses, alarms, and data processing [59].

Based on the evolution of manufacturing technology, some authors pointed to the start of Industry 5.0 in 2020 [60]. Industry 5.0 focuses on the interaction of

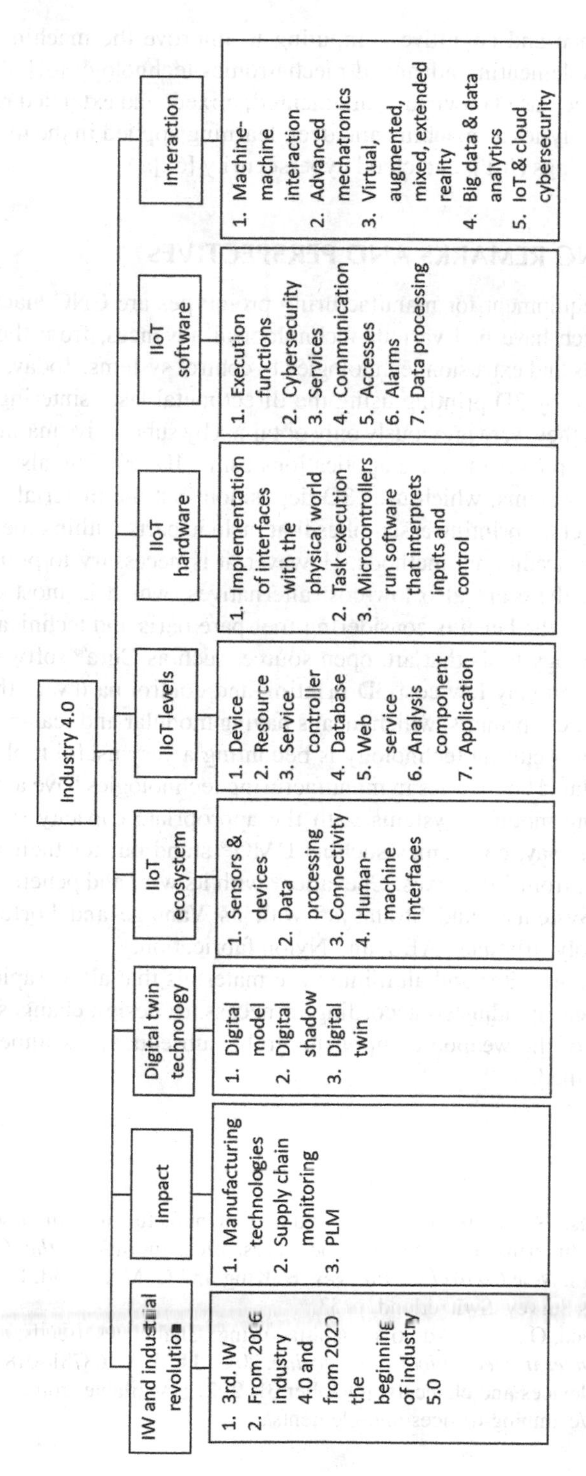


**FIGURE 5.9** Characteristics of industry 4.0.

human intelligence and cognitive computing to improve the machine-machine interaction in implementing advanced mechatronics technology [61]. This interaction is developed based on virtual, augmented, mixed, and extended reality, big data, and data analytics as machine and deep learning applied in the manufacturing, internet of things (IoT), and cloud cybersecurity [62].

## 5.11 CLOSING REMARKS AND PERSPECTIVES

The most used equipment for manufacturing prototypes are CNC machines and 3D printers, which have had various technological advances, from the range of printing materials and extrusion technologies to control systems. Today, some barrels are fabricated by 3D printing using the direct metal laser sintering (DMLS) process, whereas they were previously only obtained by subtractive manufacturing.

Currently, it is possible to have applications using 4D printing, also known as shape-morphing systems, which uses 3D deposition, but the material is deformable; this is perfect for printing assemblies that include parts within others, impossible to obtain by traditional methods. However, it is necessary to point out that with technology, there are also low-cost alternatives, which in most cases it is advisable to verify the benefits considering the spare parts and technical support. There are technology tools that are open source, such as Cura® software, which is compatible with many low-cost 3D printing and control hardware that can be implemented in these printers, which means having modular and scalable systems.

Fiber-optic laser cutting technology is becoming a very useful tool for prototyping. Consolidated companies in manufacturing technologies have also focused on manufacturing modular systems with the appropriate capacity to manufacture RPs. In this way, companies such as DMG® stand out for their small-size CNC machinery, from 3 to 5 axes, Chevalier® with its wire and penetration EDM machines, 3D Systems®, and Stratasys® with its Vantage and Fortus models, which include polycarbonate, ABS, and Nylon fabrication.

Clay-based composites and aluminum are materials that allow rapid machining that can be easily adjusted according to necessary design changes based on the functioning of the weapon components to be subsequently scanned, parameterized, and refined.

## REFERENCES

1. Jenzen-Jones, N., Small arms and additive manufacturing: An assessment of 3D-printed firearms, components, and accessories, in *Behind the Curve, New Technologies, New Control Challenges*, B. King and G. McDonald, Editors. 2015, Small Arms Survey: Switzerland, p. 33.
2. Abdulhameed, O., et al., Additive manufacturing: Challenges, trends, and applications. *Advances in Mechanical Engineering*, 2019. **11**(2): p. 1687814018822880.
3. Clamping devices and elements. [October 31, 2021]; Available from: https://unity-group.co.in/clamping-devices-and-elements/.

4. Boboulos, M.A., *CAD-CAM & Rapid Prototyping Application Evaluation*. 2010, Ventus Publishing ApS: London.
5. Xu, X., *Integrating Advanced Computer-Aided Design, Manufacturing, and Numerical Control: Principles and Implementations*. 2009, IGI Global: Hershey, PA.
6. Noor Hatem, Y.Y., A.K. Aini Zuhra, and M.A. Mohammed, Reviewing of STEP-NC standards related to manufacturing industries. *International Journal of Scientific & Technology Research*, 2020. **9**(4): p. 6.
7. Understanding laser automation axis controls. [October 31, 2021]; Available from: https://www.controllaser.com/blog/2018/12/11/understanding-laser-automation-axis-controls/.
8. Autodesk, fundamentals of CNC Machining: A Practical Guide for Beginners. Autodesk Inc. 2014, United States of America, p. 256.
9. T.R. Technology, 3D Printing: Thematic Research Reports. 2019, GlobalData: London, UK, p. 57.
10. Zhang, Y., X. Xu, and Y. Liu, Numerical control machining simulation: A comprehensive survey. *International Journal of Computer Integrated Manufacturing*, 2011. **24**(7): pp. 593–609.
11. Yudianto, Y., An overview of direct or distributed Numerical Control in Computer Numerical Control Applications. *Journal of Mechanical Science and Engineering*, 2020. **7**(2): pp. 25–29.
12. İç, Y.T. and M. Yurdakul, Development of a decision support system for machining center selection. *Expert Systems with Applications*, 2009. **36**(2): pp. 3505–3513.
13. Control sinumerik. [October 31, 2021]; Available from: https://new.siemens.com/global/en/products/automation/systems/cnc-sinumerik/automation-systems/sinumerik-840.html.
14. Types of CNC machines. [October 31, 2021]; Available from: https://www.cnclathing.com/guide/classification-of-cnc-machine-system-what-are-the-types-of-cnc-machines-cnclathing.
15. Fitzpatrick, M., *Machining and CNC Technology*. 2013, McGraw Hill Higher Education: New York.
16. Latif, K., et al., A review of G code, STEP, STEP-NC, and open architecture control technologies based embedded CNC systems. *The International Journal of Advanced Manufacturing Technology*, 2021. **114**: pp. 1–18.
17. Creation of G-code. [October 31, 2021]; Available from: https://www.ajpdsoft.com/modules.php?name=News&file=article&sid=664.
18. Rodriguez, E. and A. Alvares, A STEP-NC implementation approach for additive manufacturing. *Procedia Manufacturing*, 2019. **38**: p. 9–16.
19. Hiemenz, J., *3D Printing with FDM: How It Works*. 2018, Stratasys: Israel, p. 7.
20. Types of FDM 3D printers. [October 31, 2021]; Available from: https://all3dp.com/2/cartesian-3d-printer-delta-scara-belt-corexy-polar/.
21. Ligon, S.C., et al., Polymers for 3D printing and customized additive manufacturing. *Chemical Reviews*, 2017. **117**(15): pp. 10212–10290.
22. How to choose the right 3D printing materials. [October 31, 2021]; Available from: https://www.emergingedtech.com/2021/02/how-to-choose-right-3d-printing-materials-for-classroom/.
23. What materials can be 3D printed. [October 31, 2021]; Available from: https://blogmech.com/flexible-3d-printing-materials-material-strength/.
24. Aboushama, M., S. Beyerlein, and M. Bednarz, Evaluation of continuous fiber reinforcement desktop 3D printers, seminar paper, Technical University of Ingolstadt, Ingolstadt, Germany, 2020.

25. McGetrick, P.J., et al., Experimental testing and analysis of the axial behavior of intermeshed steel connections. *Proceedings of the Institution of Civil Engineers-Structures and Buildings,* 2020, paper No. 1900181, pp. 1–21.
26. Olivero, M., et al. Measurement techniques for the evaluation of photodarkening in fibers for high-power lasers, Proc. SPIE 7914, Fiber Lasers VIII: Technology, Systems, and Applications. 79142U 2011. International Society for Optics and Photonics, San Francisco, CA, USA.
27. Raycus laser. [October 31, 2021]; Available from: https://en.raycuslaser.com/.
28. IPG photonics. [October 31, 2021]; Available from: https://www.ipgphotonics.com/en/products/lasers/high-power-cw-fiber-lasers/1-micron-1/yls-sm-1-10-kw.
29. Dobelis, J. and V. Beresnevich, Fundamental precision dependencies of a CNC laser cutter. *Transport & Engineering*, 2015. **36**: 1–7.
30. Simatic technology. [October 31, 2021]; Available from: https://new.siemens.com/global/en/products/automation/systems/industrial/simatic-technology.html.
31. Ballistic advantage. [October 31, 2021]; Available from: https://www.ballisticadvantage.com/about.
32. Firearm gauges. [October 31, 2021]; Available from: https://www.brownells.com/gunsmith-tools-supplies/measuring-tools/firing-pin-protrusion-gauges/firing-pin-protrusion-gauge-prod26477.aspx.
33. Equbal, A., A.K. Sood, and M. Shamim, Rapid tooling: A major shift in tooling practice. *Manufacturing and Industrial Engineering*, 2015. **14**(3–4): 1–9.
34. Stereolithography. [October 31, 2021]; Available from: https://www.3dsystems.com/stereolithography.
35. 3D printing metal. [October 31, 2021]; Available from: https://www.eos.info/en/additive-manufacturing/3d-printing-metal.
36. Smith, M.F., *Additive Manufacturing at Sandia.* 2015, Sandia National Lab (SNL-NM): Albuquerque, NM.
37. Laser cladding. [October 31, 2021]; Available from: https://www.laserline.com/en-int/laser-cladding/.
38. 3D print. [October 31, 2021]; Available from: https://prometal3d.com/en/content/8-3d-print.
39. Solid ground curing. [October 31, 2021]; Available from: https://galraz.wixsite.com/il-am-industry/solid-ground-curing-sgc-by-cubital.
40. Laserform. [October 31, 2021]; Available from: https://www.3dsystems.com/materials/laserform-ni718.
41. Selective laser sintering. [October 31, 2021]; Available from: https://www.3dsystems.com/selective-laser-sintering.
42. Methods of rapid tooling worldwide. [October 31, 2021]; Available from: https://wohlersassociates.com/Oct00MMT.htm.
43. Khushairi, M.T.M., et al., Effects of metal fillers on properties of epoxy for rapid tooling inserts. *International Journal on Advanced Science, Engineering and Information Technology*, 2017. **7**: p. 1155–1161.
44. Metal sprayed moulds. [October 31, 2021]; Available from: https://www.ronald-simmonds.de/technologie/metal-sprayed-moulds/.
45. Cast Kirksite re-emerges as RT approach for molding plastics. [October 31, 2021]; Available from: https://www.armstrongrm.com/pages/kirksitearticle.html.
46. 3D Keltool. [October 31, 2021]; Available from: http://www.3dsystems.ru/products/productiontooling/3dkeltool/products_3dkel_howitworks.asp.htm.
47. Nagahanumaiah, R.B. and N. Mukherjee. Tool path planning for investment casting of functional prototypes/production molds. In *National Conference on Investment Casting*, Durgapur, India, September 22-23, 2003.

48. Silicone rubber tooling technology. [October 31, 2021]; Available from: https://www.simtec-silicone.com/silicone-rubber-tooling-technology/.
49. Levy, G.N., R. Schindel, and J.-P. Kruth, Rapid manufacturing and rapid tooling with layer manufacturing (LM) technologies, state of the art and future perspectives. *CIRP Annals*, 2003. **52**(2): pp. 589–609.
50. Industry 4.0 and the fourth industrial revolution explained. [October 31, 2021]; Available from: https://www.i-scoop.eu/industry-4-0/.
51. Finance, A., *Industry 4.0 Challenges and Solutions for the Digital Transformation and Use of Exponential Technologies*. 2015, Finance, Audit Tax Consulting Corporate: Zurich, Swiss, pp. 1–12.
52. Boschert, S. and R. Rosen, *Digital Twin: The Simulation Aspect, in Mechatronic Futures*. 2016, Springer: Berlin, Heidelberg, pp. 59–74.
53. Fuller, A., et al., Digital twin: Enabling technologies, challenges and open research. *IEEE Access*, 2020. **8**: pp. 108952–108971.
54. Understanding the digital twin. [October 31, 2021]; Available from: https://www.chemengonline.com/understanding-the-digital-twin/?printmode=1.
55. Digital twins: The Doppelgänger approach to digital success. [October 31, 2021]; Available from: https://www.devopsonline.co.uk/digital-twins-the-doppelganger-approach-to-digital-success/.
56. Petrik, D. and G. Herzwurm. Platform ecosystems for the industrial internet of things-A software intensive business perspective. in SiBW, 2018.
57. Smart sensors for industry 4.0. [October 31, 2021]; Available from: https://www.te.com/usa-en/industries/sensor-solutions/applications/iot-sensors/industry-4-0.html.
58. Industrial internet of things. [October 31, 2021]; Available from: https://internetofthingsagenda.techtarget.com/definition/Industrial-Internet-of-Things-IIoT.
59. Dhirani, L.L., E. Armstrong, and T. Newe, Industrial IoT, cyber threats, and standards landscape: Evaluation and roadmap. *Sensors*, 2021. **21**(11): p. 3901.
60. Di Nardo, M. and H. Yu, *Special Issue "Industry 5.0: The Prelude to the Sixth Industrial Revolution"*. 2021, Multidisciplinary Digital Publishing Institute: Basel.
61. George, A.S. and A.H. George, Industrial revolution 5.0: The transformation of the modern manufacturing process to enable man and machine to work hand in hand. *Journal of Seybold Report*, 2020. **1533**: p. 9211.
62. Maddikunta, P.K.R., et al., Industry 5.0: A survey on enabling technologies and potential applications. *Journal of Industrial Information Integration*, 2021: p. 100257.

# 6 Experimental Physical Models, Test Benches, and Prototypes

## 6.1 INTRODUCTION

The scheduling of the product development tasks includes critical activities, among which is the assessment of problems in the subsystems design, the evaluations of the solution proposals, and the validation of such proposals, which allows the approval of each phase of the firearm design. The scientific rigor and the precision of the results of the experimentation are fundamental in the decisions taken to increase the technological maturity of the firearm design.

The experimental physical model is subject to the evaluation of the operating principle of the firearm, for which a test protocol is established to carry out the experimentation in the test benches (TBs) defined by experimental batches. There are companies specialized in developing TBs with which the evaluations of the behavior of the experimental physical model are programmed, including the certifications of the approved tests.

The prototype is subject to the validation of a replicable product, which can be used for demonstration purposes with users and in later stages for industrial demonstration, using pilot batches to verify its behavior with users in different real operating environments.

This chapter addresses the design of TBs, features, outlining of testing protocol, and experimentation of both subsystems and the entire firearm, considering the laboratory environment to evaluate the experimental physical model (EPM) and the real environment to evaluate the prototype.

## 6.2 TESTS PROTOCOLS AND PRODUCT VALIDATION PROCESS

While previous chapters explain the design process of small weapons including the design methodology (Chapter 2), the custom and product requirements (Chapter 3), the CAD modeling and CAE simulation (Chapter 4), and the CAM assessment and rapid prototypes (Chapter 5), this chapter focuses on the validating of the firearm design. Likewise, Chapters 7–9 describe the materials used in the production of small weapons, heat treatments and surface hardening of small weapon components, and manufacturing processes for small weapon components, respectively.

DOI: 10.1201/9781003196808-6

The validation of a firearm is carried out through EPM in TBs, whereas that of prototypes is achieved through emulation and testing in a real environment using test protocols and usability metrics. Although there is a lot of information in the literature on firearm design, a common mistake is to experiment without a test protocol or expected error metrics. A testing protocol is a guide to structure the experimentation according to case studies and tasks concerning design parameters (DPs). Planning the study objectives, cases, tasks, materials, methods, and result reports will help get reliability metrics and mitigate risk during design and manufacture. An external laboratory (outsourcing with a trusted partner) specialist in ballistic tests can be a good option to increase the readability of the test and to avoid the workshop blind. The design team task will be focused on the analysis of reported data.

A test protocol serves for repeatability expectations, acceptance criteria, regulatory requirements, and applicable standards. Typical protocol sections are outlined as follows: scope, purpose, reference documents (test standards, regulatory guidance, previous test reports, published literature), test samples, materials and calibrated equipment, methods, acceptance criteria, exception conditions, data analysis and documentation requirements, references, traceability, specimen preparation, test configurations, testing frequency, target cycles, target load data or method for load selection, testing environment, testing rates, revision history, and signatures to ensure that all parties are aware of any changes and to get everyone on the same page. Any exceptions, changes, or modifications to the test protocol should also be noted in order to prevent confusion, set clear expectations, and preserve the necessary information for future reference and use.

The accomplishment of end-user requirements is verified in the validation stage. As shown in Figure 6.1, the end user can be classified into an expert user who aims the tactical issues which are the procedures to the correct use of a firearm; an operational execution user who performs the shooting activities continually; and a client who is responsible for strategic planning and use policy.

Nowadays, the product approval process regarding end-user requirements involves techniques to perform the verification, validation, and qualification of firearms. Verification shows the right development of firearm-related requirements. Validation focuses on the right firearm to end-user expectation in an operative environment [1]. Firearm qualification or certification is a part of product validation that proves a firearm has passed performance and quality assurance test according to lineaments, regulations or conditions of certification bodies, accreditation bodies, and contracts, emitting the declaration of conformity (DoC).

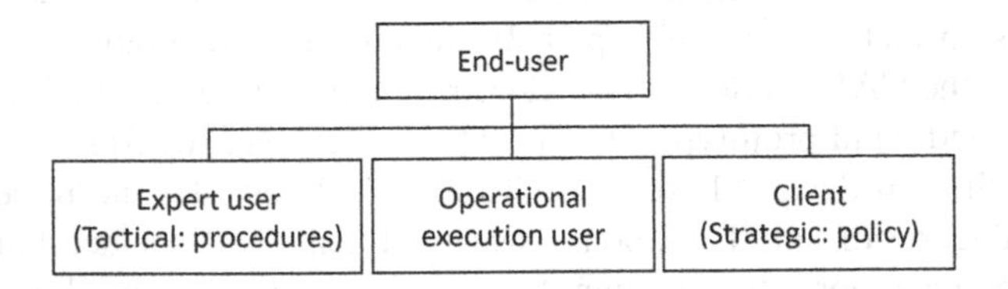


**FIGURE 6.1** Classification of end user to firearm design.

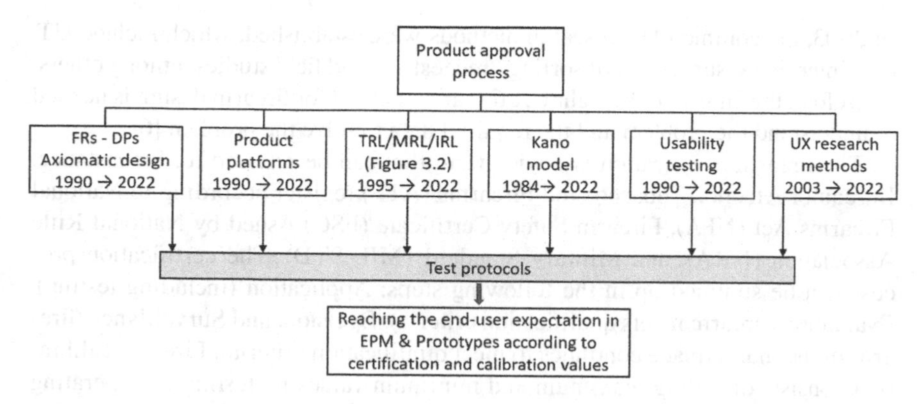


**FIGURE 6.2** Framework of the product approval process.

The product approval process (Figure 6.2) considers the functional requirements (FRs) and DPs based on axiomatic design analyzed in Chapter 3, the requirements for product platform architecture presented in Chapter 2, requirements to reach the technology readiness level (TRL)/manufacturing readiness level (MRL)/investment readiness level (IRL) established in Figure 3.2, Kano model, usability testing (UT), and user experience (UX) research methods. In the framework of Figure 6.2, the starting dates are considered from the wide use of the methodologies, which have been maintained, updated, and increased in impact until now. Through test protocols, the product approval methodologies (PAMs) are implemented, reaching end-user expectations. Suh developed the axiomatic design based on the independence (coupling, decoupling, and uncoupled FRs) and information (minimizing the information content of the design) axioms [2]. In 1990, several companies popularized the concept "product platform" and researchers such as Meyer, Lenherd, Utterback, Cusumano, Ulrich, Sanderson, and Uzumeri further developed this concept. In 1995, Mankins created the TRL scale, which was later adopted by the U.S. Department of Defense (DOD) [3]. In 1984, Kano published the Kano model of product development and customer satisfaction, assigning the threshold (expected by end user), performance (increasing the enjoyment but are not essential to end user), and excitement (surprise elements and delight end user) attributes [4]. Usability has its roots in 1911 with the description of times and motion, later reducing the work motion into smaller steps; this method was used to teach soldiers how to assemble and disassemble weapons in the dark. In 1936, the usability as features of the new Frigidaire refrigerator was posted. In 1947, the User Preference Department, later renamed the Human Factors Department, was formed at Bell labs. In 1957, the Human Factors Society was formed. In 1981, the usability and usability tests were described. In 1985, the design for usability was disclosed. In 1986, the system usability scale (SUS) became the most widely used questionnaire for usability evaluation, so that in 1990 Shackel publishes Human Factors and Usability, which defined the usability as a function of efficiency, effectiveness, and satisfaction (ISO 9241 part 11); from these facts, the usability widely matures and consolidates, then giving way to UX [5].

In 2003, the common UX research methods were established, which include UT, user interviews, surveys, card sorting, tree testing, and field studies, among others. Therefore, the choice of the right UX research method for firearm design is needed to understand the problem and the required data to solve the problem [6].

The firearm certification is a wide term that can be referred to, for instance, Bureau of Alcohol, Tobacco, and Firearms (ATF) register according to National Firearms Act (NFA), Firearm Safety Certificate (FSC) issued by National Rifle Association (NRA), and Military Standard (MIL-STD). The certification process can be summed up in the following steps: Application (including testing), Evaluation (firearm meets qualification criteria), Decision, and Surveillance (firearm in the marketplace continues to meet qualification criteria). Firearm calibration consists of setting maximum and minimum values by testing the operating principle systematically concerned to a reference standard.

## 6.3 FIREARM USABILITY AND UX

At present, terms such as Usability, UX, and customer experience (CX) are often used, and the design of firearms is not the exception. Usability refers to the ease of use of a firearm, considering effectiveness (learnability, memorability, and error frequency), efficiency, and satisfaction [7]. UX covers a wider aspect of end-user interaction with the firearm, including functionality, findability, reliability, value, accessibility, and delight [8]. CX is based on attraction, awareness, advocacy, purchase, discovery, and cultivation [9].

The usability and UX testing are oriented to improve the firearm. In the case of usability, the aim is to make solutions for developing the user tasks easier and more intuitive, minimizing steps, removing roadblocks, and finding how and what the shooter does during the firearm operation. In the case of UX, the aim is to increase the meaning and value of tasks, emotional connection, and knowing of user feel [10].

The goals of UT usually include identifying problems in the firearm design, uncovering opportunities to improve, and learning about the target user behavior and preferences. The elements of UT are the facilitator, the tasks, and the participant [11].

Aspects that add cost include competitive testing of multiple designs, international testing in multiple countries, testing with multiple user groups, quantitative studies, detailed analysis and report about the findings, and participants recruiting costs based on the requirements. Expert review is another general method of UT. As the name suggests, this method relies on bringing in experts with experience in the field (possibly from companies that specialize in UT) to evaluate the usability of a product [12].

## 6.4 FIREARM EPMs AND DEMONSTRATION PROTOTYPES

The design team previously defines the solution formulation, reaching the design concept through mockups. At this stage, the ALPHA firearm is conformed in the

TRL3, which requires a test protocol that includes evaluating the following: operating principle, which is the inventive activity that refers to novelty and industrialization, but it is not easily deductible by an expert in the field; initial materials and manufacturing implications; and the product feasibility regarding market demand, competitors, differential attributes, suppliers, and price estimate. The test protocol for TRL4 is focused on laboratory testing using an EPM experimental pilot, considering high-safety conditions because performance is unknown due to uncertain variables. The validation of TRL4 firearm results in obtaining the EPM approval.

The BETA firearm is constituted in TRL5, which requires a test protocol that includes evaluating the following: prototype with parts of high reliability and experimental lot included into the demonstration pilot using the optimization due to communality and DFA index; analysis of manufacturing process developed with limited production; and competitive analysis.

## 6.5 TEST BENCHES AND STANDARDS TO FIREARM PERFORMANCE

TBs are experimentation platforms for assessment, evaluating, verifying, and validating firearm systems, providing rigorous, transparent, and repeatable analysis of ballistic phenomena according to test protocols of standards and testing methods. A TB can be purchased or designed to specific testing requirements. The design of a TB encompasses the use of instrumentation parts as sensors, piezoelectric, strain gauges, servomotors, and data acquisition cards (DACs) controlled by programmable logic controller (PLC) or embedded systems by programming, for instance, in LabVIEW®. When a TB design is required, there are two alternatives, either it is planned in parallel to firearm design, or it is planned once an EPM is obtained. During the testing of a new firearm, unexpected results may occur, in which case a hypothesis is proposed to find the cause and repeat the critical result through destructive proofs.

For the correct operation of a firearm TB, the following should be considered (Figure 6.3): ballistic laboratory accreditation and certification bodies; firearm, ammunition, and protection standards, as well as testing methods; and indoor and outdoor shooting ranges.

The ballistic laboratory accreditation bodies comply with the following: ISO/IEC-17020, which specifies requirements for the competence of body performing inspection, as well as for the impartiality and consistency of the inspection activities; ISO/IEC-17025, which enables all types of laboratories that performs testing, sampling, or calibration and providing reliable results; EN ISO-9001 (Quality management systems) [13], Home Office Scientific Development Branch (HOSDB), which is responsible for UK body armor standards [14], Allied Quality Assurance Publications (AQAP) such as AQAP-2110 NATO, which covers the quality assurance requirements for design, development, and production; and codes for entities located outside the United States, which are called NATO Commercial and Government Entity (NCAGE) codes [15]. The accreditation

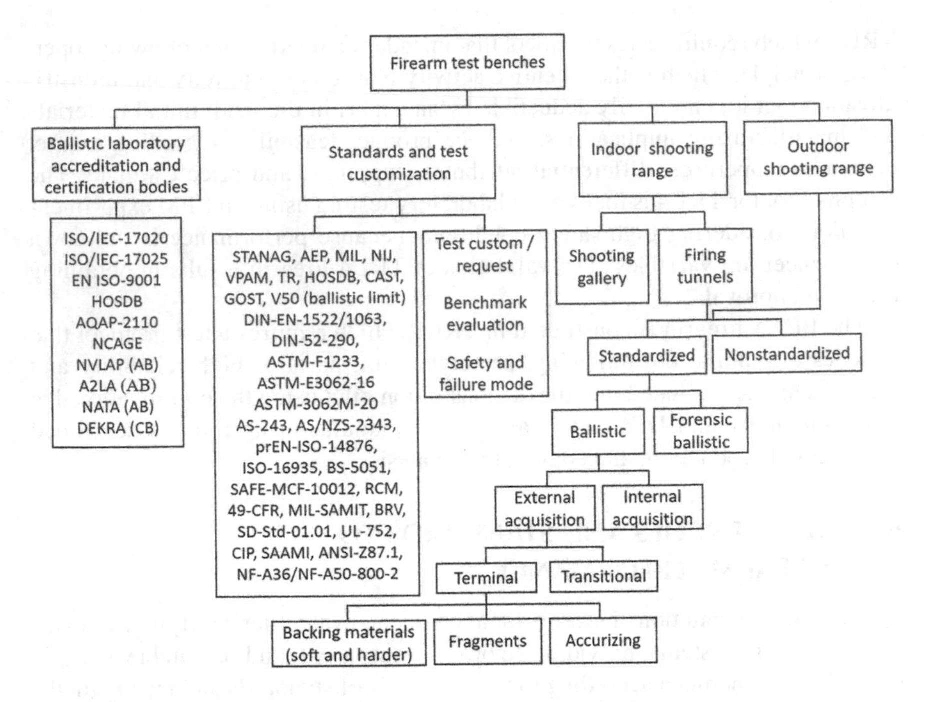


**FIGURE 6.3** Firearms test benches to complement standards.

bodies are the National Voluntary Laboratory Accreditation Program (NVLAP) [16], The American Association for Laboratory Accreditation (A2LA) [17], and the National Association of Testing Authorities (NATA). One of the ballistic laboratory certification bodies is DECKRA [18].

Standards and test customization are classified into regulations from organizations and test customization. Standards include the following: Standardization Agreement (STANAG), which is a NATO standardization document that specifies the agreement of member nations to implement a standard; Allied Engineering Publication (AEP) [19]; MIL (for instance, MIL-STD and MIL-H); National Institute of Justice (NIJ) [20]; Association of test centers for Anti-attack Materials and Constructions (VPAM, German abbreviation for Vereinigung der Prüfstellen für angriffshemmende Materialien und Konstruktionen) [21]; technical guidelines (TR, German abbreviation for Technische Richtlinien); HOSDB; Centre for Applied Science and Technology (CAST, standards for portable ballistic protection for UK police) [22]; Governmental Standard (GOST, Russian acronym for gosudarstvennyy standart) [23]; V50 ballistic limit, which is the velocity at which a specific bullet is expected to penetrate the armor half of the time [24]; DIN-EN-1522 (Windows, doors, shutters and blinds—Bullet resistance—Requirements and classification) [25]; DIN-EN-1063 (Glass in building, security glazing, testing and classification of resistance against bullet attack) [26]; DIN-52-290 (security glazing) [27]; ASTM-F1233, which defines three factors to determine the success or failure of any attempt to forcefully enter (or exit) the glazing or system; ASTM-E3062/E3062M-20 (Standard Specifications for Indoor Ballistic Test Ranges for Small Arms and Fragmentation Testing of Ballistic-resistant Items) [28]; AS-243 (Australian Ballistic Standard); AS/NZS-2343, which is an armor spec for Australia and New Zealand that specifies the ballistic requirements for bullet resistant panels and elements including transparent, opaque, and translucent components [29]; prEN-ISO-14876 (Protective clothing) [30]; ISO-16935, which sets forth test procedures to evaluate resistance of security glazing materials and products against ballistic impact with classification by weapon and ammunition [31]; BS-5051 (Specification for glazing for interior use) [32]; SAFE-MCF-10012 (Brunswick Ballistic Standards) [33]; RCM (Canadian ballistic standard); Code of Federal Regulations (CFR, 49-CFR hazardous materials transportation regulations and federal motor carrier regulations [34]; MIL-SAMIT (ballistic resistance) [35]; BRV (ballistic vehicle resistance standard developed in Germany for armored vehicle testing against various caliber ammunition) [36]; SD-Std-01.01 (Certification Standard Forced Entry and Ballistic Resistance of Structural Systems) [37]; UL-752 (Standard for Bullet-Resisting Equipment) [38]; Permanent International Commission (CIP, French abbreviation, Commission internationale permanente) [39]; Sporting Arms and Ammunition Manufacturers' Institute (SAAMI) [40]; ANSI-Z87.1 (This standard sets criteria to requirements, testing, permanent marking, selection, care, and use of protectors to minimize injuries) [41]; NF-A36-800-2 (Hot-rolled weldable steel sheets for armoring: firing test method) [42]; and NF-A50-800-2 (Hot-rolled weldable aluminum alloy sheets for armoring: ballistic testing method). The test customization considers the test custom or test request, benchmark evaluation, and safety and failure mode.

The indoor shooting range is classified into shooting gallery and firing tunnels. Firing tunnels can be standardized or nonstandardized. The former is classified into ballistic and forensic ballistic. Ballistic can be of external or internal acquisition. The external acquisition considers the terminal and transitional ballistic. The terminal ballistic analyzes the backing materials, both the soft (gelatin) and harder, fragments, and accurizing; accurizing is the process of improving the accuracy and precision of a firearm [43]. The outdoor shooting range is classified in the same way as firing tunnels into standardized and nonstandardized tests.

As can be seen in Figure 6.4, ballistic tests consider test criteria, analysis, type, and techniques and methods. Test criteria include Army, Navy, and protection ballistic limits, critical depth of intrusion, hard armor plate arrest, back-face signature (BFS), shooting cycle synchro, felt recoil, interchangeability, good trigger that should not have any roughness or hesitation in its stroke, regardless of pull weight [44], suitable weight (weapon between 2.8 and 4 kg should be loaded with 30 rounds; pistol between 1.3 and 2.3 kg should be loaded with magazine), and permissible noise.

Test analysis includes, among others, ballistic resistance, environmental conditioning, vibration, salt fog, humidity, full-scale vehicle, impact, simulated blast, lot acceptance evaluations, mud exposure, and extreme temperature [45]. Test types consider the armor (body, vehicle, and structural), helmets, firearms,

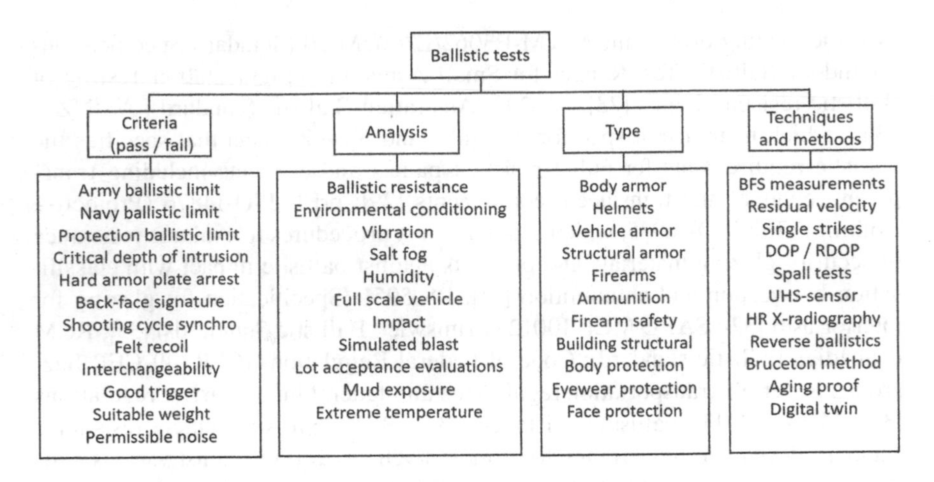


**FIGURE 6.4** Framework of ballistic testing.

ammunition, firearm safety, building structural, and protection (body, eyewear, and face) [45].

Techniques and methods include the BFS measurements, which include the amount of deformation or damage created on the backside of a piece of armor after an impact [46,47], residual velocity, single strikes, depth of penetration (DOP)/residual depth of penetration (RDOP), spall tests [48], ultra high speed (UHS) sensor, high-resolution (HR) X-radiography, reverse ballistics, Bruceton method (sensitivity analysis and sensitiveness tests of explosives) [49], aging proof, and digital twin (Real-time communication between the virtual TB and physical TB).

## 6.6 FIREARM PROTOTYPES IN REAL ENVIRONMENT

Firearm prototype version 1 is reached in TRL6, which requires a test protocol that includes evaluating the following: prototypes validated in entirely firearm testing indoor and outdoor shooting ranges, including industrial demonstration using selected and created manufacturing technologies, producibility assessments, and identification of required facilities and skills.

Firearm prototype version 2 is reached in TRL7, which requires a test protocol that includes evaluating assembly group validation in outdoor shooting range using refined and integrated manufacturing with risk management plan, preliminary processes, materials, and tools. The validation of TRL7 firearm prototype results in obtaining a low-fidelity valuable minimum product (VMP) approved with feedback of end users and potential clients in field testing with objects in motion, under water, sand, snow, and mud.

Later, a transition phase in TRL7 corresponds to a validation of the production by assembly groups according to the capability to produce the prototypes version 2, with which the MRL6 is reached. A certified firearm (TRL8) requires a test protocol that includes evaluating the performance of certification bodies, pilot line, final materials, user manual, technical support, and maintenance organization. Thus, a high-fidelity VMP is reached with the declaration of conformity NATO and commercial clearances.

## 6.7 POLYMER FIREARMS AND 3D PRINTED PROTOTYPES

New firearm design has been tested using novel materials based on cost reduction by homogenizing the component materials. Figure 6.5 shows a timeline of technologies used in the manufacture of firearms. Initially, the parts were made of steel and wood covers, as was the first commercial repeating firearm, the Colt Paterson revolver, patented as "Improvement in firearms" (USX9430I1) by Samuel Colt in 1836 [50]. Later, in 1958 Remington® patented (US3023527) a firearm having receiver bearing surfaces of synthetic resinous material, which shows the incorporation of Nylon from Dupont® into the receiver [51]. Since then, several manufacturing processes have been integrated into firearms, achieving today the functioning using most components made of polymers, maintaining the critical parts as breech bolt group, hammer, and barrel made of steel. The first stereolithography (SLA) 3D printed parts were developed between 1983 and 1986 in the patent EP0535720B1 by Charles W. Hull, who cofounded 3D Systems® in 1986 [52,53]. This fact allowed complement the manufacturing of firearm RPs, so firearm parts such as covers and receivers were obtained, helping to design complex components and more attractive nonfunctional prototypes of firearms due to low mechanical resistance of SLA material. Later, in the period from 1988 to 1989, the fused deposition modeling (FDM) technology was developed, as described in the patent US00521329A by Steven Scott Crump, who cofounded Stratasys® in 1989 [54,55]. Since then, the firearm RP manufacturing has been hugely enhanced, which allowed manufacturing functional prototypes of firearms with better mechanical properties, in such a way that the design of new firearms using FDM and steel parts can be visualized and operated without ammunition and fired on TBs. Today, the EPM and RP manufacturing is done using first SLA in the research stage, later FDM, and finally the rapid tooling in order to the firearm prototype is integrated by steel, FDM (covers and parts except for mechanisms), and polyamide (ASTAMID® and VESTAMID®) parts. The polyamide

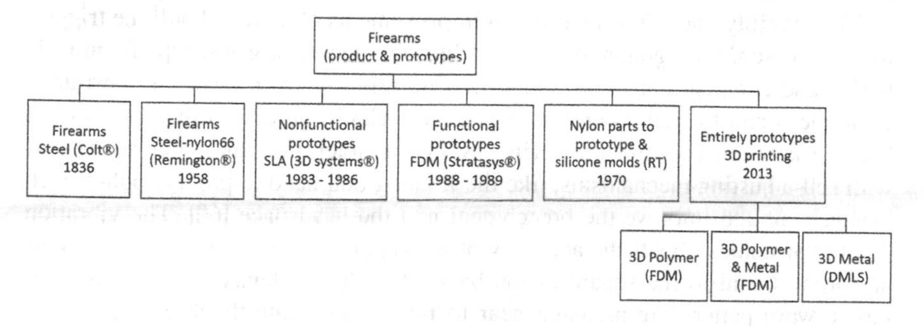


**FIGURE 6.5** Timeline of technologies for firearm manufacturing.

parts obtained from RT are possible by the use of vacuum casting through the silicone mold technology, which surged in 1970 [56].

The FX05 rifle, invented in 2005, was the first weapon in using SLA, FDM (Stratasys FDM Vantage S), rapid tooling by vacuum casting, investment casting, sintering, and machining parts in its development; this sped up the times of development, usability, and UX.

Entirely prototype 3D printing was created in 2013, being the Liberator pistol the first weapon made from a 3D polymer by FDM using a Stratasys 3D printer [57]. The pistol Ruger Charger was created using FDM 3D printers to get polymer and metal parts using plastic filament infused with metal powder or nearly full-metal filaments [58]. The company Solid Concepts (acquired by Stratasys® in 2014) created the pistol Browning 1911 using DMLS technology in EOSINT M270 Direct Metal 3D Printer manufactured in 34 3D printed components [59]. As can be seen in Figure 6.5, RP and RT technologies have been improving over the years.

## 6.8 FIREARM PERFORMANCE CASE STUDIES

There are several case studies in scientific communications, patents, and especially in reports made during the experience of this book's authors. Hence, the framework on firearm performance case studies (FPCS) shown in Figure 6.6 has been consolidated to guide readers to have an overview of the evaluation of EPMs and prototypes. The FPCS are classified into the following: usability and UX to get user expectative metrics; accurizing consists in enhancing the accuracy (hit exactly when aiming) and precision (hit on the same place repeatedly, being the standard deviation the precision value); ballistic, which is divided into ballistic data by machine learning, pattern identification, firing effects, rifling, and ballistic coefficient, and ballistic comparison between types of barrels using the same ammunition; clearances and tolerances; recoil dynamics and vibration; materials, which include tribology analysis, heat treatment of additive manufacturing materials, and fatigue life analysis; impact penetration and perforation; safety, which considers drop testing, cook-off, selective fire, overloaded ammo, and calibration; kinematics and dynamics, which include firearm platforms and scaling, trigger mechanism, breech bolt and headspace, feed and eject, and momentum and impulse.

The usability and UX is focused on improvements of hold and pull the trigger; for this reason, an ergonomic design, adjustable features, grips, bipods, muzzle brake, and compensator to counteract the muzzle rise from recoil are desirable. Clearances and tolerances are an engineering challenge due to the search for a balance between cost manufacturing by closest tolerances and open tolerances with self-adjusting mechanisms, like the floating bolt head to provide bolt-breech engagement and improve the breech seal and the headspace [60]. The vibration due to harmonics affects the accuracy of a weapon long barrel (harmonic effects are proportional to the square of the barrel length), so devices to alter the harmonic wave pattern are mounted near to muzzle to locate the sweet point [61]; patent US005423145A describes a harmonic vibration tuning device mounted in

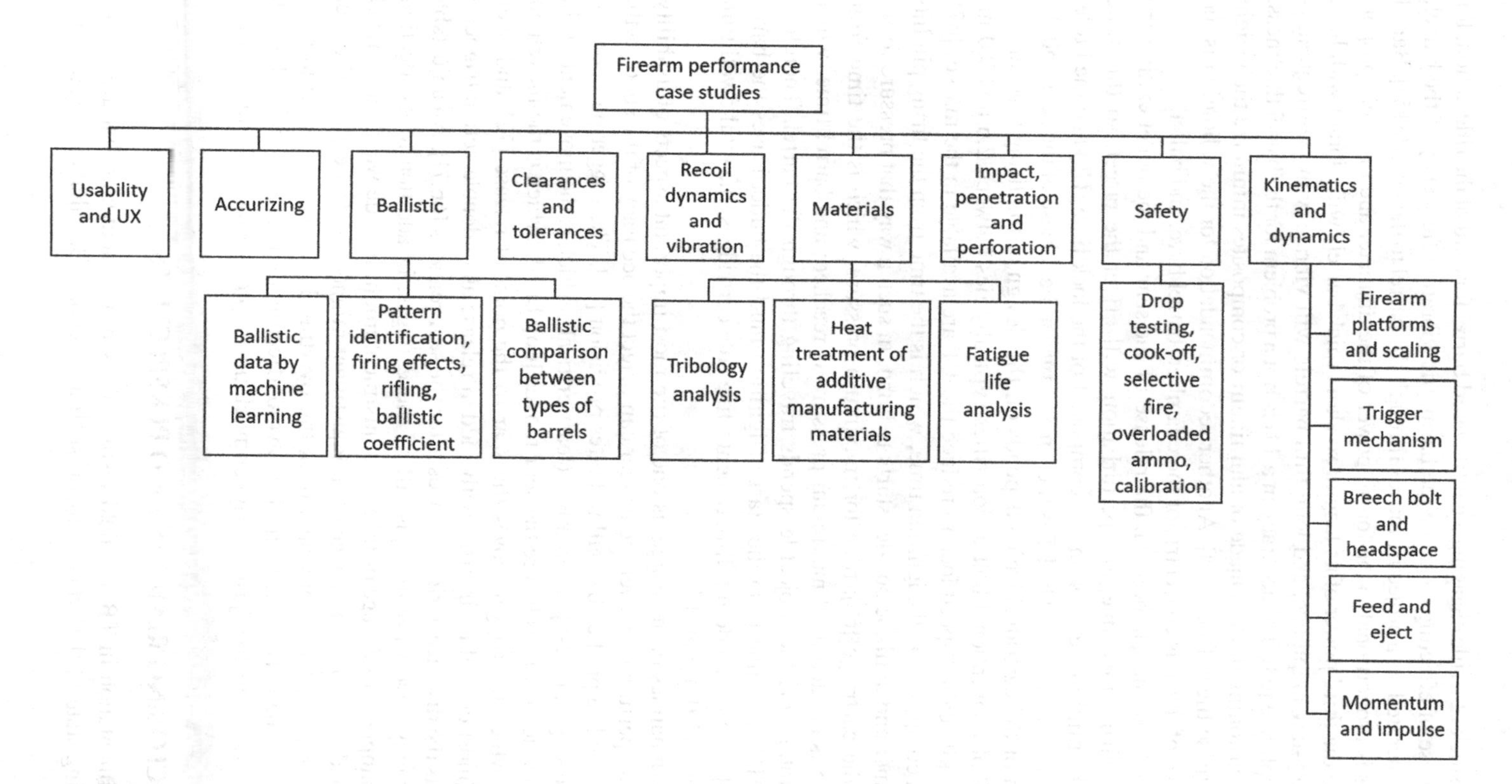


**FIGURE 6.6** Framework of firearm performance case studies.

the forestock of a rifle engaging a cantilevered barrel forward of the rifle action, for the purpose of effecting a change the harmonic vibration frequency of the barrel, that is generated, as a projectile moves through the barrel to the muzzle [62]. Barrel stiffness is proportional to the fourth power of the diameter and inversely proportional to the third power of its length, so short thick barrels will vibrate with high frequency-low amplitude, and long thin barrels will vibrate with low frequency-high amplitude [63]. The accuracy in a TB is not a problem for the barrel thickness. Barrel tensioning devices made of aluminum or composites minimize the weight and increase the stiffness [64]. Another recommendation for the validation is the assurance of the axis concentric of the chamber, barrel bore, and bullet.

There is a time gap between the release of the sear and the bullet exiting the barrel. During this time, any perturbation will affect the target, so this time should be minimized; this time is composed of the lock time (LT) and the bullet dwell time (BDT) [65]. The LT is the time between the release of the sear and the ignition of the cartridge, which depends on the design of the firing mechanism. The LT of conventional bolt action rifles typically lies between 2.6 and 9.0 ms [65,66]; kits called speed lock to reduce the LT are available in the market [67]. BDT is composed by the ignition time, which is the time from the firing pin hits the primer and until the powder starts to burn, in such a way that pressure forms inside the cartridge [68]; time for maximum pressure, which is the time from pressure is created until maximum pressure is reached; and barrel time, which is the total time the projectile spends traveling through the barrel. Therefore, the BDT is the time from the start of ignition until the bullet leaves the barrel [68–70]. In most modern full-bore centerfire rifle cartridges, the total dwell time lies around 1.0–1.5 ms [71].

The dynamics of the trigger is one of the most important aspects of usability, since any perturbation over the trigger can affect the accuracy of the shot, so the trigger needs a predictable pull to better precision [72]. The trigger pull consists of three stages: take-up or pretravel (least critical), which is the movement of the trigger which happens before the sear moves [66]; break, which is the movement during which the trigger moves the sear to the point of release [72]; and overtravel (most critical in firearms with sudden release as in double-action triggers), which is the distance a trigger moves after the sear releases [66,71]. An adjustable trigger may have ways to adjust all of these stages, in addition to its location. For example, a first stage or take-up adjustment might include weight and travel. A second stage or sear engagement adjustment might include weight and travel. A trigger stop adjustment would limit the overtravel [70]. One consideration for design and manufacturing teams is to avoid affecting user safety, so it is advisable not to allow users to adjust the trigger mechanism, preventing law liability issues.

## 6.9 CLOSING REMARKS AND PERSPECTIVES

Experimentation in TB and field testing are sensitive activities that are carried out using state-of-the-art equipment, such as a recoil TB, bullet trap system with deceleration chamber, a device consisting of a non-destructive method for capturing bullets and accurately analyzing view of land and groove impression test barrel system, and target testing. Developers of testing equipment have modernized their technology; for instance, the equipment of Oehler® and Kistler® for ballistics labs, transducers according to SAAMI method, and electronic pressure velocity and action time (EPVAT) according to NATO method.

The setup and design of a TB should be viewed as a collaborative activity between vendors, scientists, and developers. Teradyne® is a company dedicated exclusively to developing test stations for many areas, including systems for the defense and aerospace industry.

There are companies as Teledyne DALSA® specializing in high-performance digital imaging and semiconductor technology that fabricate and customize ultra-high-speed imaging systems with frame rates up to 100 million fps. The UHS technology is suitable for analyzing the behavior of case studies with ballistic systems and explosives. Also, this technology enables stress analysis of firearm components.

The instrumentation equipment has evolved to be friendly to the software with industrial-strength code, libraries, software development kit (SDK), high-performance 1D/2D CMOS and CCD cameras, 3D sensors of high-accuracy laser profiling, stereo imaging and time of flight, image acquisition boards, smart cameras including vision tools with embedded software, uncooled long-wave infrared sensors for industrial and defense applications, and X-ray generators for non-destructive testing (NDT).

## REFERENCES


1. Product validation. [October 31, 2021]; Available from: https://www.nasa.gov/seh/5-4-product-validation.
2. Suh, N.P., Axiomatic design theory for systems. *Research in Engineering Design*, 1998. **10**(4): pp. 189–209.
3. Mankins, J.C., Technology readiness assessments: A retrospective. *Acta Astronautica*, 2009. **65**(9–10): pp. 1216–1223.
4. Kano model analysis. [October 31, 2021]; Available from: https://www.mindtools.com/pages/article/newCT_97.htm.
5. Sauro, J., A brief history of usability. 2013 [October 31, 2021]; Available from: https://measuringu.com/usability-history/.
6. UX research methods. [October 31, 2021]; Available from: https://maze.co/guides/ux-research/ux-research-methods/.
7. The difference between usability and user experience. [October 31, 2021]; Available from: https://fuzzymath.com/blog/difference-between-usability-and-user-experience/.
8. User experience basics. [October 31, 2021]; Available from: https://www.usability.gov/what-and-why/user-experience.html.
9. CX vs. UX: What's the difference? [October 31, 2021]; Available from: https://www.usertesting.com/blog/cx-vs-ux.
10. Usability vs. user experience: What's the difference? [October 31, 2021]; Available from: https://pt.slideshare.net/domain7/ux-vs-usability/9.

11. Usability testing. [October 31, 2021]; Available from: https://www.usability.gov/how-to-and-tools/methods/usability-testing.html.
12. Sauro, J., How much does a usability test cost? 2018 [October 31, 2021]; Available from: https://measuringu.com/usability-cost/.
13. Standards. [October 31, 2021]; Available from: https://www.iso.org/standards.html.
14. CAST (HOSDB) UK Home Office Body Armour Standards. [October 31, 2021]; Available from: https://www.bodyarmour.uk/cast-hosdb-uk-home-office-body-armour-standards/.
15. NATO Standardization Office. [October 31, 2021]; Available from: https://nso.nato.int/nso/nsdd/main/standards.
16. National Voluntary Laboratory Accreditation Program (NVLAP). [October 31, 2021]; Available from: https://www.nist.gov/nvlap/personal-body-armor-lap.
17. American Association for Laboratory Accreditation (A2LA). [October 31, 2021]; Available from: https://a2la.org/about/.
18. DEKRA large-scale test site. [October 31, 2021]; Available from: https://www.dekra.us/en/process-safety-testing/large-scale-test-site/.
19. NATO standardization. [October 31, 2021]; Available from: https://www.nato.int/cps/en/natohq/topics_69269.htm.
20. National Institute of Justice. [October 31, 2021]; Available from: https://nij.ojp.gov/about-nij.
21. Association of test centers for anti-attack materials and constructions (VPAM). [October 31, 2021]; Available from: https://www.vpam.eu/.
22. Tichler, C., Portable Ballistic Protection for UK Police, 2011, Centre for Applied Science and Technology United Kingdom, p. 32.
23. RussianGost. [October 31, 2021]; Available from: https://www.russiangost.com/t-about.aspx.
24. Frank, D.E., Ballistic tests of used soft body armor. Vol. 86, 1986, US Department of Commerce, National Bureau of Standards.
25. DIN EN 1522. [October 31, 2021]; Available from: https://www.en-standard.eu/din-en-1522-windows-doors-shutters-and-blinds-bullet-resistance-requirements-and-classification/.
26. DIN EN 1063. [October 31, 2021]; Available from: https://www.en-standard.eu/din-en-1063-glass-in-building-security-glazing-testing-and-classification-of-resistance-against-bullet-attack-english-version-of-din-en-1063/.
27. DIN Standards. [October 31, 2021]; Available from: https://www.din.de/en/about-standards/din-standards.
28. Standard specification for indoor ballistic test ranges for small arms and fragmentation testing of ballistic-resistant items. [October 31, 2021]; Available from: https://www.astm.org/Standards/E3062.htm.
29. AS/NZS 2343:1997 bullet-resistant panels and elements. [October 31, 2021]; Available from: https://www.standards.govt.nz/shop/asnzs-23431997/.
30. ISO/TR 11610:2004 protective clothing. [October 31, 2021]; Available from: https://www.iso.org/obp/ui/#iso:std:iso:tr:11610:ed-1:v1:en.
31. ISO 16935:2007 glass in building, bullet-resistant security glazing, test and classification. [October 31, 2021]; Available from: https://www.iso.org/standard/38168.html.
32. Bullet-resistant glazing. [October 31, 2021]; Available from: https://www.architecturalarmour.com/tech-spec/bullet-specs/bs5051.
33. Brunswick ballistic standards. [October 31, 2021]; Available from: http://www.closefocusresearch.com/brunswick-ballistic-standards.

34. 49 CFR hazmat transportation regulations. [October 31, 2021]; Available from: https://www.labelmaster.com/49-cfr.
35. Ballistic testing, product qualification, engineering & research. [October 31, 2021]; Available from: http://ballistic-testing.com/.
36. VPAM BRV armored vehicle testing and certification. [October 31, 2021]; Available from: https://www.interarmored.com/certified-toyota-landcruiser/.
37. USACE, Unified facilities guide specification. 2009: USA, p. 25.
38. Standard for bullet-resisting equipment. [October 31, 2021]; Available from: https://standardscatalog.ul.com/ProductDetail.aspx?productId=UL752.
39. Origin of the C.I.P. [October 31, 2021]; Available from: https://www.cip-bobp.org/en.
40. The Sporting Arms and Ammunition Manufacturers' Institute (SAAMI). [October 31, 2021]; Available from: https://saami.org/about-saami/.
41. ANSI/ISEAZ87.1-2020AmericanNationalStandardforOccupationalandEducational Personal Eye and Face Protection Devices. [October 31, 2021]; Available from: https://webstore.ansi.org/Standards/ISEA/ANSIISEAZ872020?gclid=Cj0KCQiA-eeMBhCpARIsAAZfxZDbSFWV9rVknQIhtSe5EnfZe0sq7IrsjLxBcrpNxN-0cyYsUbpsp3GwaApozEALw_wcB.
42. NF A36-800-2 hot-rolled weldable steel sheets for armouring - Part 2: firing test method. [October 31, 2021]; Available from: https://www.boutique.afnor.org/en-gb/standard/nf-a368002/hotrolled-weldable-steel-sheets-for-armouring-part-2-firing-test-method/fa177889/44472.
43. SAAMI glossary. [October 31, 2021]; Available from: https://web.archive.org/web/20070918032334/http://www.saami.org/Glossary/display.cfm?letter=A.
44. Good trigger. [October 31, 2021]; Available from: https://www.grantcunningham.com/2006/04/what-is-a-good-trigger/.
45. Ballistics laboratory. [October 31, 2021]; Available from: https://www.contractlaboratory.com/labclass/industries/industry-details.cfm?ballistics-testing&industry_id=208.
46. BackFace Signature (BFS) [October 31, 2021]; Available from: https://citizenarmor.com/blogs/blog/what-is-backface-signature-bfs-blunt-force-trauma#:~:text=Backface%20signature%20is%20the%20amount,and%20caliber%20of%20the%20projectile.
47. Pirvu, C., L. Deleanu, and C. Lazaroaie. Ballistic tests on packs made of stratified aramid fabrics LFT SB1. 7th International Conference on Advanced Concepts in Mechanical Engineering (ACME 2016), Iasi, Romania, In *IOP Conference Series: Materials Science and Engineering*, 2016, IOP Publishing.
48. Crouch, I. and B. Eu, Ballistic testing methodologies, in *The Science of Armour Materials,* I. Crouch, Editor. 2017, Elsevier: Amsterdam, Netherlands, pp. 639–673.
49. Dixon, W.J. and A.M. Mood, A method for obtaining and analyzing sensitivity data. *Journal of the American Statistical Association*, 1948. **43**(241): pp. 109–126.
50. Colt, S., Improvement in firearms, U.S.P. Office, Editor. 1836, Colt company: United States, p. 7.
51. Leek, W.E., Firearm having receiver bearing surfaces of synthetic resinous material, U.S.P. Office. 1958, Remington Arms Company, p. 10.
52. Hull, C.W., Method and apparatus for production of three-dimensional objects by stereolithography, E.P. Office. 1986, 3D Systems, Inc., p. 19.
53. 3D systems our story. [October 31, 2021]; Available from: https://www.3dsystems.com/our-story.
54. Crump, S.S., Apparatus and method for creating three-dimensional objects, U.S. Patent. 1988, Stratasys, Inc., p. 15.

55. The history of 3D printing. [October 31, 2021]; Available from: https://www.stratasys.com/explore/article/3d-printing-history.
56. The history of vacuum casting. [October 31, 2021]; Available from: https://www.mk-technology.com/?pageID=185.
57. Droege, J., 3D printers (weapons). 2019 [October 31, 2021]; Available from: https://slideplayer.com/slide/12857508/.
58. 3D Printed Semi-automatic Ruger Charger Pistol. [October 31, 2021]; Available from: https://3dprint.com/8398/3d-printed-gun-semi-automatic/.
59. Stratasys completes acquisition of solid concepts. [October 31, 2021]; Available from: https://investors.stratasys.com/news-events/press-releases/detail/226/stratasys-completes-acquisition-of-solid-concepts.
60. Clearances, tolerances and other errors. [October 31, 2021]; Available from: http://www.epi-eng.com/mechanical_engineering_basics/clearance_and_tolerance.htm.
61. Barrel tuner vibration analysis: Effects of tuner adjustments on vibration frequency and the shift in barrel node points. [October 31, 2021]; Available from: https://web.archive.org/web/20070831000414/http://www.varmintal.com/atune.htm.
62. Nasset, J.L., Rifle-Barrel harmonic vibration tuning device, U.S. Patent, 1995.
63. Lilja, D., A look at the rigidity of benchrest barrels. [October 31, 2021]; Available from: https://web.archive.org/web/20070927212954/http://www.riflebarrels.com/articles/barrel_making/rigidity_benchrest_rifles.htm.
64. Ruger Mini 14/30 Harmonic Stabilizer or NEW Tuneable Boss System™. [October 31, 2021]; Available from: https://web.archive.org/web/20070214010556/http://accuracysystemsinc.com/ruger_mini_14_30.html.
65. Wakeman, R., Locktime. [October 31, 2021]; Available from: http://www.chuckhawks.com/locktime.htm.
66. Tubb Precision Speedlock Firing Pin - Rem 700 Short Action. [October 31, 2021]; Available from: http://www.davidtubb.com/firingpin-remington-700.
67. The mini-14 rifle. [October 31, 2021]; Available from: https://outlands.tripod.com/armory/mini-14.htm.
68. Practical accuracy in the field: Shooting from unsupported positions. [October 31, 2021]; Available from: http://www.chuckhawks.com/shooting_unsupported_positions.htm.
69. Anschutz trigger. [October 31, 2021]; Available from: https://web.archive.org/web/20071005110928/http://www.davidtubb.com/tcom_images/t2k_images/t2k_manual_8.html.
70. Adjustable sights. October 31, 2021]; Available from: https://web.archive.org/web/20041121235152/http://www.findarticles.com/p/articles/mi_m0BTT/is_2000_Annual/ai_61620977.
71. Wallack, L.R., *Encyclopedia of American Gun Design and Performance: Rifles*. 1983, Winchester Press: Winchester, VA.
72. Why a custom Bullseye gun is important. [October 31, 2021]; Available from: http://www.bullseyepistol.com/cstomgun.htm.

# 7 Materials Used in the Production of Small Weapons

## 7.1 INTRODUCTION

The materialization and overall performance of a firearm depend largely on the materials selected to make its components. One could have a firearm with the best mechanical design, but if the materials with which its components are made are not suitable, its behavior will not be ideal, or, even worse, it could fail catastrophically. The selection of a material, together with the manufacturing process, cannot be separated from the mechanical design. In fact, the choice of a material is dictated by the mechanical design, although sometimes the design of a product must be manufactured based on a new material. The amount of materials available to the designer is enormous, and databases with their properties are usually offered. In the early stages of design, the range of materials can be vast, having many possible candidates. As the design progresses, the number of materials is reduced until having, in the last stage of design, a few options or even only one, where precise and detailed data are needed. These data are usually provided by the material producer, usually in the form of a data sheet. It should be noted that a material delivered by one producer may have slightly different properties from those delivered by another producer. Therefore, it is recommended to select a specific supplier in the last stage of the design.

New materials have been developed rapidly in recent decades, so most modern small firearms are made of steel, aluminum, polymers, and composites. Each material provides attributes to the firearms, such as mechanical strength, corrosion resistance, wear resistance, thermal resistance, weight, safety, aesthetics, and shooting performance.

Most of the books, and even diverse courses offered in some universities, dealing with the design of mechanical components, including firearms, do not take into account the selection of materials. For this reason, the intent of this chapter is to give the reader knowledge about the main materials used in the manufacture of firearms, the fundamentals for their selection based on their properties and specific application in firearms, the designation systems for steel and aluminum alloys, the description of the main engineering polymers, the basis of metal (MMCs)- and polymer-matrix composites (PMCs), and the emerging research on ceramic materials for the manufacture of small weapon components.

DOI: 10.1201/9781003196808-7

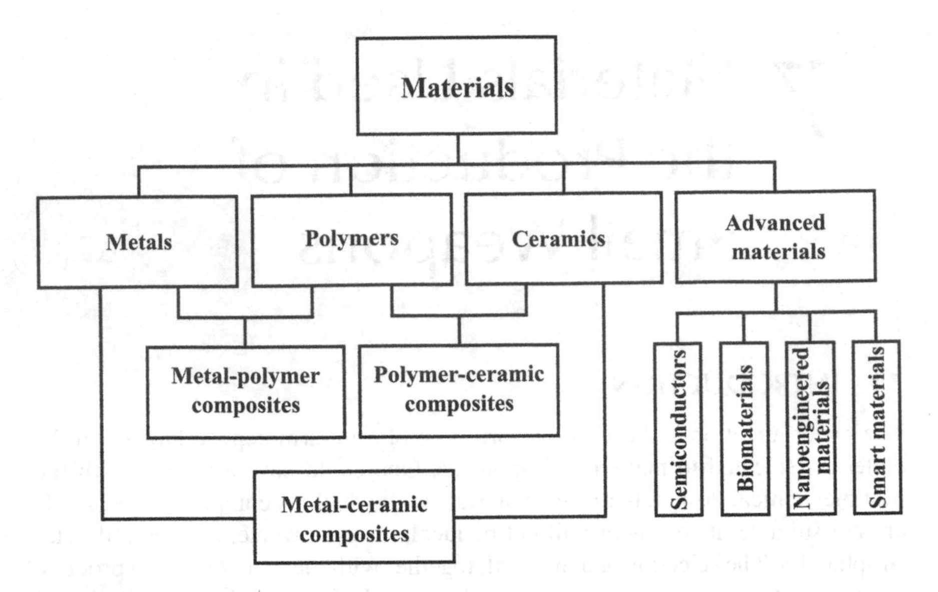


**FIGURE 7.1** Classification of materials.

## 7.2 CLASSIFICATION OF MATERIALS

Based on their chemical makeup and atomic structure, solid materials are classified into metals, polymers, and ceramics [1]. The combination of two or more of these materials forms a composite. It should be noted that there exist advanced materials, also made of metals, polymers, or ceramics, which are used in high-tech applications. Figure 7.1 shows the complete classification, but this chapter will not cover advanced materials.

### 7.2.1 Metals

Metals consist of atoms linked by metallic bonds. Most of the elements in the periodic table are metals; examples of metals are iron, aluminum, titanium, copper, and nickel. Most pure metals possess low mechanical properties, the reason why they are usually combined with other metals to form alloys. Metallic materials are usually divided into two large classes, ferrous and nonferrous, depending on whether they have iron or another element as the main constituent, respectively.

Metals and their alloys are ductile, malleable, good conductors of heat and electricity, and they have relatively high melting points, density, strength, hardness, and stiffness. One disadvantage of many metallic materials, especially ferrous ones, is their reactivity with the environment, causing them to rust and corrode; consequently, it is necessary to protect them with some surface treatment.

### 7.2.2 Polymers

Polymers are organic solids made up of long chains of carbon atoms linked through covalent bonds. By their origin, polymers are divided into natural polymers, which are produced by living organisms, and synthetic polymers, which are made by man [2]. By their thermomechanical properties, polymers are classified into thermoplastics (e.g., PET and Nylon 6,6), elastomers (e.g., natural and synthetic rubbers), and thermosets (e.g., Bakelite and epoxy resins) [3].

Polymers possess low densities, which are lower than those of metals and ceramics. Their strength-to-weight ratio is comparable to that of metals. Most polymers are poor conductors of heat and electricity, although some of them can be conductive through doping or the manufacture of PMCs. Polymers offer excellent corrosion resistance, impact resistance, and thermal insulation.

### 7.2.3 Ceramics

Ceramics are inorganic compounds made up of metallic and non-metallic elements from opposite ends of the periodic table, whose interatomic bonding is predominantly ionic [1]. Examples of ceramics are aluminum oxide (alumina, $Al_2O_3$), silicon carbide (SiC), tungsten carbide (WC), titanium nitride (TiN), and silicon dioxide (silica, $SiO_2$). Ceramics have low density compared to metals.

Most ceramics are poor conductors of heat and electricity; hence, they are used as thermal and electrical insulators. They are hard, stiff, and resistant to heat, wear, and corrosion. The main disadvantage of ceramics is their brittleness, making it difficult to design and manufacture components in comparison with metals and polymers.

### 7.2.4 Composites

Composites are produced when two or more materials are joined to give a combination of properties, which cannot be obtained in the original materials. Most composites are made up of two phases: a continuous phase named matrix, which surrounds and binds a dispersed phase named reinforcement. The properties of composites depend on the properties of the constituent phases, their relative content, and the reinforcement geometry [1]. Composites are classified according to the nature of the matrix: metal-matrix, polymer-matrix, and ceramic-matrix composites (CMCs). The reinforcing phase may be particles, fibers, or structural, whose concentration, size, shape, distribution, and orientation influence the properties of composites. An example of PMC is the Nylon family reinforced with glass fibers, with which different firearm components are manufactured. Such components are lightweight and possess good strength, stiffness, and impact resistance.

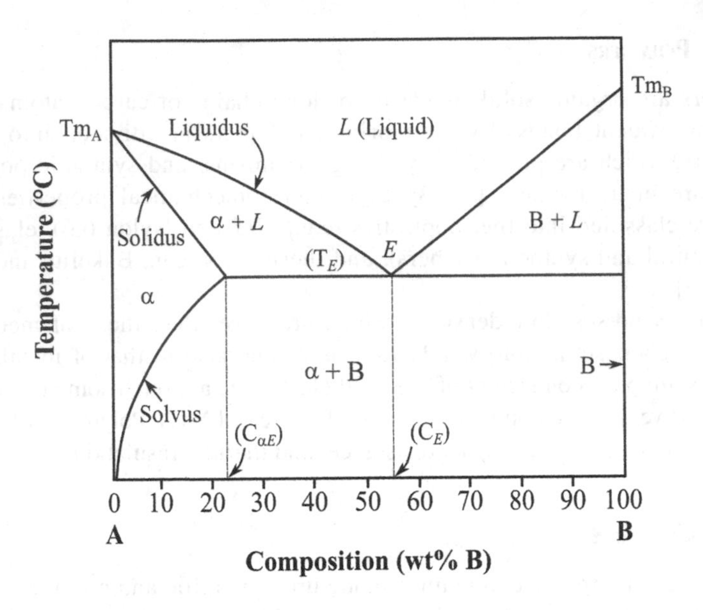


**FIGURE 7.2** Hypothetical binary phase diagram.

## 7.3 BINARY PHASE DIAGRAMS

A binary phase diagram is a graph of temperature versus composition, where the equilibrium phases present at a given temperature and composition are indicated. Figure 7.2 presents a schematic of a binary phase diagram, indicating several terms that will be explained below.

Phase diagrams provide valuable information on which phases are thermodynamically stable in an alloy and whether they could be present over a long time when the component is subjected to a particular temperature, for example, in a gun barrel made of steel. Other information obtained from phase diagrams is about melting, solidification, solubility limits, the presence of solid solutions, and phase transformations. These diagrams are useful to metallurgists, materials engineers, and materials scientists in different areas [4]:

- Development of new alloys for specific applications
- Fabrication of alloys into useful configurations
- Design and control of heat treatment procedures for specific alloys to produce the required properties
- Solving performance problems with specific alloys in commercial applications, thus improving product predictability.

Understanding the phase diagram of an alloy system is of utmost importance since there is a strong correlation between the mechanical properties and the microstructure, being this related to the characteristics of the phase diagram [1].

The hypothetical binary phase diagram of Figure 7.2 consists of two components, A and B, being pure metals of which an alloy is composed. The melting point of pure component A (0 wt% B) is $Tm_A$, while that of pure component B (100 wt% B) is $Tm_B$. This type of phase diagram has a special point called eutectic, *E*, whose composition $C_E$ and temperature $T_E$ are invariant. The alloy corresponding to the composition $C_E$ is called eutectic alloy. $T_E$ is the lowest temperature at which a mixture of A and B, at the fixed composition $C_E$, can melt.

Different phases can be distinguished: α, B, *L*. A phase is a physically homogeneous and separable portion of a system that has uniform physical and chemical characteristics; every pure material and every solid, liquid, and gaseous solution are considered to be a phase. When the components of a system are completely miscible in the solid form, a solid solution is formed, whose representation is made with a Greek letter, such as α in Figure 7.2. The α phase is a solid solution rich in component A and its solubility is limited to the concentration $C_{\alpha E}$ at the temperature $T_E$.

In addition, different boundaries can be distinguished in the phase diagram of Figure 7.2. The boundary between the liquid field (*L*) and the two-phase fields (α+*L* and B+*L*) is called the liquidus. The boundary between the two-phase field (α+*L*) and the solid field (α) is the solidus. The boundary between the α and α+B regions is termed the solvus, which corresponds to the solid solubility limit of B in A to form the α solid solution.

## 7.4 THE Fe-C PHASE DIAGRAM

Ferrous alloys are based on iron-carbon alloys, so that the Fe-C phase diagram provides the basis for understanding the different phases, heat treatments, and properties of cast irons and steels. The Fe-C phase diagram is somewhat complex, but only up to 6.7 wt% C is considered; compositions between 6.7 and 100 wt% C are not of interest since, in practice, all cast irons and steels have carbon contents <6.7 wt%.

Figure 7.3 presents the Fe-$F_3C$ phase diagram, where it can be seen that the vertical line at the composition 6.7 wt% C represents the iron carbide ($F_3C$) phase. It is worth mentioning that although the phase diagram is based on $F_3C$, the content is always measured in terms of carbon (wt% C).

Pure iron (Fe) (0 wt% C) can have different forms or crystal structures when heated before it melts; this is known as polymorphism in materials science. The stable form at room temperature is α-iron (α-Fe), named ferrite, which has a body-centered cubic (BCC) structure. α-Fe is also a solid solution of carbon in BCC Fe, with a maximum content of 0.022 wt% C at 727°C. Figure 7.4 shows a close-up of the Fe-$F_3C$ phase diagram to observe these details. Ferrite is relatively soft, and it is magnetic at temperatures below 768°C.

α-Fe transforms to the γ-iron (γ-Fe) phase, named austenite, at 912°C. γ-Fe has a face-centered cubic (FCC) structure. It is also a solid solution of carbon in FCC Fe, with a maximum content of 2.14 wt% C at 1147°C. Austenite is a nonmagnetic

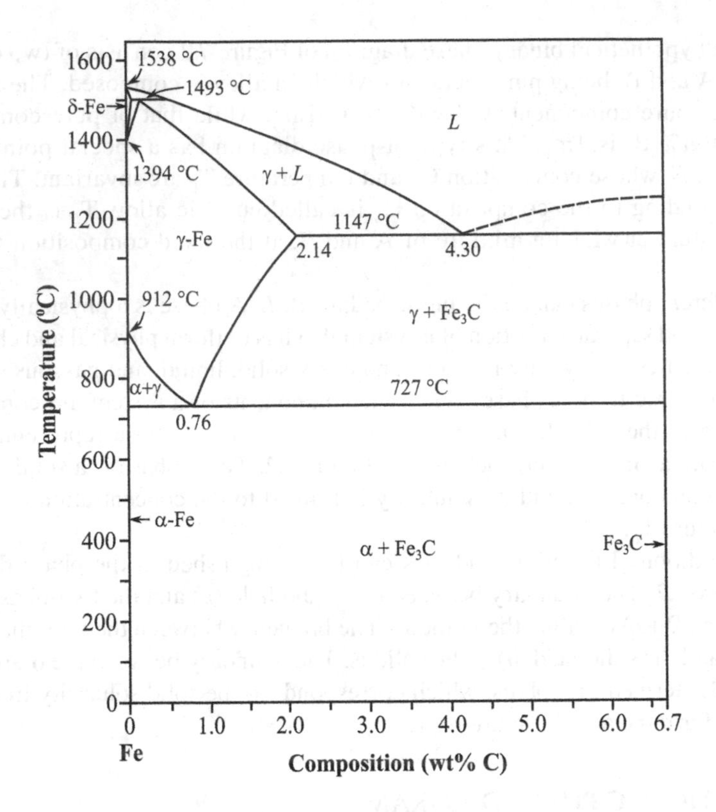


**FIGURE 7.3** Schematic of the Fe-$F_3C$ phase diagram.

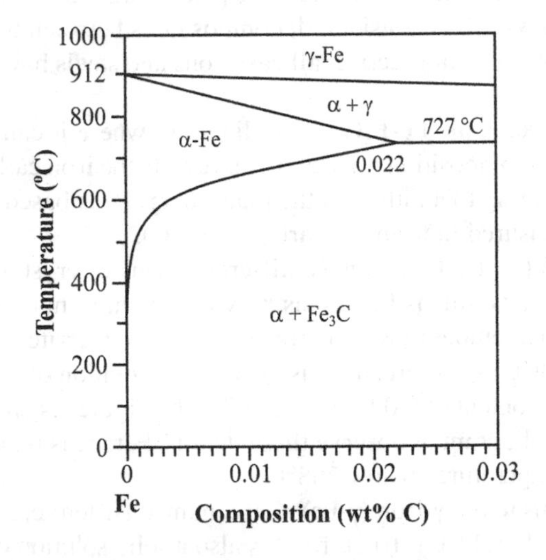


**FIGURE 7.4** Schematic of the Fe-$F_3C$ phase diagram showing the Fe-rich zone.

phase. γ-Fe transforms to δ-iron (δ-Fe), named δ-ferrite, at 1394°C. δ-Fe becomes BCC structure again. The maximum content of carbon in δ-ferrite is 0.09 wt% at 1493°C and it melts at 1538°C. This phase has a similar structure as that of α-Fe, but it is of no technological importance due to the high temperature at which it occurs. Iron carbide ($Fe_3C$), named cementite, is a very hard and brittle ceramic phase. It is because of these properties that cementite is used to strengthen cast irons and steels. It has an orthorhombic crystal structure. Cementite is accountable for the great variety of microstructures and properties produced in steels [5]. This phase coexists with the α phase ($\alpha+Fe_3C$) below 727°C and with the γ phase ($\gamma+Fe_3C$) between 727°C and 1147°C.

In the Fe-$F_3C$ phase diagram, the dividing point between steels and cast irons is 2.14% C. Steels are considered in the range from 0.008 to 2.14 wt% C, while cast irons in the range from 2.14 to 6.7 wt% C. A feature of the Fe-$Fe_3C$ phase diagram is the existence of a eutectoid point at 0.76 wt% C and 727°C, with which steels may be classified as hypoeutectoid, eutectoid, and hypereutectoid steels. Another feature is the existence of a eutectic point at 4.3 wt% C and 1147°C, with which cast irons may be referred to as hypoeutectic, eutectic, and hypereutectic cast irons. It should be pointed out that cast irons are outside the scope of this chapter because they are not used in the manufacture of modern small firearm components.

## 7.5 ALUMINUM ALLOYS PHASE DIAGRAMS

Figure 7.5 presents phase diagrams of aluminum with four of the most common alloying elements.

Pure aluminum (Al) has a FCC crystal structure. The formation of a solid solution (α) is possible with each alloying element, which preserves the FCC phase. The solid solubility of Cu, Mg, Mn, and Zn in aluminum is 5.65, 17.1, 1.82, and 83.1 wt% at 548°C, 450°C, 658°C, and 443°C, respectively.

## 7.6 MECHANICAL PROPERTIES OF MATERIALS

The components of small weapons are exposed to different stresses during shooting, transport, handling, and even at the time of storage. Therefore, the designers and engineers must ensure that the components, and the weapon as a whole, will have enough strength, ductility, stiffness, toughness, and resistance to wear, heat, and corrosive environments, among other properties and characteristics. Some of the most important properties of materials, especially metals, are those obtained from tension and hardness tests. These properties are commonly reported in the technical data sheets when raw materials are supplied by a vendor.

The tension test consists in applying a load uniaxially along the longitudinal axis of a specimen, usually until its complete fracture. The test is commonly carried out using standard specimens and testing machines, under certain specific conditions. For instance, the American Society for Testing and Materials (ASTM) covers the

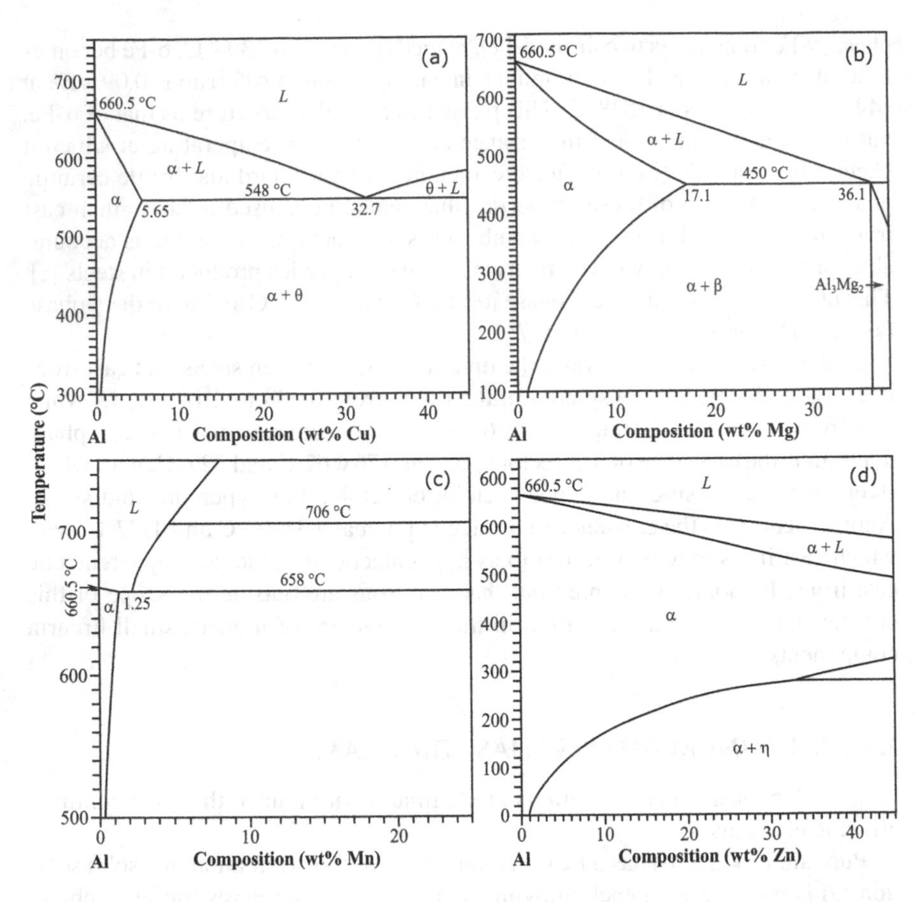


**FIGURE 7.5** Schematic of (a) Al-Cu, (b) Al-Mg, (c) Al-Mn, and (d) Al-Zn phase diagrams showing the Al-rich zones.

tension testing of metallic materials in its ASTM E8/E8E standard [6], which is equivalent to other standards issued by different organizations around the world. Examples of these organizations are the European Committee for Standardization (CES), British Standards Institution (BSI), German Institute for Standardization (DIN), French Standardization Association (AFNOR), International Organization for Standardization (ISO), Japanese Industrial Standards (JIS), Bureau of Indian Standards (BIS), and Official Mexican Standards (NOM).

Figure 7.6 illustrates the tensile stress-strain curve corresponding to a typical ductile metal, where the main mechanical properties acquired from a tension test are defined; the inset corresponds to a virtual round tension test specimen. The slope of the linear elastic region of the curve, which obeys Hooke's law, is the modulus of elasticity or Young's modulus; this property is a measure of the material stiffness. The yield strength is the stress at which plastic deformation begins; by convention, this property is determined using the 0.002 strain offset

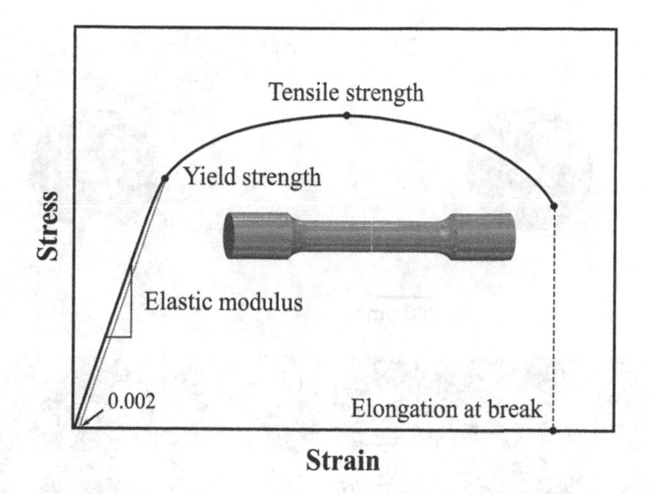


**FIGURE 7.6** Schematic of a tensile stress-strain curve of a ductile metal.

technique. The tensile strength is the maximum tensile stress that a specimen is capable of sustaining. The elongation at break or fracture strain is the strain where a specimen breaks; this property is an indication of the material ductility. The area under the stress-strain curve indicates the toughness of a material, i.e., its ability to absorb energy and plastically deform before breaking. This means that the greater the deformation of the material, the more energy will be required to break it.

It is important to note that some weapon components must withstand compressive stresses rather than tensile stresses, especially those that absorb loads during recoil. For such components, the materials must be tested in compression to ensure that they will resist these mechanical stresses. The compression test is conducted in a manner similar to the tension test, except that the specimen is subjected to an increasing axial compressive load [7]. The mechanical properties obtained in compression are the same as indicated in Figure 7.6.

The hardness is the ability of a material to resist localized plastic deformation. There exist various hardness-testing techniques, such as Brinell [8], different Rockwell scales [9], Vickers and Knoop [10], and Shore [11]. Figure 7.7 shows micrographs of indentations with some of these techniques. The selection of the technique depends on the material features, i.e., its microstructure, type, size of the specimen, and its condition. A conversion among different scales is available for specific materials [12]. Hardness tests are simple, inexpensive, and nondestructive techniques, with which it is possible to estimate the tensile strength of materials. In fact, there are tables and graphs in the literature that relate to both properties [13]. Note that there is no general method to accurately convert the hardness values among the different scales, or even to tensile strength values. The conversions are approximations and should be treated with caution.

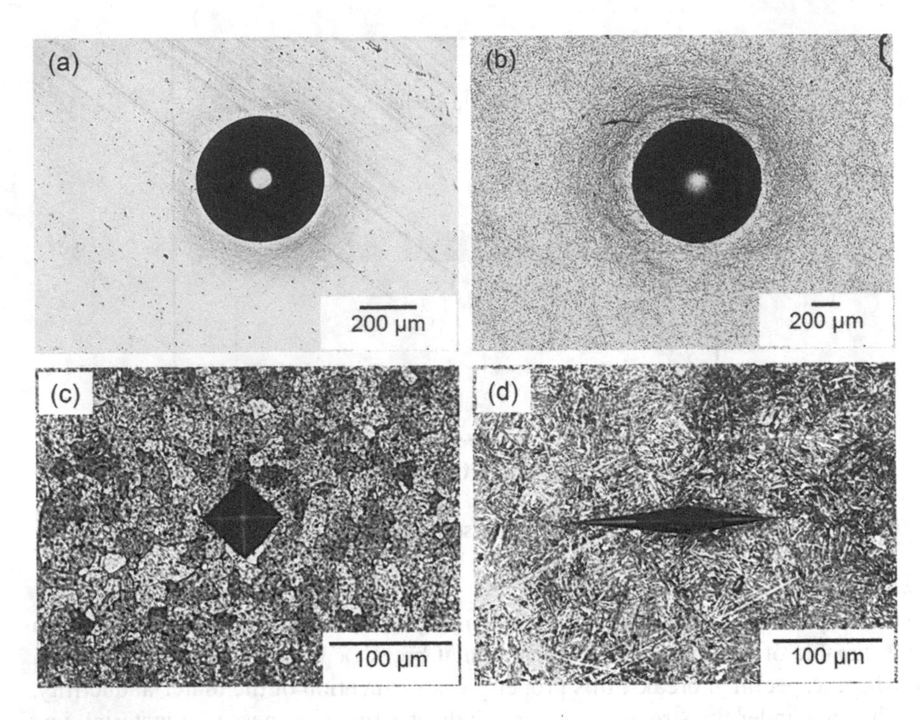


**FIGURE 7.7** Micrographs showing the indentation of (a) Brinell, (b) Rockwell C, (c) Vickers, and (d) Knoop. (Images courtesy of Dr. Caleb Carreño Gallardo, CIMAV.)

## 7.7 STEELS

Steels are the materials most used in multiple applications worldwide due to different aspects like the abundance of iron ore on Earth; the relatively low cost for their making, forming, and processing; and the wide range of mechanical properties that can be obtained with them. The development of firearms was due to the advancement in steel metallurgy, and, at present, different steels are the base materials in any firearm.

### 7.7.1 Classification of Steels

Steels can be classified by a variety of different systems, depending on [5]:

- **The chemical composition:** plain carbon, low-alloy, or high-alloy steels (Figure 7.8).
- **The manufacturing methods:** open hearth, basic oxygen process, or electric furnace methods.
- **The finishing method:** hot rolling or cold rolling.
- **The product form:** bar, plate, sheet, strip, tubing, structural shape, etc.
- **The deoxidation practice:** killed, semikilled, capped, or rimmed steel.

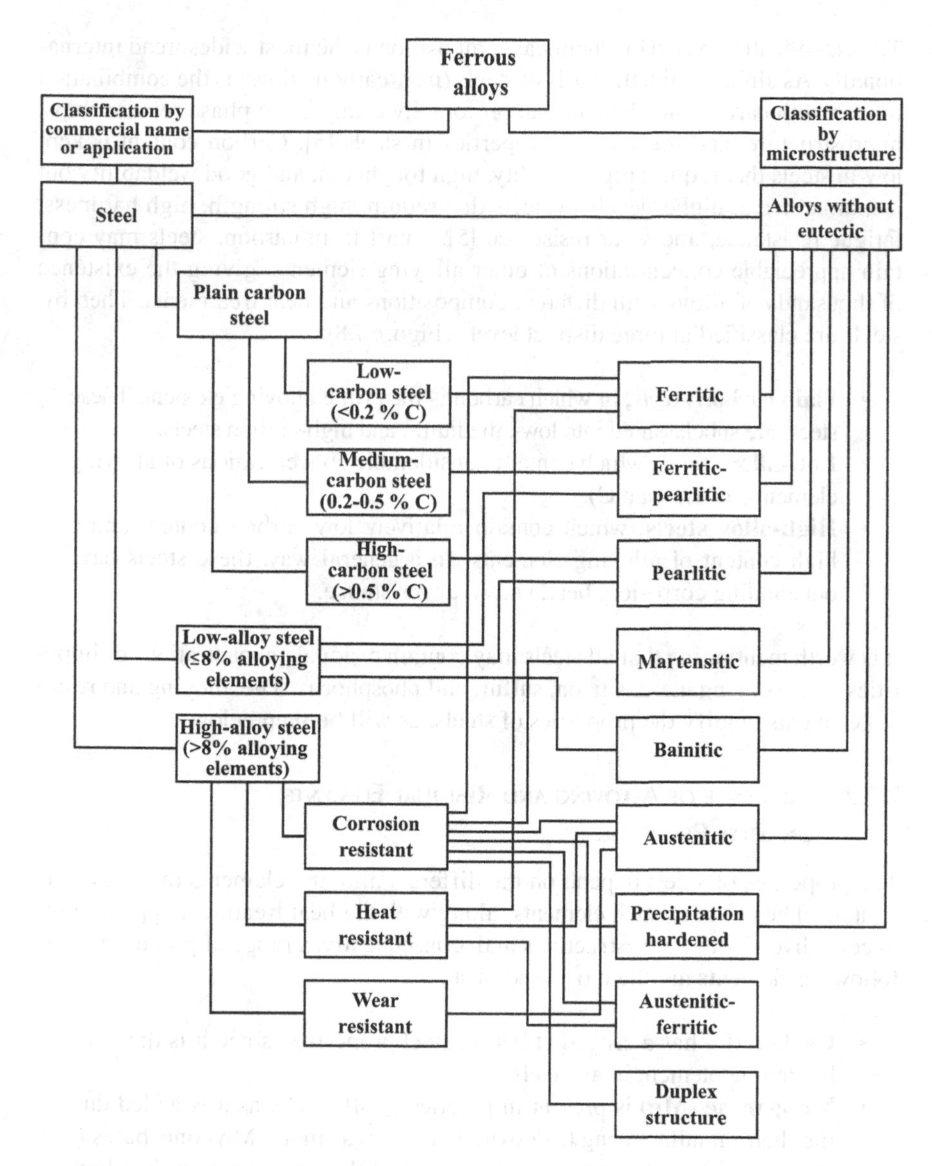


**FIGURE 7.8** Classification of steels. (Courtesy of Prof. Doru Michael Stefanescu, The University of Alabama and The Ohio State University.)

- **The microstructure:** ferritic, pearlitic, martensitic, etc. (Figure 7.8).
- The required strength level, as specified in different standards.
- **The heat treatment:** annealing, quenching and tempering, thermomechanical processing, etc.
- **Quality descriptors:** forging quality, commercial quality, etc.

The classification based on chemical composition is the most widespread internationally. As already said, the basis of steels (iron-carbon alloys) is the combination of iron with carbon, an element that profoundly changes the phase relationships, microstructure, and mechanical properties in steels [5]. Carbon content is kept low in steels that require high ductility, high toughness, and good weldability but is maintained at higher levels in steels that require high strength, high hardness, fatigue resistance, and wear resistance [5]. Apart from carbon, steels may contain appreciable concentrations of other alloying elements, giving the existence of thousands of alloys with different compositions and heat treatments. Thereby, steels are classified at three distinct levels (Figure 7.8):

- **Plain carbon steels,** in which carbon is the prime alloying element. These steels are subclassified into low-, medium-, and high-carbon steels.
- **Low-alloy steels,** which contain considerable concentrations of alloying elements (other than C).
- **High-alloy steels,** which contain relatively low carbon content and a high content of alloying elements. In a general way, these steels have outstanding corrosion, heat, and wear resistance.

It is worth mentioning that all steels may contain residual concentrations of impurities such as manganese, silicon, sulfur, and phosphorus. The alloying and residual elements modify the properties of steels, as will be seen below.

### 7.7.2 Influence of Alloying and Residual Elements on Steel Properties

The properties of steels depend on the different alloying elements that they may contain. The effect of such elements, along with the heat treatment applied, produces a diversity of microstructures and, consequently, a range of properties. The following elements are the most used in steels:

- **Carbon (C)** has a crucial effect on steel properties, since it is the main hardening element in all steels.
- **Manganese (Mn)** is present in practically all steels, as it is added during their manufacturing to deoxidize and degas them. Mn contributes to strength and hardness but to a lesser extent than does carbon. It is beneficial for the hardenability and surface quality of steel. If the steels did not have manganese, they could not be rolled or forged.
- **Phosphorus (P)** is usually considered deleterious to the mechanical properties of steels. It segregates but to a lesser degree than sulfur. It increases strength and hardness but decreases ductility and toughness, which can produce brittleness. This decrease in ductility and toughness is greater in quenched and tempered higher-carbon steels. Low-carbon free-machining steels are often specified with higher phosphorus content to improve machinability.

- **Sulfur (S)** is very detrimental to surface quality, particularly in low-carbon and low-manganese steels, so that a maximum limit is specified for most steels. In the case of free-machining steels, sulfur is added to improve machinability. Sulfur decreases weldability and toughness. Sulfur has a greater segregation tendency in the form of sulfide inclusions.
- **Silicon (Si)** is intentionally added during the manufacturing process as a deoxidizer and degasser; its role is to prevent pores and other internal defects from appearing in steels. It is less effective than Mn in increasing as-rolled strength and hardness and has a slight tendency to segregate. In low-carbon steels, Si is usually detrimental to surface quality.
- **Chromium (Cr)** increases the resistance to oxidation and corrosion, as well as the hardness and tensile strength of steels. It improves hardenability and high-temperature strength, prevents deformation by quenching, and increases wear resistance. When used with a toughening element such as nickel, it produces superior mechanical properties. In conjunction with molybdenum, chromium increases the strength of steels.
- **Molybdenum (Mo)** notably improves tensile strength, hardness, toughness, and creep resistance of steels. It retains the hardness at elevated temperatures.
- **Nickel (Ni)** increases tensile strength and hardness without sacrificing toughness. It increases corrosion resistance when introduced in enough amounts (at least 8 wt%). When used in constructional steels, it strengthens the ferrite phase. Further, Ni lowers the critical cooling rate during the heat treatments. Along with Cr, Ni improves the mechanical properties.
- **Vanadium (V)** tends to refine the grain and decrease the hardenability. It is a very strong deoxidizing element and tends to form carbides. Vanadium, like titanium and zirconium, is an effective grain growth inhibitor.

### 7.7.3 Designation of Steels

The American Iron and Steel Institute (AISI) and the Society of Automotive Engineers (SAE) provide a basic four-digit system to designate the chemical composition of carbon steels and alloy steels (Table 7.1). The first digit indicates the main alloying element, the second digit specifies the approximate weight percentage of the alloying element, and the third and fourth digits refer to the weight percentage in hundredths. Thus, a 1060 steel is a plain carbon steel with 0.60 wt% C, while a 4140 steel is a chromium-molybdenum steel containing 1.0 wt% Cr and 0.40 wt% C. The stainless steels have assigned a three-digit number that identifies the basic family and the particular alloy within each family (Table 7.1).

Table 7.2 presents the composition of some carbon steels and alloy steels used in the manufacture of small weapon components. When the content of an alloying element is specified as a range, the steelmaker adjusts its content to the average

**TABLE 7.1**
**General AISI/SAE Designation System**

| Specification | Classification |
|---|---|
| 1xxx | Carbon steels |
| 2xxx | Nickel steels |
| 3xxx | Nickel-chromium steels |
| 4xxx | Chromium-molybdenum steels |
| 5xxx | Chromium steels |
| 6xxx | Chromium-vanadium steels |
| 7xxx | Tungsten steels |
| 8xxx | Nickel-chromium-molybdenum steels |
| 9xxx | Silicon-manganese steels |
| 200 | Austenitic stainless steels with chromium, nickel, manganese, or nitrogen |
| 300 | Austenitic stainless steels with chromium and nickel |
| 400 | Ferritic or martensitic stainless steels with chromium and possibly carbon |
| 500 | Martensitic stainless steels with low chromium (<12%) and possibly carbon |

**TABLE 7.2**
**Composition (wt%) of Selected Carbon Steels and Alloy Steels [14,15]**

| Alloy | C | Mn | P | S | Si | Cr | Mo | Ni |
|---|---|---|---|---|---|---|---|---|
| 1015 | 0.13–0.18 | 0.30–0.60 | 0.04 | 0.05 | | | | |
| 1020 | 0.18–0.23 | 0.30–0.60 | 0.04 | 0.05 | | | | |
| 1045 | 0.43–0.50 | 0.60–0.90 | 0.04 | 0.05 | | | | |
| 1060 | 0.55–0.65 | 0.60–0.90 | 0.04 | 0.05 | | | | |
| 1080 | 0.75–0.88 | 0.60–0.90 | 0.04 | 0.05 | | | | |
| 1140 | 0.37–0.44 | 0.70–1.00 | 0.04 | 0.08–0.13 | | | | |
| 1215 | 0.09 | 0.75–1.05 | 0.04–0.09 | 0.26–0.35 | | | | |
| 1330 | 0.28–0.33 | 1.60–1.90 | 0.035 | 0.04 | 0.15–0.35 | | | |
| 1524 | 0.19–0.25 | 1.35–1.65 | 0.04 | 0.05 | | | | |
| 4140 | 0.38–0.43 | 0.75–1.00 | 0.035 | 0.04 | 0.15–0.35 | 0.80–1.10 | 0.15–0.25 | |
| 4340 | 0.38–0.43 | 0.60–0.80 | 0.035 | 0.04 | 0.15–0.35 | 0.70–0.90 | 0.20–0.30 | 1.65–2.00 |
| 8620 | 0.18–0.23 | 0.70–0.90 | 0.035 | 0.04 | 0.15–0.35 | 0.40–0.60 | 0.15–0.25 | 0.40–0.70 |
| 416 | 0.15 | 1.25 | 0.60 | ≤0.15 | 1.00 | 12.00–14.00 | 0.60 | |

*The balance of the composition is iron.*

between both values of the range; a single value means a maximum content. From Table 7.2, it follows that 1015, 1020, 1045, 1060, and 1080 alloys correspond to carbon steels, which have certain content of manganese, phosphorus, and sulfur. The 1140 alloy is a resulfurized carbon steel, in which the sulfur content was increased; observe the range in which it should be. The 1215 alloy is a rephosphorized and resulfurized carbon steel; note the phosphorus and sulfur ranges and, in addition, the low carbon content. Although the 1330 alloy is from the 1xxx series, it is considered an alloy steel in which silicon is added, in addition to having considerably increased the range in the manganese content. The 1524 alloy is a high-manganese carbon steel. Both the 4140 and 4340 alloys are chromium-molybdenum steels, although the latter also contains nickel. The 8620 alloy is a nickel-chromium-molybdenum steel. Finally, the 416 alloy is a martensitic stainless steel that, like all steels of this type, is more sensitive to heat treatment variables than are carbon and low-alloy steels. One characteristic of stainless steels is their high content of chromium (at least 12 wt%), which makes them resistant to oxidation and corrosion by forming a passive layer of chromium oxide (passivation) on the surface, which seals iron from oxidation.

The AISI/SAE designations for steels are equivalent in other regions and countries. Take, for example, a 4340 steel in bar presentation. The ASTM covers this alloy in the A29/A29M standard [14], which is equivalent to the 41NiCrMo7-3-2 (DIN 1.6563) designation in the EN 10263-4:2001 standard [16], applicable in European countries such as Germany, UK, and France. The corresponding alloy in Japan is the SNCM439 designation, which is covered by the JIS G4053-2008 standard [17]. The equivalent alloy under ISO is the 36CrNiMo4 designation, covered by the ISO 683-18:1996 standard [18].

In order to observe the effect of carbon and alloying elements on the mechanical properties of steels, Table 7.3 displays some of the alloys used in the manufacture of small weapon components. Note that these properties are presented as a matter of general information, since they can be modified depending on the heat treatment conditions, as will be seen in Chapter 8.

Compare the 1060 alloy with the 1080 alloy, which are in the hot rolled condition. Both alloys are carbon steels, the former with 0.60 wt% C and the latter with

**TABLE 7.3**
**Mechanical Properties of Selected Carbon Steels and Alloy Steels**

| Alloy | Condition | Yield Strength (MPa) | Tensile Strength (MPa) | Modulus of Elasticity (GPa) | Elongation at Break (%) | Brinell Hardness (HBW) | References |
|---|---|---|---|---|---|---|---|
| 1015 | Hot rolled | 190 | 345 | 200 | 28 | 101 | [19] |
| 1020 | Hot rolled | 205 | 380 | 186 | 25 | 111 | [20] |
| 1045 | Hot rolled | 310 | 565 | 206 | 16 | 163 | [21] |
| 1060 | Hot rolled | 370 | 660 | 200 | 12 | 201 | [22] |
| 1080 | Hot rolled | 425 | 772 | 205 | 10 | 229 | [23] |
| 4140 | Annealed | 415 | 655 | 205 | 25 | 197 | [24] |
| 4340 | Annealed | 470 | 745 | 192 | 22 | 217 | [25] |
| 8620 | Annealed | 385 | 530 | 205 | 31 | 149 | [26] |
| 416 | Annealed | 275 | 515 | 200 | 30 | 262 | [27] |

0.80 wt% C. This difference in carbon content causes the yield strength, the tensile strength, and the hardness to increase by 15%, 17%, and 14%, respectively; however, the elongation at break decreases by 17%. Now compare the 4140 and 4340 alloys in the annealing condition. Both of them are alloy steels with 0.40 wt% C but with different content of the alloying elements, particularly the nickel in the 4340 alloy. This difference in the alloying elements causes the yield strength, the tensile strength, and the hardness of the 4340 alloy to increase by 13%, 14%, and 10%, respectively, and that its elongation at break decreases by 13%. In a general way, increasing strength occurs at the expense of elongation properties.

Steels listed in Table 7.2 are suitable for the manufacture of small weapon components thanks to the properties that can be achieved with each of them [28,29]. Note that the terms related to heat and surface treatments will be addressed in Chapter 8.

- **1015 alloy** is considered a borderline grade of steel, since its carbon content is high for best formability and slightly low compared with grades of carbon steels used for most carburizing hardening treatment. This alloy has excellent forgeability, reasonably good cold formability, and excellent weldability. Machinability is relatively poor compared with other grades such as 1100 and 1200. This steel can be used in forged, cold-headed, or cold-formed parts, which are low strength with wear-resistant and hard surfaces. Pins and bolts for small weapons are manufactured from 1015 alloy.
- **1020 alloy** is the most widely used steel. It is available in a variety of product forms. It has excellent forgeability and weldability but a notably poor machinability. This steel is widely used as a carburizing steel for simple structural applications such as cold-headed bolts. It is suitable for manufacturing axles, general engineering and machinery parts and components, shafts and camshafts, gudgeon pins, ratchets, light duty gears, and spindles. Trigger guards, floorplates, sights, sling swivels, and other steel hardware are made of 1020 steel.
- **1045 alloy** is specified as a medium-carbon steel. It is available in a variety of product forms, mainly as stock for forging, as well as special quality grades of steel compositions. This steel has excellent forgeability, good weldability, fair machinability, and high-strength and impact properties. It responds readily to heat treatment, being its as-quenched hardness at least 55 HRC (hardness Rockwell C). It is widely used for all industrial applications requiring more wear resistance and strength. Gears, pins, rams, shafts, rolls, sockets, axles, spindles, worms, bolts, ratchets, light gears, studs, crankshafts, guide and connecting rods, torsion bars, and hydraulic clamps are made of 1045 alloy.
- **1060 alloy** is a versatile high-carbon steel. Product forms include various thicknesses of flat stocks for fabricating parts to be tempered. This steel has good forgeability. It is not recommended for welding. Its hardness is near 65 HRC in the as-quenched condition, which is close to the

maximum Rockwell hardness. Firearms guide pins are manufactured from 1060 steel.

- **1080 alloy** is a high-carbon steel highly hardenable, especially if its manganese content is close to the upper limit (0.90 wt% Mn). Its as-quenched hardness is near 65 HRC. Thanks to its high carbon content, it may have free carbide, which enhances the abrasion resistance but decreases ductility. This steel has good forgeability but poor weldability. 1080 steel is one of the common spring steel grades. It finds applications in piano wires, springs, shafts, and in suspension parts.
- **4140 alloy** is among the most widely used medium-carbon alloy steel. It is relatively inexpensive, considering its high hardenability. Fully hardened 4140 ranges from about 54 to 59 HRC, depending upon the exact carbon content. This steel has very good forgeability but fair machinability and poor weldability because of susceptibility to weld cracking. It has high ductility and can be formed using conventional techniques in the annealed condition. It requires more pressure for forming because it is tougher than plain carbon steels. 4140 alloy is used to manufacture components subjected to stress and impact loads. Barrels, bolt receivers, and high-stress items like muzzle brakes are made of 4140 steel.
- **4340 alloy** is a high-hardenability steel, more than any other standard AISI grade. Depending on the precise carbon content, as-quenched hardness ranges from 54 to 59 HRC. Hardened 4340 is not suitable for welding by conventional methods. This steel has good toughness, fatigue resistance, and wear resistance. It can be forged without difficulty, although its hot strength is considerably higher than that of carbon or lower alloy grades, requiring more powerful forging machines. Its machinability is relatively poor. 4340 steel is widely used in the aerospace industry because of its great strength. Power transmission gears and shafts, aircraft landing gear, and other structural parts are made of 4340 steel. It is also used for manufacturing firearm barrels.
- **8620 alloy** is used extensively as a case hardening steel for both carburizing and carbonitriding. As-quenched surface hardness usually ranges from around 37 to 43 HRC. This steel has reasonably high hardenability, excellent forgeability, and weldability, although alloy steel practice should be used in welding to minimize susceptibility to weld cracking. It has good machinability and cold strain plasticity. 8620 steel can be used for a number of medium-strength applications such as camshafts, fasteners, gears, and chains/chain pins. Some receivers are made of this 8620 alloy.
- **416 alloy** is a low-cost, free-machining stainless steel whose machinability is 85%, the highest of all stainless steels. It has good nonseizing and nongalling properties, as well as good corrosion resistance. This steel is highly resistant to acids, alkalis, fresh water, and dry air. It resists oxidation up to 760°C. It exhibits poor weldability and is not recommended for forging operations requiring severe deformation. 416 steel is capable

of hardening to 42 HRC. Some of its applications include the manufacture of valves, pump and motor shafts, parts of washing machines, gears, bolts, nuts, and automatic screw-machined components. Some buffer tubes and barrels are made of 416 steel due to its good machining properties and the best precision that can be achieved with it, compared to components made of Cr-Mo steels; these features are exploited in target shooting weapons. Nonetheless, this steel is more expensive, and due to its lower mechanical properties, the life of barrels is slightly less than that of Cr-Mo steel barrels, which are preferred for military weapons.

There are some steels whose composition includes vanadium, which are selected to produce hammer-forged barrels. Examples of these steels are AISI 41V45 and DIN 32CrMoV12-10. Other stainless steels can be found in the manufacture of weapon components; for instance, AISI 316 (austenitic stainless steel) is used to produce trigger guards and floorplates and AISI 17-4 PH (precipitation-hardened steel) is used to produce barrels, bolts, gas blocks, and receivers. It is to be noted that there have been efforts to manufacture most of the weapon components in stainless steel, as was the case with the extinct Arcadia Machine & Tool company; however, some problems encountered were poor wear and galling resistance of both slide and receiver in automatic pistols, and loss of cylinder timing in revolvers under heavy use, compared to standard carbon steels. Cases of success have been some Smith & Wesson firearms, such as the Model 60 revolver [30] and the SW22 Victory pistol [31], whose frame, barrel, and cylinder/slide are made of stainless steel.

Table 7.4 makes a comparison of the previous steels with other designations. It should be noted that only a few examples are named, since different steels may be considered under the same designation. For more details, go to Ref. [32].

## 7.8 ALUMINUM ALLOYS

Aluminum is the most abundant metal in the earth's crust and the third most abundant, after oxygen and silicon [33]. In terms of volumes used, aluminum alloys rank second after steel, which is due to their versatility derived from their excellent and diverse range of physical, chemical, and mechanical properties. Although aluminum and its alloys generally have low mechanical resistance compared to steels, they have other advantages such as low density, which is around one-third of that of steels. They are notable for their ability to resist corrosion due to the passivation phenomenon, are nonmagnetic, noncombustible, and nontoxic. They have high reflectivity and rapid heat dissipation. Aluminum and its alloys are malleable, ductile, and easily worked by different manufacturing processes. Structural components made of aluminum and its alloys are of great importance in industries such as aerospace, transportation, and construction. The weapon industry is not alien to the use of aluminum and its alloys because, thanks to the aforementioned properties, different weapon components are manufactured with them.

**TABLE 7.4**
**Comparison of AISI Steels with Other Designations**

| AISI | DIN | AFNOR | BSI | CEN | JIS |
|---|---|---|---|---|---|
| 1015 | C15 | CC12 | 080M15 | C15 | S15C |
| | Cq15 | C15D | 080A15 | C15D | S15CK |
| | 1.0401 | FR15 | C15D | 1.0401 | SWRM15 |
| | 1.1141 | XC15 | C15E2C | 1.1141 | SWRCH15K |
| 1020 | 1.0044 | S275JR | 43/25HR | 1.0044 | SS400 |
| | C22 | C22 | 43B | C22 | STAM390G |
| | Ck22 | C20D | 070M20 | S275JR | SNR400A |
| | St 44-2 | C20D2 | C22 | C20D | SNR400B |
| 1045 | C45 | C45 | 080M46 | C45 | S45C |
| | Ck45 | XC45 | CFS8 | 2C45 | S45CM |
| | 1.0503 | C45E | C45E | 1.0503 | S48C |
| | 1.1191 | XC48H1 | AW2 | 1.1191 | |
| 1060 | C60 | CC55 | 080A62 | C60 | S60C |
| | Cm60 | C60 | 60CS | 1.0601 | S58C |
| | 1.0601 | C58D | 60HS | C58D | SWRH62B |
| | 1.1223 | C60D | En9 | 1.1223 | S65C |
| 1080 | 1.0616 | C86D | C86D | C86D | CS85 |
| | D 85-2 | C86D2 | 80CS | 1.0616 | |
| | C86D | | 80HS | C86D2 | |
| | 1.1265 | | C86D2 | 1.1265 | |
| 4140 | 1.7223 | 42CD4TS | 708M40 | 1.7223 | SNB22 |
| | 41CrMo4 | 42CD4 | 709M40 | 42CrMo4 | SNB22-1 |
| | 42CrMo4 | | EN19 | 42CrMo4V | SNB22-2 |
| | 42CrMo4V | | | 41CrMo4V | SNB22-3 |
| 4340 | 1.6563 | 35NCD6 | S 149 | 1.6563 | SNB24 |
| | 41NiCrMo7-3-2 | 34CrNiMo8 | 817M40 | 41NiCrMo7-3-2 | SNCM439 |
| | 1.6565 | | 816M40 | 1.6565 | SNB24-1 |
| | 40NiCrMo6 | | EN24 | 34CrNiMo6V | SNB24-2 |
| 8620 | 1.6523 | 20NCD2 | 805M20 | 21NiCrMo2 | SNCM220H |
| | 21NiCrMo2 | 20NiCrMo2-2 | 21NiCrMo2-2 | 20NiCrMo2-2 | SNCM220WCH |
| | 20NiCrMo2-2 | | | 1.6523 | SNCM220RCH |
| | 1.7334 | | | 1.7334 | 20NiCrMo2 |
| 416 | 1.4005 | Z11CF13 | 416S21 | 1.4005 | SUS416 |
| | X12CrS13 | X12CrS13 | X12CrS13 | X12CrS13 | |

It is worth mentioning that the name of aluminum is spelled differently in American English than it is in British English. This is because when this metal was first isolated from alum, a hydrated double sulfate salt of aluminum, it was named as "aluminum," which was later changed as "aluminium" for consistency with most chemical element names that use the suffix "ium." Now, "aluminum" is the name still used in the United States, whereas "aluminium" is used in Europe and other countries [33].

**TABLE 7.5**
**Classification of Aluminum Alloys [34]**

| Wrought Alloys | | Cast Alloys | |
|---|---|---|---|
| **Alloy Series** | **Principal Alloying Element** | **Alloy Series** | **Principal Alloying Element** |
| 1xxx | ≥99.000% aluminum | 1xx.x | ≥99.000% aluminum |
| 2xxx | Copper | 2xx.x | Copper |
| 3xxx | Manganese | 3xx.x | Silicon plus copper and/or magnesium |
| 4xxx | Silicon | 4xx.x | Silicon |
| 5xxx | Magnesium | 5xx.x | Magnesium |
| 6xxx | Magnesium and silicon | 6xx.x | Unused series |
| 7xxx | Zinc | 7xx.x | Zinc |
| 8xxx | Other elements | 8xx.x | Tin |
| 9xxx | Unused series | 9xx.x | Other elements |

### 7.8.1 Classification of Aluminum Alloys

The International Alloy Designation System (IADS), based on the classification developed by the Aluminum Association (AA) of the United States, classifies aluminum alloys according to their chemical composition. IADS classification is used in most countries, and it differentiates two types of alloys (Table 7.5):

- **Wrought aluminum alloys**, which are mainly intended for the production of wrought products in the solid form by hot and/or cold working, such as rolling, forging, extrusion, drawing, bending, shearing, pressing, and stamping. These alloys are designated by a four-digit number, whose first digit indicates the alloy group, which is determined by the major alloying element. The second digit specifies the variations in the original alloy, such as impurity limits; thus, a "0" will indicate the original alloy, whereas "1" to "9" will indicate differences in the composition. The third and fourth digits designate a specific alloy or the alloy purity.
- **Cast aluminum alloys**, which are primarily intended for the production of castings by different casting processes, such as sand casting, permanent mold casting, die casting, investment casting, centrifugal casting, squeeze casting, and continuous casting. These alloys are designated by a four-digit number with a decimal point between the last two digits. The first digit indicates the alloy group, which is determined by the major alloying element. The second and third digits designate a specific alloy or the alloy purity. The last digit indicates the product form: "0" for casting alloys and "1" or "2" for ingot compositions.

It is worth mentioning that only wrought aluminum alloys are used in the manufacture of small weapons; hence, cast aluminum alloys will not be covered in this book.

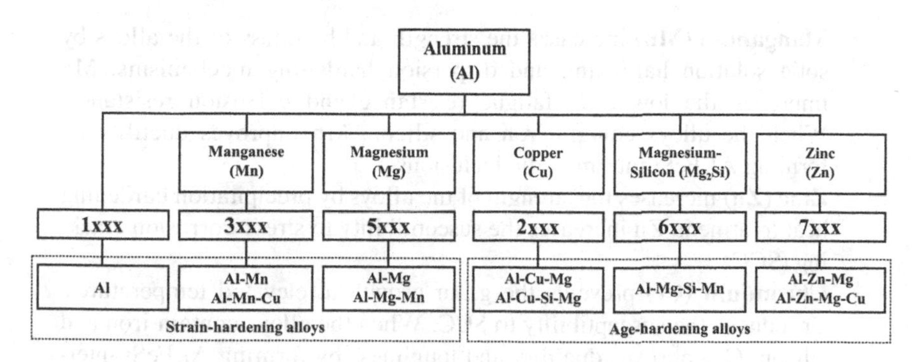


**FIGURE 7.9** Designation of aluminum alloys.

### 7.8.2 Designation of Wrought Aluminum Alloys

Wrought aluminum alloys can be divided into two groups as can be seen in Figure 7.9. One group contains alloys whose mechanical properties are controlled by strain hardening and annealing; 1xxx, 3xxx, and 5xxx series belong to this group. The second group comprises alloys that respond to age- or precipitation hardening, such as 2xxx, 6xxx, and 7xxx series. These treatments will be discussed in Chapter 8. Note that 4xxx and 8xxx series are not considered in such a classification. The former are used mainly for welding and brazing electrodes and brazing sheets rather than for structural purposes, while the latter, which do not respond to heat treatment, are used for specific applications such as bearings and bottle caps [33].

### 7.8.3 Influence of Alloying Elements and Impurities on Aluminum Alloy Properties

The addition of alloying elements and impurities affects the properties of aluminum and its alloys, especially during their different heat treatments. The following elements are the most used or found in aluminum and its alloys:

- **Silicon (Si)** increases the alloy strength and improves the resistance to wear. In combination with Mg, Si allows strengthening the alloys by precipitation hardening heat treatment.
- **Copper (Cu)** increases the hardness and tensile and fatigue strength by the effect of solid solution hardening. It strengthens the alloys by precipitation hardening heat treatment. However, it decreases the ductility and corrosion resistance.
- **Magnesium (Mg)** strengthens and hardens the alloys by the solid solution hardening mechanism without a considerable decrease of ductility. Mg helps retain the resistance to corrosion, particularly to seawater and alkaline solutions. In combination with silicon or zinc, Mg allows strengthening the alloys by precipitation hardening heat treatment.

- **Manganese (Mn)** increases the strength and hardness of the alloys by solid solution hardening and dispersion hardening mechanisms. Mn improves the low-cycle fatigue resistance and corrosion resistance. When the alloys contain iron and silicon, Mn improves ductility by forming $Al_5FeSi$ intermetallic inclusions.
- **Zinc (Zn)** increases the strength of the alloys by precipitation hardening heat treatment. Zn increases the susceptibility to stress-corrosion cracking (SCC).
- **Chromium (Cr)** prevents the grain growth at elevated temperatures. Cr reduces the susceptibility to SCC. When the alloys contain iron and silicon, Cr improves ductility and toughness by forming $Al_5FeSi$ intermetallic inclusions.
- **Nickel (Ni)** increases hardness and strength at elevated temperatures and reduces the coefficient of thermal expansion. In combination with Fe, Ni forms the $Al_9FeNi$ intermetallic, which causes dispersion hardening and assists in stabilizing the microstructure.
- **Titanium (Ti)** is commonly added to refine the grain size by the formation of $Al_3Ti$. In combination with boron, the grain refining effect is increased due to the formation of $TiB_2$.
- **Boron (B)**, in combination with titanium, refines the grain size due to the formation of $TiB_2$.
- **Iron (Fe)** increases the strength but decreases the ductility of alloys due to the formation of Al-Fe intermetallics. This element is an undesirable impurity in most aluminum alloys.

Table 7.6 presents the composition of some aluminum alloys used in the manufacture of small weapon components. 1100 alloy is basically pure aluminum with some amount of alloying elements. 6061 alloy contains magnesium and silicon as major alloying elements. 7075 alloy has zinc as the main alloying element. Each of these dominant alloying elements confers different mechanical properties, as can be seen in Table 7.7. However, such properties can be enhanced by specific strain hardening and heat treatments, as will be seen in Chapter 8.

Aluminum alloys listed in Table 7.6 are appropriate for the manufacture of small weapon components due to the properties that can be attained with each of

**TABLE 7.6**
**Composition (wt%) of Selected Wrought Aluminum Alloys [35,36]**

| Alloy | Si | Fe | Cu | Mn | Mg | Cr | Zn | Ti |
|---|---|---|---|---|---|---|---|---|
| 1100 | 0.95 (Si+Fe) | | 0.05–0.20 | 0.05 | | | 0.10 | |
| 6061 | 0.40–0.80 | 0.70 | 0.15–0.40 | 0.15 | 0.80–1.20 | 0.04–0.35 | 0.25 | 0.15 |
| 7075 | 0.40 | 0.50 | 1.20–2.00 | 0.30 | 2.10–2.90 | 0.18–0.28 | 5.10–6.10 | 0.20 |

The balance of the composition is aluminum.

**TABLE 7.7**
**Mechanical Properties of Selected Heat-Treatable Wrought Aluminum Alloys**

| Alloy | Condition | Yield Strength (MPa) | Tensile Strength (MPa) | Modulus of Elasticity (GPa) | Elongation (%) | Brinell Hardness (HBW) | References |
|---|---|---|---|---|---|---|---|
| 1100 | Annealed | 35 | 90 | 69 | 35 | 23 | [37] |
| 6061 | Annealed | 55 | 124 | 69 | 25 | 60 | [38] |
| 7075 | Annealed | 103 | 228 | 72 | 17 | 60 | [39] |

them [1,33,40]. Aluminum framing is one of the most important components in modern small weapons, with which it is possible to save around 30% in weight compared to steel. Compared with steel components, some downsides of aluminum components are their low durability, a greater felt recoil, and the difficulty to be repaired/welded. Accordingly, it is important to select the suitable alloy, taking into account the weight-performance ratio of the weapon.

- **1100 alloy** has limited application in forging because it cannot be strengthened by heat treatment. This alloy can be produced in the form of sheet, plate, tube, wire, and hollow tableware. Its applications include food, chemical, and storage equipment, light reflectors, and heat exchangers. In the weapon industry, the 1100 alloy is used for the production of heat gun diffusers and heat shields.
- **6061 alloy** is also considered as an aircraft aluminum. It is used for general applications from flashlight bodies to canoes, railroad cars, furniture, pipelines, bicycle frames, automotive parts, and welded structures. In weapon applications, it is used to manufacture receivers, hunting rifle floorplates, scope rings, handguards, and buffer tubes.
- **7075 alloy** is one of the highest strength alloys among aluminum alloys. Its strength is comparable to that of many steels. It has good fatigue resistance and very good corrosion resistance. Its machinability properties are not as good as those of the 6061 alloy are and its cost is relatively higher. Its applications comprise aircraft structural parts. Concerning weapons, 7075 aluminum is used for producing rifle receivers, pistol frames, buffer tubes, port doors, barrel nuts, and railed handguards, among others.

## 7.9 TITANIUM ALLOYS

A wide range of materials that combine low density and high strength or durability has been created in the past few decades. The aerospace, automotive, and biomedical industries are examples where lightweight and high-strength structural materials are required. Titanium (Ti) presents several advantages for these purposes, such as high strength-to-weight ratio and good corrosion resistance. The density of pure titanium is 4.51 g/cm$^3$, which represents about 57% of that of iron. The corrosion resistance of titanium is based on the formation of a passive protective oxide layer (mainly of $TiO_2$), similar to the passivation phenomenon that takes place in aluminum alloys and stainless steels.

Titanium alloys were initially developed for aeronautical and aerospace industries and almost immediately for orthopedic applications. Now they continue to be widely used for the manufacture of different components of such industries. The weapon industry has not been left behind, and today, some components are being manufactured from Grade 5 titanium alloy (Ti6Al4V), which contains 6 wt% aluminum and 4 wt% vanadium; its chemical composition (wt%) specified by the ASTM B265 standard is 5.5-6.75 Al, 3.5–4.5 V, ≤0.08 C, and ≤0.4 Fe [41]. The mechanical properties of this alloy in the annealed condition are: yield strength of 828 MPa, tensile strength of 895 MPa, elongation of 10%, elastic modulus of 114 MPa, and hardness of 36 HRC (334 Brinell) [29,41]. These properties can be enhanced through different heat treatments. Examples of components made of grade 5 titanium alloy are gas blocks, shot selectors, pins, barrel nuts, muzzle brakes, clamps, and bolt carrier groups, among others [42].

## 7.10 SYNTHETIC POLYMERS

Synthetic polymers are used worldwide in multiple applications at industrial, technological, and scientific levels. Different synthetic polymers have been applied in the small weapon industry; therefore, it is necessary to explain their basic concepts so that the reader understands why they are used in one or another weapon component.

There are several reasons why various weapon components are manufactured from polymers. Without a doubt, one of them is saving weight; indeed, the density of polymers is around 85% and 55% less than that of steels and aluminum alloys, respectively. Another reason is, of course, that polymers are resistant to damage and corrosion, are thermal insulators, and the component color can be customized. One more reason is the manufacturing economy; with processes such as injection molding, the production of plastic components is much faster compared to machining processes, even if these are made by computer numerical control technology.

### 7.10.1 Glass Transition Temperature

While melting temperature or melting point ($T_m$) matters for metals and their alloys, the glass transition temperature ($T_g$) is taken into account for polymers. $T_g$ is the temperature at which an amorphous polymer becomes soft, turning into a rubbery state. In other words, $T_g$ is the temperature at which the transition from the glass state to the rubber-like state occurs. There are various standards for the assignment of the $T_g$ of polymers by specific methods, such as the ASTM E1356 [43].

It must be noted that $T_m$ is taken into account for crystalline polymers, but also $T_g$ because they usually have an amorphous portion, i.e., they can actually be semicrystalline. It is common to take both $T_g$ and $T_m$ as a reference during the processing and handling of polymers, as well as to see them reported in technical data-sheets.

### 7.10.2 Nomenclature of Synthetic Polymers

The International Union of Pure and Applied Chemistry (IUPAC) is the universally recognized authority on chemical nomenclature and terminology, which has established rules for assigning systematic names to polymers [44]. However, it is very common to name polymers based on the monomers from which they are derived, or even by their trademark or generic name. There are some standards where the terminology of polymers is given, such as ASTM D1600 [45].

Take as an example a polymer whose IUPAC name is "benzene-1,4-diamine, terephthalic acid," its name based on the monomers is "poly(p-phenylene terephthalamide)," its generic name is para-aramid or simplify aramid, and its trademark names are Kevlar® (DuPont de Nemours, Inc.) and Twaron® (Teijin Aramid). Another example is "poly(azanediyladipoylazanediylhexane-1,6-diyl)" based on IUPAC, "poly(hexamethylene adipamide)" based on its monomers, and "polyamide 66," "PA66," or "Nylon 66," according to its generic names.

### 7.10.3 Classification of Polymers by Properties

According to their thermomechanical properties, synthetic polymers are traditionally classified into thermoplastics, elastomers, and thermosets [2,3]. However, thermoplastic elastomers are recently being considered, as will be seen below.

- **Thermoplastics**, commonly called "plastics," are polymers that harden when cooled, say at room temperature, but soften when heated; this process is said to be totally reversible. This kind of polymers exhibits a high $T_g$. Thermoplastic components are generally manufactured by the simultaneous application of heat and pressure, through different manufacturing processes such as injection molding, extrusion, casting, and blow molding. Nylon 66, Nylon 6, polypropylene, and polybutylene terephthalate are examples of thermoplastic polymers used in the manufacture of weapon components, although most of them are structurally reinforced with fibers, as will be seen in the Composites section. Frames, handguards, grips, stocks, forearms, Picatinny rails, trigger guards, head guards, magazines, and some components of the trigger mechanism are made from thermoplastics.
- **Elastomers** are rubbery materials that can be stretched many times their original dimension and that recover their initial dimensions when

the applied stress is released. According to IUPAC, an elastomer is a polymer that displays rubber-like elasticity [46]. Elastomers are typically amorphous polymers with a certain degree of crosslinking. A common example of crosslinking is that carried out in the vulcanization process, where natural rubber is heated with sulfur to form crosslinks between long rubber molecules, resulting in increased rigidity and durability. Styrene-butadiene rubber (SBR) and nitrile-butadiene rubber (NBR) are examples of elastomers.

- **Thermosets** are highly crosslinked polymers that are generally very hard, insoluble, and do not soften when heated. Thermosets are often formed by resins, such as epoxy adhesive, which, mixed with hardener, becomes solid ("sets"). These polymers are generally harder and stronger than thermoplastics and have better dimensional stability, but they present brittle fractures under stress. Epoxies, phenolics, vulcanized rubbers, and Bakelite are thermosets.
- **Thermoplastic elastomers (TPEs)** are the combination of a thermoplastic with an elastomer, i.e., an elastomer comprising a thermoreversible network [46]. They are usually referred to as "thermoplastic rubbers." TPEs combine the properties of thermoplastics with the softness and flexibility of elastomers. There are different types of TPEs, such as thermoplastic polyether block amides (TPAs), thermoplastic copolyester elastomers (TPCs), thermoplastic polyurethane elastomers (TPUs), thermoplastic polyolefins (TPOs), thermoplastic vulcanizate elastomers (TPVs), and thermoplastic styrenic block copolymers (TPSs), among others. Shock absorbers, vibration dampers, and recoil pads, as well as the soft touch of handguards, forearms, Picatinny rails, grips, and magazines are made from TPEs.

## 7.11 COMPOSITES

A composite consists of at least two constituent phases, which are chemically dissimilar and separated by a distinct interface. One of these constituents is the matrix, which is a continuous phase that surrounds and binds the reinforcement, which is dispersed within the matrix. To guarantee an adequate interfacial bond between the matrix and the reinforcement, there must be a good wettability and an appropriate chemical stability between both constituents. These aspects, in addition to a good dispersion and a suitable orientation of the reinforcement, ensure a good load transfer from the matrix to the reinforcement. The mechanical behavior of the composite will depend on the resistance of the matrix-reinforcement interface; a weak interface will result in a material with low mechanical properties.

Composite materials are generally optimized to achieve a particular balance of properties, depending on their specific applications. Thus, the choice of a suitable matrix and reinforcing material will be necessary (Figure 7.10).

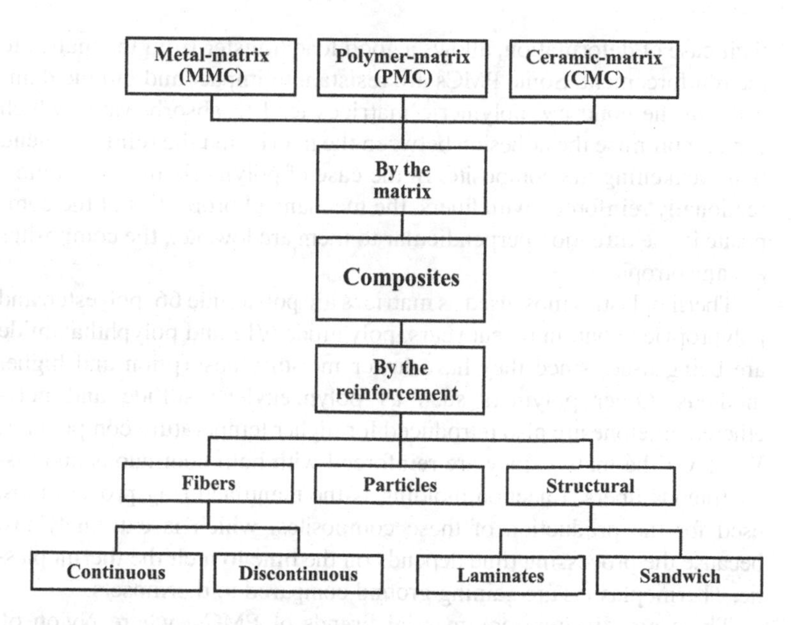


**FIGURE 7.10** Classification of composites.

### 7.11.1 Matrices

A level of classification of composite materials is according to the type of matrix (Figure 7.10):

- **MMCs** consist of a ductile metal or alloy into which a reinforcement is embedded. The reinforcing materials are usually ceramic particles or fibers. MMCs have some advantages over PMCs, including higher operating temperatures, nonflammability, excellent abrasion resistance, and greater resistance to degradation by solvents. However, some of their disadvantages are their higher weight and cost, which may limit their use. An example of this kind of composites is aluminum reinforced with SiC particles; SiC provides better strength, stiffness, hardness, and wear resistance to the aluminum matrix. In the case of weapon components, research is being done on gun barrels, whose metallic matrix is reinforced structurally with both a metal [47] and a ceramic [48].
- **PMCs** are used in a wide range of applications to a greater extent than MMCs and CMCs. Polymeric matrices are attractive due to different aspects [49]. One of them is their low density, which is often something higher than that of water. Polymeric matrices can be used both in solution and in the molten state to impregnate the reinforcing material, applying pressures and temperatures much lower than those necessary for metallic or ceramic matrices. Polymers are often very resistant to corrosive environments, resulting in useful properties for the composites produced. The low modulus of elasticity of most polymers, together with

their ease of deformation, allows a good load transfer from the matrix to the reinforcement. Some PMCs are resistant to impact and fatigue damage. On the contrary, polymeric matrices tend to absorb water, which can compromise the adhesion between the matrix and the reinforcement, thus weakening the composite. In the case of polymeric matrices unidirectionally reinforced with fibers, the mechanical properties of the composite in the direction perpendicular to them are low, i.e., the composites are anisotropic.

Thermoplastics most used as matrices are polyamide 66, polyester, and polypropylene but, in recent years, polyamide 6/12 and polyphthalamide are being used, since they have lower moisture absorption and higher modulus. Other polymers such as polyphenylene sulfide and polyetheretherketone are also introduced for higher temperature components. Whatever the matrix, they are reinforced with both continuous and discontinuous fibers. Injection molding is the manufacturing process most used for the production of these composites, which is extremely fast because the processing time depends on the time to melt the thermoplastic. Thermoplastics are gaining ground compared to thermosets.

There are different commercial brands of PMCs, where Nylon 66 and polyphthalamide reinforced with both short and long glass fibers, in percentages of up to 40%, are being used for manufacturing weapon components. These PMCs have demonstrated good durability, strength, stiffness, impact and wear resistance, and low- and high-temperature resistance.

- **CMCs:** The progress in materials has allowed improving the brittle behavior of ceramics by adding ceramic particles or fibers, thus developing CMCs. The improvement in the fracture properties results when the dispersed particles or fibers hinder the crack propagation in different ways: deflecting the crack tips, bridging across the crack faces, absorbing energy during pullout as they debond from the matrix, and/or causing a redistribution of stresses in regions adjacent to the crack tips [1].

  The matrices and fibers most commonly used for CMCs are carbon (C), silicon carbide (SiC), alumina ($Al_2O_3$), and mullite ($Al_2O_3$-$SiO_2$). The ceramic type of the matrix can be the same or different from that of the reinforcing material; i.e., a carbon matrix can be reinforced with carbon fibers (C/C composite), but it can also be reinforced with SiC (C/SiC composite).

  There is an investigation on a hybrid material consisting of a CMC (SiC matrix/$Al_2O_3$ fibers) and an MMC (aluminum matrix/$Al_2O_3$ fibers), which is being developed for barrels [50].

### 7.11.2 Reinforcements

Another level of classification of composite materials is based on the reinforcement form (Figure 7.10); Figure 7.11 schematizes some of these forms.

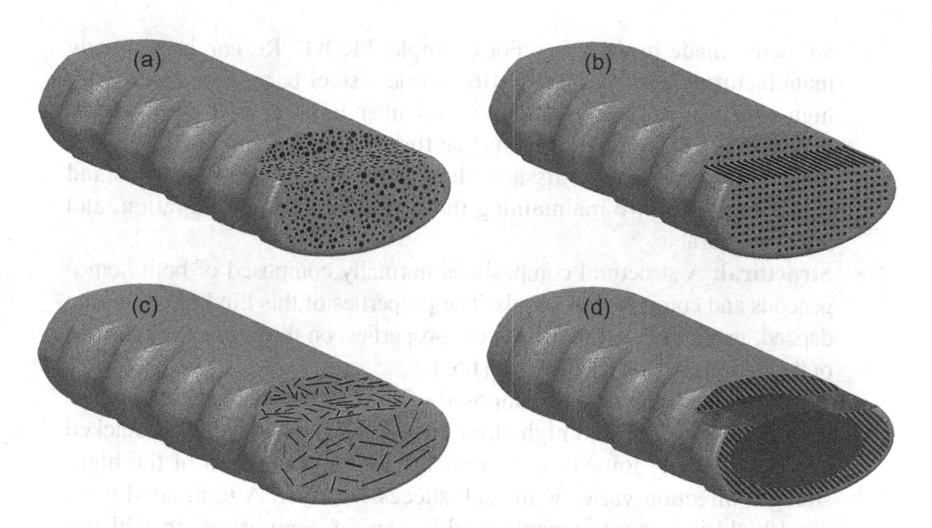


**FIGURE 7.11** Types of reinforcements: (a) particles, (b) continuous or long fibers, (c) discontinuous or short fibers (whiskers), and (d) structural.

- **Particles** are reinforcements whose dimensions are roughly equal. Thus, spheres, rods, and flakes, among others, are considered particles. The use of particles as reinforcing material is more widely accepted in MMCs, since they are associated with lower costs and allow obtaining a greater isotropy of properties in the product. However, strict control of the size and purity of the particles is essential. Typical reinforcements most commonly used in particle form are carbides (TiC, $B_4C$), oxides ($SiO_2$, $TiO_2$, $ZrO_2$, MgO), and silicon nitride ($Si_3N_4$).
- **Fibers** are defined as thin and long filaments of matter, with a diameter generally on the order of a few microns and with a length-to-diameter ratio between 100 for discontinuous fibers and virtually infinite for continuous fibers [51]. Fibers can be discontinuous (short fibers or whiskers) and continuous (long fibers). The former have a random orientation and are used when it is necessary for the component to have isotropic mechanical properties, i.e., its mechanical resistance will be the same in all directions. The latter have a preferred orientation and are used to reinforce the matrix in the direction in which the component will be subjected to a certain stress. Fibers can be metallic (steel, W, Mo, Nb, etc.), polymeric (Nylon, PET, aramid, etc.), and ceramic (C, B, $Al_2O_3$, SiC, glass, etc.) in nature.

Almost three-quarters of PMCs are reinforced with glass fibers; nevertheless, carbon, aramid, and plant fibers are increasing their market share each year [49]. The fact that glass fibers are the most used to reinforce polymeric matrices is due to their low cost and ease of use, which facilitates manufacturing. PMCs reinforced with carbon fibers have become a standard choice for the aerospace industry, but some efforts are being made in weapons. For example, PROOF Research is already manufacturing and marketing 416 stainless steel barrels wrapped with high-strength, aerospace-grade carbon fiber impregnated with a resin matrix [52]. The company claims that these barrels are up to 64% lighter than traditional barrels, dissipate heat in an improved way, withstand high volumes of fire maintaining their accuracy, reduce vibration, and are more durable.

- **Structural:** A structural composite is normally composed of both homogeneous and composite materials. The properties of this kind of composite depend, in addition to the constituent properties, on the geometrical design of the various structural elements [1,53].

  A laminar composite is composed of two-dimensional sheets or panels that have a preferred high-strength direction. The layers are stacked and subsequently joined together such that the orientation of the high-strength direction varies with each successive layer. A laminated glass windshield is a good example of this type of composites, in which a plastic adhesive is placed between two layers of glass; upon impact, the adhesive prevents glass fragments from flying away. Some laminates are often made by stacking single sheets of continuous fibers in different orientations to obtain the desired mechanical properties.

  A sandwich composite is designed to be lightweight beams or panels having relatively high stiffness and strength. They are composed of thin layers of a facing material joined to a lightweight filler material, such as a polymer or metal foam or a metallic honeycomb. Neither the filler nor the facing material is strong or rigid, but the composite possesses both properties. A familiar example is corrugated cardboard, where a corrugated core of the paper is bonded on either side to flat, thick paper; neither the corrugated core nor the facing paper is rigid, but the combination is.

## 7.12 CERAMICS

As already mentioned, due to some disadvantages of ceramics, their use is limited in the manufacture of weapon components. Diverse research efforts have been made on advanced ceramics for potential application in gun barrels, such as those of Swab et al. [54], who analyzed the mechanical and thermal properties of alumina ($Al_2O_3$), zirconia ($ZrO_2$), silicon aluminum oxy-nitride (SiAlON), silicon carbide (SiC), and silicon nitride ($Si_3N_4$), with the aim of identifying the optimum material for further evaluation. Then, they studied the effective volume and area of ceramic c-ring test specimen geometries, making use of experimental and numerical analyses [55]. They continued with the evaluation of the previous ceramics by probabilistic modeling tools in order to employ them as liner materials for barrels [56]. They were able to manufacture a small rifled gun barrel tube made of alumina by means of powder injection molding, which was promising to continue testing with other advanced ceramic materials and longer barrels [57].

Following these works, Carter applied a probabilistic model, which verified experimentally, for ceramic-lined gun barrels to investigate optimal geometries and prestress levels for different liner and sheath materials across various caliber systems [58]. After all this series of investigations, these researchers conclude that $Si_3N_4$ and SiAlON are the top-performing ceramics, since they exhibited the best combination of thermal and mechanical properties and proved to be the most capable of surviving the interior ballistic conditions and functioning as a barrel liner [59,60]. The analysis of these materials continued through comparative modeling between a 4340 steel barrel and a SiAlON lined gun barrel in order to predict the thermal behavior for single and multiple shots [61]. In addition, Grujicic et al. [50,62,63] have conducted various studies on hybrid materials, specifically focused on a SiC lining/CrMoV steel jacket gun barrel.

Although mass production of barrels made of ceramic has not yet been achieved, all these studies are undoubtedly laying the groundwork for the potential application of ceramic materials in weapon components in the near future.

## 7.13 CLOSING REMARKS

The mechanical design should comprise the selection of materials. Different materials have been emerging over the years, and the weapon industry has taken advantage of this. Most modern small weapons are made from improved steels, and the use of light metals such as aluminum- and titanium-based alloys is already a reality. Efforts are being made with MMCs and hybrid materials to enhance the performance of firearm components. Synthetic polymers and PMCs have been applied in the manufacture of different components. Ceramic materials have been in constant investigation for their possible application in barrels. Each type of material provides certain attributes to the firearm components, which contribute to achieving a safe, reliable, durable, and efficient firearm.

## REFERENCES

1. Callister, W.D. and D.G. Rethwisch, *Materials Science and Engineering: An Introduction*. 10th ed, 2018, John Wiley & Sons: New York.
2. Klein, D.R., *Organic Chemistry*. 3rd ed., 2017, John Wiley & Sons: New York.
3. Hamley, I.W., *Introduction to Soft Matter: Synthetic and Biological Self-Assembling Materials*. 2013, John Wiley & Sons: New York.
4. Okamoto, H., M.E. Schlesinger, and E.M. Mueller, *ASM Handbook Volume 3: Alloy Phase Diagrams*. 2016, ASM International: Materials Park, OH.
5. ASM Handbook Program, *ASM Handbook Volume 1: Properties and Selection: Irons, Steels, and High-Performance Alloys*. 1990, ASM International: Materials Park, OH.
6. ASTM E8/E8M-21 Standard test methods for tension testing of metallic materials. 2021, ASTM International: West Conshohocken, PA.
7. ASTM E9-19 Standard test methods of compression testing of metallic materials at room temperature. 2019, ASTM International: West Conshohocken, PA.
8. ASTM E10-18 Standard test method for Brinell hardness of metallic materials. 2018, ASTM International: West Conshohocken, PA.

9. ASTM E18-20 Standard test methods for Rockwell hardness of metallic materials. 2020, ASTM International: West Conshohocken, PA.
10. ASTM E384-17 Standard test method for Microindentation hardness of materials. 2017, ASTM International: West Conshohocken, PA.
11. UNE-EN ISO 868:2003 Plastics and ebonite: Determination of indentation hardness by means of a durometer (Shore hardness) (ISO 868:2003). 2003, UNE-EN ISO.
12. ASTM E140-12B(2019)e1 Standard hardness conversion tables for metals relationship among Brinell hardness, Vickers hardness, Rockwell hardness, Superficial hardness, Knoop hardness, Scleroscope hardness, and Leeb hardness. 2019, ASTM International: West Conshohocken, PA.
13. ASTM A370-20 Standard test methods and definitions for mechanical testing of steel products. 2020, ASTM International: West Conshohocken, PA.
14. ASTM A29/A29M-20 Standard specification for general requirements for steel bars, carbon and alloy, hot-wrought. 2020, ASTM International: West Conshohocken, PA.
15. ASTM A582/A582M-12 Standard Specification for Free-Machining Stainless Steel Bars. 2017, ASTM International: West Conshohocken, PA.
16. EN 10263-4:2001 Steel rod, bars and wire for cold heading and cold extrusion. Technical delivery conditions for steels for quenching and tempering. 2001, European Committee for Standardization: Brussels, Belgium.
17. JIS G4053-2008 Low-alloyed steels for machine structural use. 2008, Japanese Industrial Standards Committee: Tokyo, Japan.
18. ISO 683-18:2014 Heat-treatable steels, alloy steels and free- cutting steels — Part 18: Bright products of unalloyed and low alloy steels. 2014, International Organization for Standardization: Geneva, Switzerland.
19. AISI 1015 Steel, hot rolled, 19–32 mm (0.75-1.25 in) round. [May 4, 2021]; Available from: http://www.matweb.com/search/DataSheet.aspx?MatGUID=8186049c0bdb42e2a683ae5809a9f9ec.
20. AISI 1020 Steel, hot rolled, 19–32 mm (0.75-1.25 in) round. [May 4, 2021]; Available from: http://www.matweb.com/search/DataSheet.aspx?MatGUID=b58ee61a3745453a9232f7864abba74f.
21. AISI 1045 Steel, hot rolled, 19–32 mm (0.75-1.25 in) round. [May 4, 2021]; Available from: http://www.matweb.com/search/DataSheet.aspx?MatGUID=4b0553daf9c245e684f2199a48179d89.
22. AISI 1060 Steel, hot rolled, 19–32 mm (0.75-1.25 in) round. [May 4, 2021]; Available from: http://www.matweb.com/search/DataSheet.aspx?MatGUID=55cb4415eb704a46b3537627e9619d52.
23. AISI 1080 Steel, hot rolled, 19–32 mm (0.75-1.25 in) round. [May 4, 2021]; Available from: http://www.matweb.com/search/DataSheet.aspx?MatGUID=5cda6846cdfb4d2383b673abb072e4c0&ckck=1.
24. AISI 4140 Steel, annealed at 815°C (1500°F) furnace cooled 11°C (20°F)/hour to 665°C (1230°F), air cooled, 25 mm (1 in.) round. [May 4, 2021]; Available from: http://www.matweb.com/search/DataSheet.aspx?MatGUID=7b75475aa1bc41618788f63c6500d36b.
25. AISI 4340 Steel, annealed, 25 mm round. [May 4, 2021]; Available from: http://www.matweb.com/search/DataSheet.aspx?MatGUID=fd1b43a97a8a44129b32b9de0d7d6c1a.
26. AISI 8620 Steel, annealed 13 mm (0.5 in.) round. [May 4, 2021]; Available from: http://www.matweb.com/search/DataSheet.aspx?MatGUID=d7f44bdac4514000a0ffe5c983485c1f.

27. 416 Stainless Steel, annealed bar. [May 18, 2021]; Available from: http://www.matweb.com/search/DataSheet.aspx?MatGUID=2d05275df2a841a1b7ddffa503f41fa1.
28. Chandler, H., *Heat Treater's Guide: Practices and Procedures for Nonferrous Alloys*. 1995, ASM international: Metals Park, OH.
29. AZO materials. [May 18, 2021]; Available from: https://www.azom.com/.
30. Model 60. [May 18, 2021]; Available from: https://www.smith-wesson.com/product/model-60.
31. SW22 victory. [May 18, 2021]; Available from: https://www.smith-wesson.com/product/sw22-victory?sku=108490.
32. Equivalent materials. [May 7, 2021]; Available from: https://equivalentmaterials.com/.
33. Polmear, I., et al., *Light Alloys: Metallurgy of the Light Metals*. 2017, Butterworth-Heinemann: Oxford.
34. ASM International, *ASM Handbook Volume 2: Properties and Selection: Nonferrous Alloys and Special-Purpose Materials*. 1990, ASM International: Materials Park, OH.
35. ASTM B209M-14 Standard specification for aluminum and aluminum-alloy sheet and plate (metric). 2014, ASTM International: West Conshohocken, PA.
36. ASTM B221M-13 Standard specification for aluminum and aluminum-alloy extruded bars, rods, wire, profiles, and tubes. 2013, ASTM International: West Conshohocken, PA.
37. Aluminum 1100-O. [May 14, 2021]; Available from: http://www.matweb.com/search/DataSheet.aspx?MatGUID=db0307742df14c8f817bd8d62207368e&ckck=1.
38. Aluminum 6061-O. [May 14, 2021]; Available from: http://www.matweb.com/search/DataSheet.aspx?MatGUID=626ec8cdca604f1994be4fc2bc6f7f63.
39. Aluminum 7075-O. [May 14, 2021]; Available from: http://www.matweb.com/search/DataSheet.aspx?MatGUID=da98aea5e9de44138a7d28782f60a8.
40. Anderson, K., J. Weritz, and J.G. Kaufman, *ASM Handbook Volume 2A: Aluminum Science and Technology*. 2018, ASM International: Materials Park, OH.
41. ASTM B265-20a Standard specification for titanium and titanium alloy strip, sheet, and plate. 2020, ASTM International: West Conshohocken, PA.
42. Vseven. The cure for the common AR. [May 19, 2021]; Available from: https://www.vsevenweaponsystems.com/.
43. ASTM E1356-08 Standard test method for assignment of the glass transition temperatures by differential scanning calorimetry. 2014, ASTM International: West Conshohocken, PA.
44. International Union of Pure and Applied Chemistry. [May 19, 2021]; Available from: https://iupac.org/.
45. ASTM D1600-18 Standard terminology for abbreviated terms relating to plastics. 2018, ASTM International: West Conshohocken, PA.
46. Alemán, J.V., et al., Definitions of terms relating to the structure and processing of sols, gels, networks, and inorganic-organic hybrid materials (IUPAC Recommendations 2007). *Pure and Applied Chemistry*, 2007. **79**(10): pp. 1801–1829.
47. Glisovic, A., S. Gravely, and J. Gravely, Enhanced metal-metal-matrix composite weapon barrels and ways of making the same. 2017, patent No. US 2017 / 0261280 A1, Google Patents, p. 31.
48. Pyka, D., et al. Concept of a gun barrel based on the layer composite reinforced with continuous filament. Proceedings of the 15th Conference on Computational Technologies in Engineering (TKI'2018), Jora Wielka, Poland. In *AIP Conference Proceedings*. 2019, AIP Publishing LLC.

49. Bunsell, A.R., S. Joannès, and A. Thionnet, *Fundamentals of Fibre Reinforced Composite Materials*. 2nd ed, 2021, CRC Press: Boca Raton, FL.
50. Katz, R., et al., Hybrid ceramic matrix/metal matrix composite gun barrels. *Materials and Manufacturing Processes*, 2006. **21**(6): pp. 579–583.
51. Herrera Ramirez, J.M., Les mécanismes de fatigue dans les fibres thermoplastiques. PhD dissertation. 2004, École Nationale Supérieure des Mines de Paris: Paris, France.
52. Carbon Fiber Barrels. For bolt-action and semi-auto rifles. [May 26, 2021]; Available from: https://proofresearch.com/barrels/.
53. Askeland, D.R. and W.J. Wright, *The Science and Engineering of Materials*. 7th ed., 2015, Cengage Learning: Boston, MA.
54. Swab, J.J., et al., *Mechanical and Thermal Properties of Advanced Ceramics for Gun Barrel Applications*. 2005, Army Research Laboratory: Research Triangle Park, NC.
55. Duffy, S., et al., Weibull analysis effective volume and effective area for a ceramic C-ring test specimen. *Journal of Testing and Evaluation*, 2005. **33**(4): pp. 233–238.
56. Emerson, R., et al., Approaches for the design of ceramic gun barrels. 2006, U.S. Army Research Laboratory. Weapons & Materials Research Directorate.
57. Bose, A., R.J. Dowding, and J.J. Swab, Processing of ceramic rifled gun barrel. *Materials and Manufacturing Processes*, 2006. **21**(6): pp. 591–596.
58. Carter, R.H., Probabilistic modeling for ceramic lined gun barrels. *Journal of Pressure Vessel Technology*, 2006. **128**: pp. 251–256.
59. Carter, R.H., et al., Material selection for ceramic gun tube liner. *Materials and Manufacturing Processes*, 2006. **21**(6): pp. 584–590.
60. Bose, A., R.J. Dowding, and J.J. Swab, Processing of rifled gun barrels from advanced materials. 2008, patent No. US 2008/0120889 A1, Google Patents, p. 12.
61. Huang, X., P. Conroy, and R. Carter. 5.56 mm ceramic gun barrel thermal analysis with cycled ammunition. In *23rd International Symposium on Ballistics Tarragona*, Spain, 2007.
62. Grujicic, M., J. Delong, and W. Derosset, Reliability analysis of hybrid ceramic/steel gun barrels. *Fatigue & Fracture of Engineering Materials & Structures*, 2003. **26**(5): pp. 405–420.
63. Grujicic, M., J. DeLong, and W. DeRossett, Probabilistic finite element analysis of residual stress formation in shrink-fit ceramic/steel gun barrels. *Proceedings of the Institution of Mechanical Engineers, Part L: Journal of Materials: Design and Applications*, 2002. **216**(4): pp. 219–231.

# 8 Heat Treatments and Surface Hardening of Small Weapon Components

## 8.1 INTRODUCTION

A heat treatment is the process that involves heating a material in the solid state up to a specific temperature for a sufficient time (usually called soaking), followed by cooling at an appropriate rate. A heat treatment is an essential stage within the entire manufacturing process of a workpiece. Most of the versatility shown by materials comes from their heat treatment, which influences on their microstructure and, finally, improves their physical and mechanical properties, such as strength, hardness, elasticity, and wear resistance, among others.

Each type of material has its own procedures to be heat-treated. Thus, certain practices will be followed for steels, which will be different for those of nonferrous alloys, ceramics, or even polymers. It is important to note that the design and control of heat treatment procedures to produce the required properties in a material are deeply based on the use of phase diagrams.

This chapter covers the basis of heat treatments of steels and aluminum alloys, focusing on those alloys used in the manufacture of small weapon components. The principles of surface hardening, which alters the metal surface properties, are also covered. A case study on the heat treatments of steel is presented here.

## 8.2 HEAT TREATMENTS OF STEELS

As stated in Chapter 7, the basis for understanding the heat treatment of steels is the Fe-$Fe_3C$ phase diagram. Figure 8.1 shows such a diagram in the range of steels (~0.008–2.14 wt% C), where the $A_1$, $A_3$, and $A_{cm}$ lines, also specified as $Ae_1$, $Ae_3$, and $Ae_{cm}$, respectively, can be seen. Their designation occurs when equilibrium conditions apply, i.e., when cooling is extremely slow. These lines correspond to critical temperatures that mark the boundaries between various phase fields:

- **$A_1$** corresponds to the eutectoid temperature (727°C), where γ-Fe begins to appear.
- **$A_3$** is the boundary between the (α+γ) region and the γ-Fe region, corresponding to hypoeutectoid steels.

DOI: 10.1201/9781003196808-8

- $A_{cm}$ is the boundary between the ($\gamma$+$Fe_3C$) region and the $\gamma$-Fe region, corresponding to hypereutectoid steels.

The letter "c" is added to the aforementioned symbols when heating conditions apply: $Ac_1$, $Ac_3$, and $Ac_{cm}$, respectively; this letter comes from the French *chauffage* (heating). The letter "r" is used when cooling conditions apply: $Ar_1$, $Ar_3$, and $Ar_{cm}$, respectively; "r" is derived from the French *refroidissement* (cooling). These distinctions are made because Ac temperatures are higher than Ae temperatures and Ar temperatures are lower than Ae temperatures, since continuous heating and cooling leave insufficient time for complete diffusion-controlled transformation at the true equilibrium temperatures.

Another aspect that the diagram in Figure 8.1 shows is the presence of the (pearlite+ferrite) mixture in the iron-rich region (hypoeutectoid steels), as well as the presence of the (pearlite+cementite) mixture in the carbon-rich region (hypereutectoid steels). Pearlite is a microconstituent with lamellar structure composed of alternating layers of ferrite ($\alpha$–Fe) and cementite ($Fe_3C$), which were described in Chapter 7 (Section 7.4, Figure 7.3). Detailed description and morphology of steel phases will be shown in subsequent sections.

Before beginning with the description of different heat treatment processes for steels, it is opportune to introduce the time-temperature-transformation (TTT) diagram to explain some important aspects.

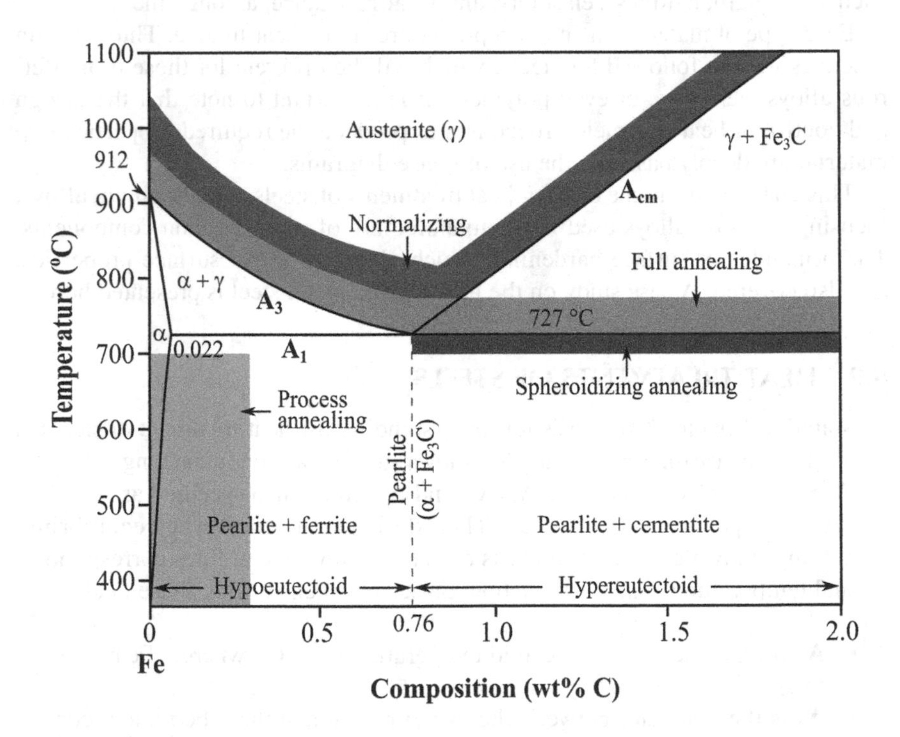


**FIGURE 8.1** Schematic of the Fe-$F_3C$ phase diagram for steel range.

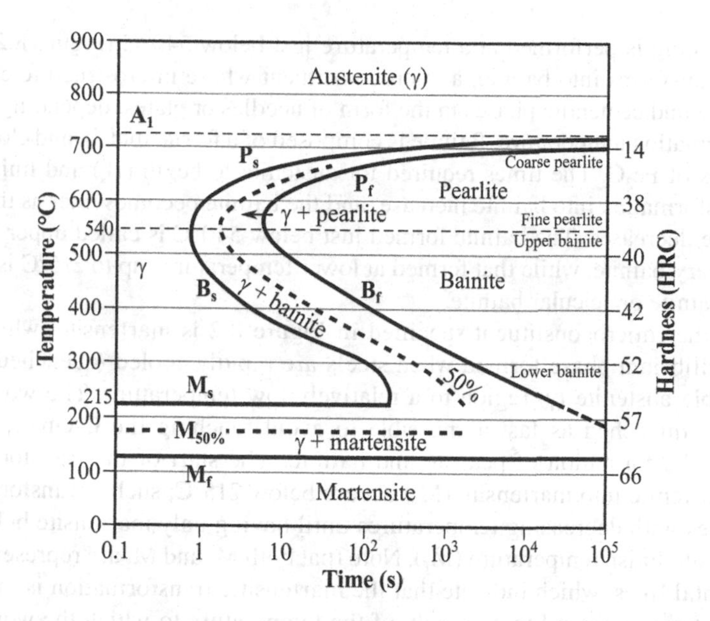


**FIGURE 8.2** Representation of the TTT diagram for a eutectoid steel (P=pearlite, B=bainite, M=martensite, s=start, f=finish).

### 8.2.1 The TTT Diagram

A TTT diagram, also called the isothermal transformation diagram or the C-curve, permits us to predict the microstructure, properties, and heat treatment required in steels. As an example, Figure 8.2 presents a schematic TTT diagram for eutectoid steel. This particular diagram is valid only for a composition of 0.76 wt% C; the curve has different configurations for other compositions. Such a diagram is a graph of temperature versus the logarithm of time, which contains (almost parallel) solid curves for representing the start of the transformation ($P_s$, $B_s$, and $M_s$) and the finish of the transformation ($P_f$, $B_f$, and $M_f$). The time required for the transformation to start and then finish depends on the temperature. The dashed curves correspond to 50% of transformation completion.

Above the $A_1$ temperature only austenite (γ) exists. The transformation from austenite to pearlite occurs when the alloy is cooled below $A_1$. To the left of the transformation start curve, only austenite is present, whereas to the right of the finish curve, only pearlite exists. In between, the austenite is in the process of transforming to pearlite, and thus, γ+pearlite are present. The thickness of pearlite depends on the temperature at which the isothermal transformation is allowed to occur. At temperatures just below $A_1$, relatively thick layers of both the α-Fe and $Fe_3C$ phases are produced; this microstructure is called coarse pearlite. As the temperature decreases, the layers become progressively thinner. At temperatures just above 540°C, the microstructure is called fine pearlite.

If cooling is performed at a temperature just below 540°C (Figure 8.2), austenite transforms into bainite, a microconstituent whose microstructure consists of ferrite and cementite phases in the form of needles or plates, depending on the transformation temperature. Bainite is composed of a ferrite matrix and elongated particles of $Fe_3C$. The times required for austenite to begin ($B_s$) and finish ($B_f$) its transformation into bainite increase, and the bainite becomes finer as the temperature decreases. The bainite formed just below 540°C is called upper bainite or feathery bainite, while that formed at lower temperatures up to 215°C is called lower bainite or acicular bainite.

Another microconstituent specified in Figure 8.2 is martensite, which is a nonequilibrium phase formed when steels are rapidly cooled (quenched) from the stable austenite (γ) region to a relatively low temperature. The workpiece must be quenched as fast as possible to avoid touching the C-curve, i.e., to prevent the formation of pearlite and bainite. The start of the transformation from austenite into martensite ($M_s$) occurs below 215°C; such a transformation increases with decreasing temperatures until having only martensite below the martensite finish temperature ($M_f$). Note that both $M_s$ and $M_f$ are represented by horizontal lines, which indicate that the martensitic transformation is independent of time, being a function only of the temperature to which the workpiece is quenched.

It is worth emphasizing that both bainite and martensite are metastable phases, so they do not appear on the Fe-$Fe_3C$ equilibrium phase diagram (Figure 8.1). Instead, they are considered on the TTT diagram (Figure 8.2).

After the description of the TTT diagram, the explanation of different heat treatments that can be applied to steels will be better understandable. The selection of the suitable heat treatment will depend on the carbon content of the steel and the desired mechanical properties of the workpiece. The most important heat treatments of steels are defined below [1–3].

## 8.2.2 Annealing

Annealing is a process involving heating and cooling, usually applied to produce softening. It is intended to make a steel workpiece easier to deform or machine. The term also refers to treatments intended to alter mechanical properties, produce a definite microstructure, or remove gases. The operation temperature and the cooling rate depend upon the material being annealed and the purpose of the treatment. The following are different types of annealing treatments.

### 8.2.2.1 Full Annealing

Full annealing, also known as critical annealing, consists of heating either hypoeutectoid steels ($C<0.76$ wt%) above the $A_3$ line to attain full austenitization (γ-Fe), i.e., the structure is fully austenitic, or hypereutectoid steels ($C>0.76$ wt%) above the $A_1$ line to attain the dual ($\gamma+Fe_3C$) region, i.e., the structure is partially austenitic. In both cases, an annealing temperature of about 50°C above such lines is suitable. The treated steel workpiece is held for a sufficient time at such temperature and then cooled slowly to room temperature usually inside the furnace. After this process, the workpiece will be in a microstructural state close to that of equilibrium, resulting in coarse pearlite with excess ferrite for hypoeutectoid steels and in coarse pearlite with excess cementite for hypereutectoid steels (Figure 8.1). Full annealing results in soft and ductile steels. This heat treatment is a time- and energy-consuming process.

#### 8.2.2.2 Process Annealing

Process annealing, frequently termed as stress-relieving annealing, is usually applied to cold-worked low-carbon steel ($C < 0.25$ wt%) workpieces to soften them sufficiently to allow further cold-working. The workpiece is usually heated to a temperature slightly below $A_1$ (Figure 8.1) for a short time and then cooled down; the cooling rate is irrelevant. If the workpiece is not to be further cold-worked, but relief of internal stresses is desired, a lower temperature range will suffice (~540°C). The workpiece is held for a sufficient time to achieve recrystallization and softening of the cold-worked ferrite grains with no phase change. This heat treatment is often applied in sheet and wire industries to workpieces cold processed by stamping, heading, extrusion, or wire drawing. Hot-worked high-carbon and alloy steels are also process annealed to prevent them from cracking and to soften them for shearing, turning, or straightening.

#### 8.2.2.3 Spheroidizing Annealing

The spheroidizing annealing is usually applied to hypereutectoid steels ($C > 0.76$ wt%), which contain a large amount of $Fe_3C$ (Figure 8.1) and therefore have poor machining characteristics. The process is carried out just below the $A_1$ temperature with long soaking time to bring the carbides into the form of spheroids rather than lamellar. Thus, the carbides are agglomerated to increase their resistance to solution in the austenite on subsequent heating. The microstructure consists of globular cementite throughout a matrix of ferrite. The purpose of this treatment is to improve machinability of workpieces and is used, for instance, to condition high-carbon steel for cold-drawing into wire.

### 8.2.3 Normalizing

Normalizing consists of heating a workpiece to a temperature above the $A_3$ or $A_{cm}$ lines, with which a full austenitization (γ-Fe) is attained (Figure 8.1). This temperature is held for no prolonged soaking times, and then the workpiece is cooled in still air. Normalizing is applied to refine and eventually make uniform the grain size of a workpiece, so that the process might be more accurately described as a homogenizing or grain-refining treatment. The resulting microstructures are pearlite or pearlite with excess ferrite or cementite, depending upon the composition of the steel and the cooling rate. Normalized steels have higher strength and ductility than those of annealed steels.

### 8.2.4 Hardening

Hardening can be accomplished by heating a workpiece above the $A_3$ or $A_{cm}$ temperatures for hypoeutectoid or hypereutectoid steels, respectively, thereby achieving complete austenitization (γ-Fe) (Figure 8.1). This temperature is held for prolonged soaking times to ensure the attainment of uniform temperature and carbon solution in the austenite. The workpiece is then rapidly cooled, much faster than in still air as in the case with normalizing heat treatment. This rapid cooling is known as quenching. Metastable phases such as martensite and bainite may be formed by quenching. Successful hardening usually means obtaining the appropriate microstructure, hardness, strength, or toughness while minimizing residual stress, distortion, and cracking.

When a steel workpiece is hardened, the medium in which it is quenched should be specified, since the cooling rate depends on the temperature and heat transfer characteristics of the quenching medium. The selection of a quenching medium depends on the hardenability of the particular alloy, the section thickness and shape involved, and the cooling rates needed to achieve the desired microstructure. The quenchants are either gases or liquids. The most common gaseous quenchants include helium, argon, and nitrogen, which are used to provide a cooling rate faster than that obtained in still air and slower than that obtained in oil. The liquid quenchants commonly used are mineral oil, water, brine, and water-based polymers:

- **Mineral oil** is used if slower quenching rates are desired; it is suitable for most alloy steels. A disadvantage of mineral oil is that it may cause water contamination, smoke, and fumes; besides, it is more expensive than water or brine quenchants and is not biodegradable.
- **Water** is a good quenching medium because of its high heat of vaporization and relatively high boiling point. This medium is suitable for some types of steel, since the absorbed gases within the water tend to bubble out, resulting in softening of the steel with cracking or warping. Therefore, when water is used as a quenching medium, it should generally be agitated. A disadvantage is that water may oxidize the workpiece.
- **Brine** consists of water in which salt is dissolved with the purpose of reducing gas absorption and preventing bubbling. A rapid cooling occurs precisely because the salt nucleates bubbles. Oxidation problems can also exist with this quenching medium.
- **Aqueous polymer solutions** have properties between mineral oil and water and brine. These media have a low cooling rate and are not compatible with some additives and antioxidants in common use. They cannot be used for steels that require quenching at high temperatures.

The severity of quenching (H coefficient) for various liquid media and agitation conditions is given in Table 8.1; air was included for comparison purposes. As can be seen, the severity increases from oil to brine and depends on the degree of agitation.

**TABLE 8.1**
**H Coefficient for Different Media and Agitation Conditions [2,4]**

| | Quenching Medium | | | |
|---|---|---|---|---|
| **Agitation** | **Oil** | **Water** | **Brine** | **Air** |
| None | 0.25–0.30 | 0.9–1.0 | 2.0 | 0.02 |
| Mild | 0.30–0.35 | 1.0–1.1 | 2.0–2.2 | |
| Moderate | 0.35–0.40 | 1.2–1.3 | | |
| Good | 0.40–0.50 | 1.4–1.5 | | |
| Strong | 0.50–0.80 | 1.6–2.0 | | 0.05 |
| Violent | 0.80–1.10 | 4.0 | 5.0 | |

### 8.2.5 Tempering

After quenching a steel workpiece, it is generally very hard and brittle and can even crack if dropped; to make the workpiece more ductile, it must be tempered. Tempering consists of reheating a quenched steel workpiece to a suitable temperature below the $A_1$ temperature (Figure 8.1) for a certain soaking time, followed by cooling at an appropriate rate. Tempering may be carried out one or more times, as in the case of tool steels that should be tempered at least twice. After tempering, the hardness of the workpiece will be reduced, but its toughness will be increased.

### 8.2.6 Martempering

Martempering, sometimes called marquenching, is a special quenching procedure to minimize problems associated with the rapid cooling (residual stress, distortion, and cracking). This heat treatment involves austenitizing (γ) followed by step quenching, at a rate fast enough to avoid the formation of ferrite, pearlite, or bainite, and soaking for long enough to ensure that the temperature is uniform but short enough to avoid the formation of bainite (Figure 8.2). The aim of this heat treatment is to obtain a martensitic microstructure with an improved ductility. A tempering treatment is not necessary because martensite is formed without the production of high-thermal stresses.

### 8.2.7 Austempering

Austempering is a special heat treatment for producing a bainitic structure in a workpiece (Figure 8.2). It comprises austenitizing (γ) followed by quenching at a certain temperature before the martensitic transformation begins ($M_s$), holding isothermally to allow the transformation of austenite into bainite, and cooling to room temperature with no specific rate. This heat treatment provides an

**TABLE 8.2**
**Characteristics of the Different Microstructures in Steels [1,2,5]**

| Phase or Microconstituent | Arrangement | Characteristics |
|---|---|---|
| Ferrite (α-Fe) | BCC | Soft, tough, magnetic, stable equilibrium phase |
| Austenite (γ-Fe) | FCC | Fair strength, nonmagnetic, stable equilibrium phase |
| Cementite ($Fe_3C$) | Orthorhombic | Hard, brittle, metastable phase |
| Coarse pearlite (lamellas ~0.4 μm) | α+$Fe_3C$, lamellar | Combination of tough ferrite and hard cementite, metastable microconstituent |
| Fine pearlite (lamellas ~0.1 μm) | α+$Fe_3C$, lamellar | Harder than coarse pearlite, metastable microconstituent |
| Spheroidite | α+globular $Fe_3C$ | Soft, ductile |
| Upper bainite | Precipitations of $Fe_3C$ on surface of α | Properties such as fine pearlite, metastable microconstituent |
| Lower bainite | Precipitations of $Fe_3C$ inside of α | Strength near martensite, but tougher than tempered martensite, metastable microconstituent |
| Martensite (α') | BCT | Hard, brittle, metastable microconstituent |
| Tempered martensite (α' -Fe) | BCT | Softer and tougher than nontempered martensite, metastable microconstituent |

alternative procedure to quenching and tempering for increasing the toughness and ductility of some steels.

As a reference of the properties that can be attained in a workpiece after undergoing a certain heat treatment procedure, Table 8.2 summarizes the features of the different phases and microconstituents described above. Furthermore, the right axis of the TTT diagram in Figure 8.2 gives an idea of the hardness that could be achieved, depending on the microconstituent into which austenite can transform.

In order to appreciate the effect of heat treatments on the mechanical properties of steels, Table 8.3 presents some steels used in the manufacture of small weapon components. These properties are given as a matter of general information, since they can be modified depending on the heat treatment conditions. In the case of 1015 and 8620 steels, they are not susceptible to hardening heat treatment due to their relatively low amount of carbon; instead, they may be cold worked or surface hardened. The rest of the steels may be hardened by heat treatment. Take the 4140 steel as an example. Comparing its mechanical properties in the annealing condition with those in the quenched and tempered condition, it is observed that the yield strength, the tensile strength, and the hardness increased by 138%, 64%, and 58%, respectively; however, the elongation at break decreases by 63%.

**TABLE 8.3**
**Mechanical Properties of Selected Steels Before and After Heat Treatment**

| Alloy | Condition | Yield Strength (MPa) | Tensile Strength (MPa) | Elastic Modulus (GPa) | Elongation at Break (%) | Brinell Hardness (HBW) | References |
|---|---|---|---|---|---|---|---|
| 1015 | Hot rolled | 190 | 345 | 200 | 28 | 101 | [6] |
| | Cold rolled | 325 | 385 | 205 | 18 | 111 | [7] |
| 1045 | Hot rolled | 310 | 565 | 206 | 16 | 163 | [8] |
| | Quenched and tempered | 1069 | 1584 | 206 | – | 450 | [9] |
| 1060 | Hot rolled | 370 | 660 | 200 | 12 | 201 | [10] |
| | Quenched and tempered | 641 | 1005 | 200 | 16 | 293 | [11] |
| 1080 | Hot rolled | 425 | 772 | 205 | 10 | 229 | [12] |
| | Quenched and tempered | 869 | 1270 | 205 | 12 | 363 | [13] |
| 4140 | Annealed | 415 | 655 | 205 | 26 | 197 | [14] |
| | Quenched and tempered | 986 | 1075 | 205 | 16 | 311 | [15] |
| 4340 | Annealed | 470 | 745 | 192 | 22 | 217 | [16] |
| | Quenched and tempered | 1145 | 1207 | 200 | 14 | 352 | [17] |
| 8620 | Hot rolled | 385 | 530 | 205 | 31 | 149 | [18] |
| | Cold rolled | 390 | 670 | 190–210 | 26 | 183 | [19] |

## 8.3 SURFACE HARDENING OF STEELS

Surface hardening, also known as case-hardening, encompasses a series of processes to substantially harden the surface of steels, while their core remains soft. These heat treatments are useful in components of small weapons where it is required to have a hard and strong surface with excellent wear and fatigue resistance but a soft, ductile, and tough core with good resistance to impact. An advantage of surface hardening over (full) hardening is that less expensive low-carbon and medium-carbon steel workpieces can be surface hardened, thus avoiding the inherent problems such as residual stress, distortion, and cracking.

Figure 8.3 displays a variety of surface-hardening methods that can be applied to steels [1–3,20]. In general, these methods are classified into those that involve the addition of layers, which increase the dimensions of the workpiece, and those that involve surface and subsurface modification without any intentional buildup or increase in workpart dimensions. The first group includes the range of hardfacing methods and coating methods. The second group comprises the range of diffusion methods and selective hardening methods.

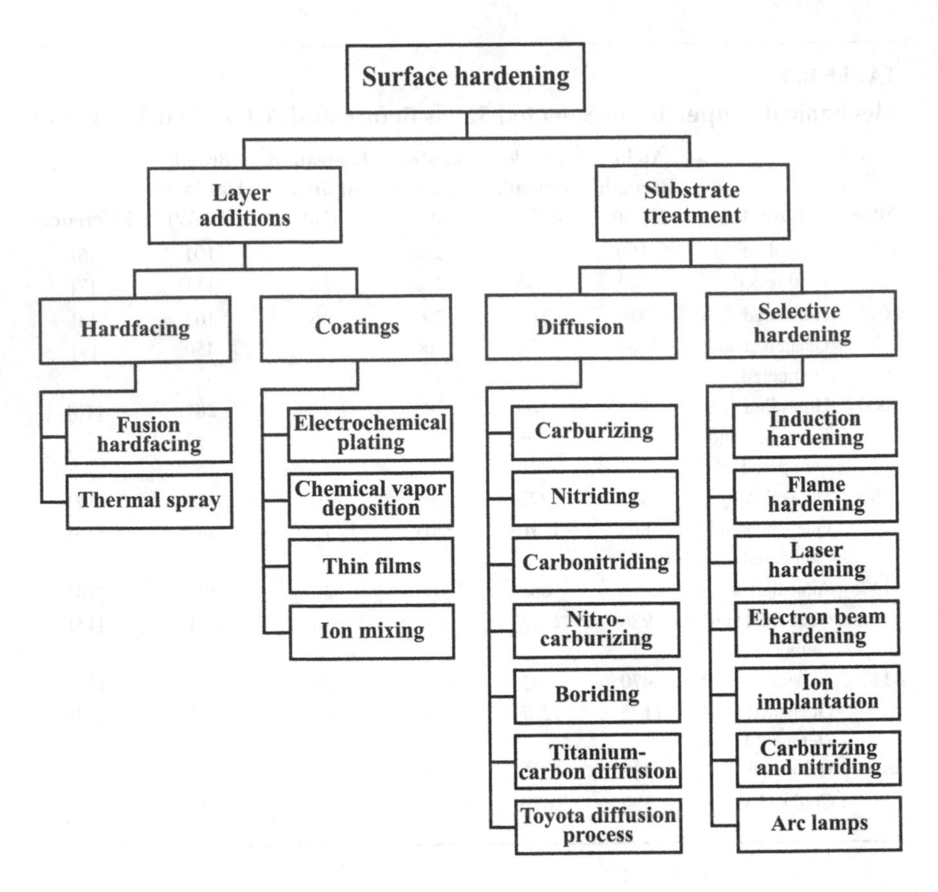


**FIGURE 8.3** Methods for surface hardening of steels.

Among the diffusion methods, carburizing and nitriding are the most widely used methods for surface hardening of small weapon components. These methods alter the chemical composition of the entire surface layer by the addition of carbon, nitrogen, or both (carbonitriding); the process involved is a thermochemical process, since heat is needed to enhance the diffusion of carbon and/or nitrogen into the surface and subsurface of a workpart. The diffusion depth of these species is dependent on time, temperature, carburizing/nitriding agent, and the chemical composition of the steel. After diffusion treatments, the workpiece will have a high-carbon/nitrogen case graduating into the low-carbon core. Table 8.4 presents the characteristics, including the range of depth and hardness reached, of different processes to perform the diffusion treatments of carburizing, nitriding, and carbonitriding [21]. Among the selective hardening methods, induction hardening and flame hardening are the most frequent methods used in the weapon industry; these methods involve heating and quenching in localized sections of a workpiece. In the following, a brief explanation of some of these methods is given [1–3,20].

**TABLE 8.4**
**Characteristics of Diffusion Treatments**

| Process | Diffusion | Process Temperature (°C) | Case Depth (μm) | Case Hardness (HRC) | Base Metal | Process Characteristics |
|---|---|---|---|---|---|---|
| **Carburizing** | | | | | | |
| Pack | Carbon | 815–1090 | 125–1500 | 50–63[a] | Low-carbon and low-carbon alloy steels | Low equipment costs, difficult to control case depth accurately |
| Gas | Carbon | 815–980 | 75–1500 | 50–63[a] | Low-carbon and low-carbon alloy steels | Good control of case depth, suitable for continuous operation, good gas controls required, can be dangerous |
| Liquid | Carbon, possibly nitrogen | 815–980 | 50–1500 | 50–65[a] | Low-carbon and low-carbon alloy steels | Faster than pack and gas processes, can pose salt disposal problem, salt baths require frequent maintenance |
| Vacuum | Carbon | 815–1090 | 75–1500 | 50–63[a] | Low-carbon and low-carbon alloy steels | Excellent process control, bright parts, faster than gas carburizing, high equipment costs |
| **Nitriding** | | | | | | |
| Gas | Nitrogen, nitrogen compounds | 480–590 | 125–750 | 50–70 | Alloy steels, nitriding steels, stainless steels | Hardest cases from nitriding steels, quenching not required, low distortion, process is slow, is usually a batch process |
| Salt | Nitrogen, nitrogen compounds | 510–565 | 2.5–750 | 50–70 | Most ferrous metals including cast irons | Usually used for thin hard cases (<25 μm), no white layer, most are proprietary processes |
| Ion | Nitrogen, nitrogen compounds | 340–565 | 75–750 | 50–70 | Alloy steels, nitriding steels, stainless steels | Faster than gas nitriding, no white layer, high equipment costs, close case control |
| **Carbonitriding** | | | | | | |
| Gas | Carbon and nitrogen | 760–870 | 75–750 | 50–65[a] | Low-carbon, low-carbon alloy, and stainless steels | Lower temperature than carburizing (less distortion), slightly harder case than carburizing, gas control critical |
| Liquid (cyaniding) | Carbon and nitrogen | 760–870 | 2.5–125 | 50–65[a] | Low-carbon steels | Good for thin cases on noncritical parts, batch process, salt disposal problems |
| Ferritic nitrocarburizing | Carbon and nitrogen | 565–675 | 2.5–25 | 40–60[a] | Low-carbon steels | Low-distortion process for thin case on low-carbon steel, most processes are proprietary |

*Source:* Adapted from Ref. [21].
[a] *Requires quenching from austenitizing temperature.*

### 8.3.1 Carburizing

Carburizing is a process that introduces carbon into the surface of low-carbon steels and low-carbon alloy steels; 1015 steel and 8620 steel in Table 8.3 are candidates for this diffusion treatment. The treatment consists of heating a workpiece in contact with a carbonaceous material to a temperature above the $A_3$ temperature (Figure 8.1), holding at that temperature for a certain soaking time and then quenching in a suitable medium.

### 8.3.2 Nitriding

Nitriding is a process in which nitrogen diffuses into the surface of a steel to form a hard case. It consists of subjecting a workpiece to the action of a nitrogenous medium at temperatures well below $A_1$ line (Figure 8.1) during a long soaking time, which may be up to 1–2 days. Because the heating does not reach the austenitizing temperature, quenching is not required. Therefore, nitriding results in minimum distortion of the workpiece.

### 8.3.3 Carbonitriding

Carbonitriding is a process in which carbon and nitrogen are supplied to the surface of a steel workpiece to form a hard case. The process is a modified form of gas carburizing, in which ammonia is introduced into the gas-carburizing atmosphere. As in gas nitriding, elemental nitrogen forms at the workpiece surface and diffuses along with carbon into the steel. Carbonitriding takes place at a lower temperature and a shorter time than in gas carburizing, producing a lower case depth. The process consists of heating the workpiece above the $A_1$ temperature (Figure 8.1), except in the case of ferritic nitrocarburizing, which is carried out below $A_1$. After a sufficient soaking time, a quenching is performed, which is typically made in oil for minimizing distortion.

### 8.3.4 Induction Hardening

Induction hardening is a selective hardening process in which a localized surface of a steel workpiece is heated by induction and then quenched. Only such a surface undergoes a martensitic transformation and is selectively hardened, while the rest of the workpiece is not affected. Carbon steels and alloy steels with a carbon content of 0.40–0.45 wt% are most suitable for this process [22].

### 8.3.5 Flame Hardening

Flame hardening is a selective hardening process that involves heating the surface of a workpiece to austenitizing temperature (Figure 8.1) by means of oxyfuel gas flames. Subsequently, the workpiece is quenched in water, with which the microstructure of the surface layer consists of hard martensite over a soft core of ferrite and pearlite. The cost of this process is considerably less than induction hardening [23]. The steel to be flame hardened must have suitable carbon and other alloy elements to produce the desired hardness, since there is no change in chemical composition.

## 8.4 HEAT TREATMENTS OF ALUMINUM ALLOYS

As mentioned in Chapter 7, wrought aluminum alloys are divided in those whose mechanical properties are controlled by strain hardening and annealing (1xxx, 3xxx, and 5xxx series) and those that respond to heat hardening (2xxx, 6xxx, and 7xxx series). Apart from the chemical composition, the microstructure and properties of aluminum workparts depend on the heat treatment and cold work they have been subjected to. Therefore, a designation of temper is necessary for the proper selection of aluminum alloys. Figure 8.4 presents the basic temper designation, which is used for both wrought and cast aluminum alloys. The meaning of each letter and digits is the following [24]:

- **F:** As fabricated, with no treatment. It is applied to workpieces shaped by cold or hot working, in which no special control over thermal conditions or strain hardening is employed.
- **O:** Annealed. It is applied to relieve strain hardening, improve ductility, and reduce strength to the lowest value.
- **H:** Strain hardened. It is applied only to wrought workpieces to increase their strength by strain hardening; a complementary thermal treatment may be performed. H is always followed by two or more digits, the first indicating a heat treatment, if any, the second indicating the degree (1–9) of strain hardening, and a third digit is assigned when the degree of control of temper is different from but close to those for the two-digit H temper designation to which it is added.
- **W:** Solution heat treated. It is an unstable temper applied only to alloys that age harden in service over months or years.
- **T:** Heat treatment to produce stable tempers. T is always followed by a digit (1–10) to indicate a specific sequence of basic treatments, as can be seen in Figure 8.4. Some of the heat treatments applied to small weapon components are the following:
    - **T4:** Solution heat treatment and natural aging. It is applied to workpieces that are not cold worked after solution heat treatment, and for which mechanical properties have been stabilized by room temperature (natural) aging.
    - **T6:** Solution heat treatment and artificial aging. It is applied to workpieces that are not cold worked after solution heat treatment, and for which mechanical properties or dimensional stability, or both, have been substantially improved by precipitation heat treatment.

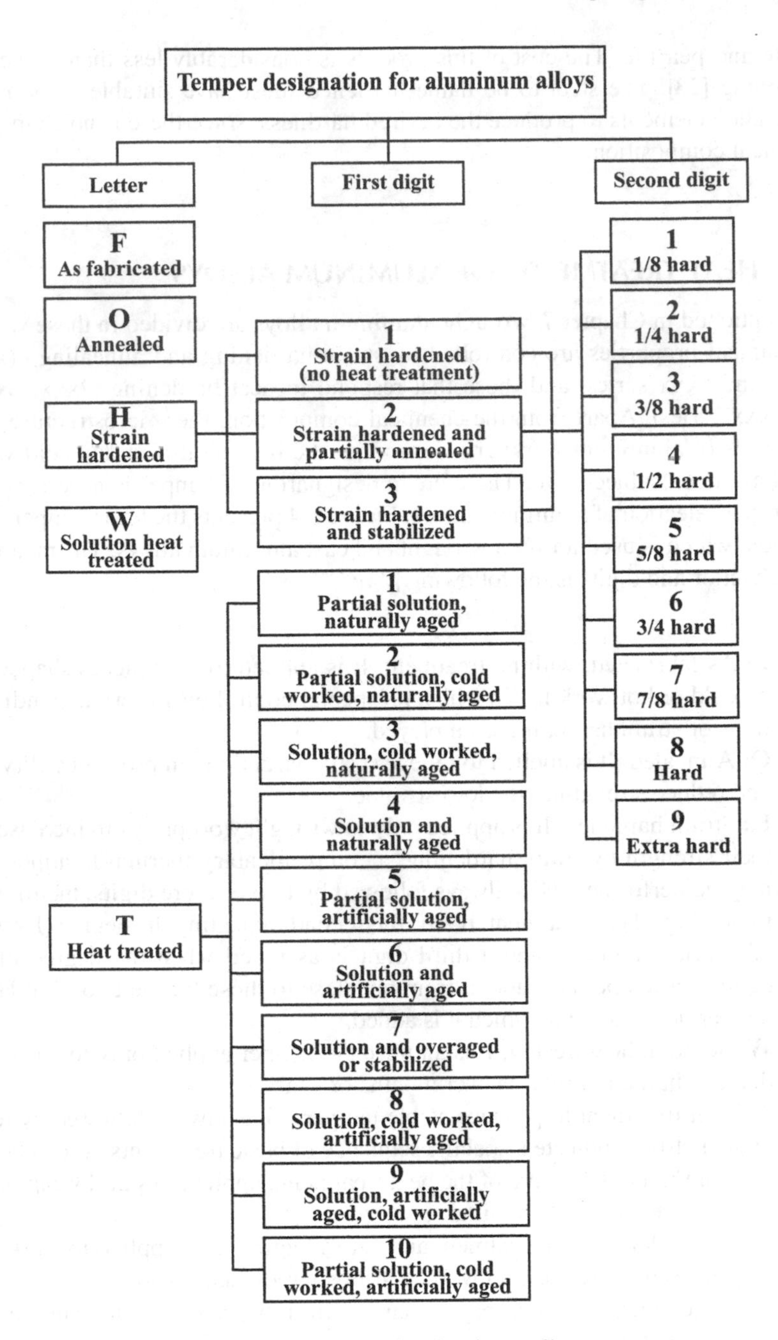


**FIGURE 8.4** Temper designation system for aluminum alloys.

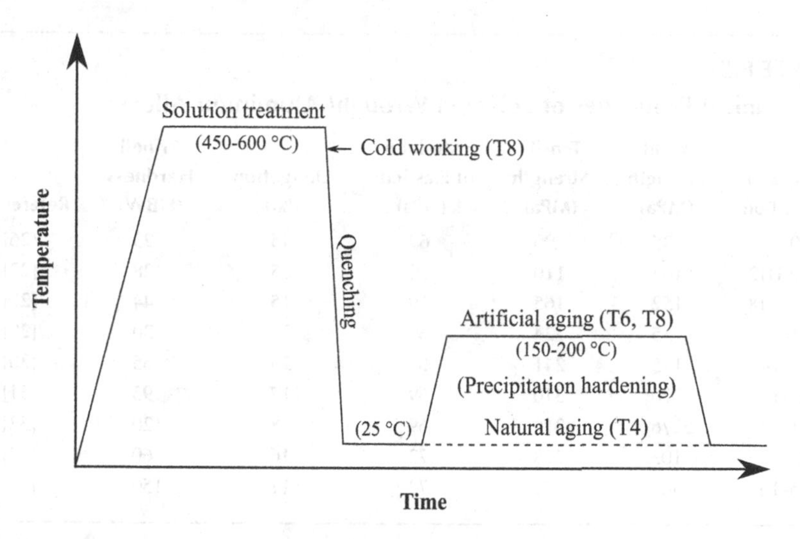


**FIGURE 8.5** Scheme of various heat treatment processes for aluminum alloys.

- **T8:** Solution heat treatment, cold working, and artificial aging. It is applied to workpieces that are cold worked specifically to improve strength after solution heat treatment, and for which mechanical properties or dimensional stability, or both, have been substantially improved by precipitation heat treatment.

The heat-treatment processes in aluminum alloys consists of three stages (Figure 8.5), which are performed in the following sequence [2,24,25]:

- **Solution treatment:** This stage involves heating the workpiece at a temperature in the range of 450°C–600°C in order to dissolve casting precipitates and disperse ("put into solution") the alloying elements through the aluminum matrix. The soaking time varies from a few minutes to one day, depending on the chemical composition and cross-section of the workpart.
- **Quenching:** This stage involves rapid cooling from the solution-treatment temperature to room temperature to avoid the reformation of precipitates and to freeze-in the alloying elements as a supersaturated solid solution in the aluminum matrix. This operation may be performed by immersing the hot workpart in cold water or by spraying it with water; the aim is to cool the workpiece in a few seconds. After quenching, the aluminum alloy is soft and ductile, being the best condition to subject the workpiece to different forming processes.
- **Aging:** In this stage, the supersaturated solid solution transforms to precipitate particles that improve the mechanical properties of the alloy. If the workpiece is reheated to a moderately high temperature

**TABLE 8.5**
**Mechanical Properties of Selected Wrought Aluminum Alloys**

| Alloy and Condition | Yield Strength (MPa) | Tensile Strength (MPa) | Modulus of Elasticity (GPa) | Elongation (%) | Brinell Hardness (HBW) | References |
|---|---|---|---|---|---|---|
| 1100-O | 35 | 90 | 69 | 45 | 23 | [26] |
| 1100-H12 | 103 | 110 | 69 | 25 | 28 | [27] |
| 1100-H18 | 152 | 165 | 69 | 15 | 44 | [28] |
| 6061-O | 55 | 124 | 69 | 30 | 30 | [29] |
| 6061-T4 | 145 | 241 | 69 | 25 | 65 | [30] |
| 6061-T6 | 276 | 310 | 69 | 17 | 95 | [31] |
| 6061-T8 | ≥276 | ≥310 | 69 | 8 | 120 | [32] |
| 7075-O | 103 | 228 | 72 | 16 | 60 | [33] |
| 7075-T6 | 503 | 572 | 72 | 11 | 150 | [34] |

(150°C–200°C), the operation is called artificial aging. If the workpiece is left at room temperature, the operation is called natural aging. Natural aging is usually a slow process, and its effect may be significant after many months or years.

The type of precipitates will depend on the alloying elements contained in the alloy. For example, the 6xxx series aluminum alloys contain mainly Al, Mg, and Si, with which the formation of $Mg_2Al_3$ and $Mg_2Si$ precipitates is possible when the alloy is age hardened [25]. The 7xxx series aluminum alloys contain mainly Al, Zn, Mg, and Cu, with which the formation of high-strength precipitates such as $CuAl_2$, $Mg_2Al_3$, $Al_{32}(Mg, Zn)_{49}$ is possible when the alloy is age hardened [25].

A 1100 alloy, one of the aluminum alloys used in the manufacture of small weapon components, is a strain-hardening alloy. This alloy can be found in the annealing condition (1100-O) and in different strain hardening conditions (1100-Hxxx). Table 8.5 displays some of these examples, where it is clear that the yield strength, tensile strength, and hardness increase as the hardening degree increases. Conversely, the elongation decreases with a higher degree of hardening.

A 6061 alloy is the most widely used aluminum alloy, although 7075 alloy is being increasingly used, in the manufacture of small weapon components. Both alloys may be subjected to aging heat treatment by different processes. Table 8.5 compares the mechanical properties of annealed 6061 alloy (6061-O) with those of the alloy under different aging conditions (6061-T4, 6061-T6, and 6061-T8), as well as of the annealed 7075 alloy (7075-O) with those of the aged condition (7075-T6). The great improvement in the properties caused by the aging heat treatments is evident, especially for the 7075 alloy. As stated above, the precipitate particles are responsible for the enhancement in the mechanical properties of the age-hardening alloys.

## 8.5 ANODIZING OF ALUMINUM ALLOYS

Anodizing is a type of surface hardening that involves the addition of a layer (Figure 8.3).

It is specifically an electrochemical process that provides the surface of the aluminum workpiece with an anodic oxide finish, which increases the wear and corrosion resistance, in addition to giving a decorative appearance. This oxide finish is aluminum oxide that, unlike a paint, is fully integrated into the underlying aluminum substrate, therefore it cannot chip or peel. Anodizing involves immersing the aluminum workpiece into an acid electrolyte bath and passing an electric current through the medium. A coating thickness of up to ~100 μm is possible to obtain [35]. The structure of the layer remains porous, thus it is necessary to apply a sealant. Further, the oxide layer can be dyed to produce attractive colors.

Small weapon components made of 6061 and 7075 aluminum alloys are usually protected by anodizing. Examples of anodized aluminum components can be found in different small weapons manufactured by Colt [36], Beretta [37], Carl Walther [38], and Smith & Wesson [39], among others.

## 8.6 A CASE STUDY BASED ON HEAT TREATMENTS OF STEELS

The case study is addressed to the heat treatment of a 4140 alloy, which is extensively used in the manufacture of small weapon components subjected to stress and impact loads, such as barrels, bolts receivers, and muzzle brakes.

A 4140 steel bar was acquired (Figure 8.6a), from which burr was produced by milling process (Figure 8.6b). The chemical composition of this burr was determined by inductively coupled plasma optical emission spectrometry; a Thermo Scientific TM iCap 6000 Series ICP spectrometer was used. The content of carbon and sulfur was determined using a Thermo Scientific FlashSmart CHNS-O elemental analyzer. Table 8.6 presents the results of these analyses, which fulfill the composition of the AISI/SAE designation.

The 4140 bar was fully annealed to obtain a microstructure close to equilibrium. For that, it was heated to 845°C for 0.5 h in a homemade furnace; this furnace was used to carry out the other heat treatments. Then the bar was cooled to room temperature inside the furnace. Round tension specimens (Figure 8.6c) were machined from the fully annealed bar in accordance with the ASTM E8/E8M standard [40]. Other samples were cut and prepared for hardness testing (Figure 8.6d).

A set of the above specimens were subjected to different heat treatments. Normalizing was carried out at 870°C, held to that temperature for 1 h, and then cooled to room temperature in still air. Hardening was performed at 855°C for 1 h and then quenched in oil with moderated agitation. Finally, quenched specimens were tempered at 205°C for 0.25 h and then cooled to room temperature in still air.

Samples with each heat treatment were mounted in a thermosetting resin (Bakelite), metallographically prepared, and etched with Nital (4% nitric acid in

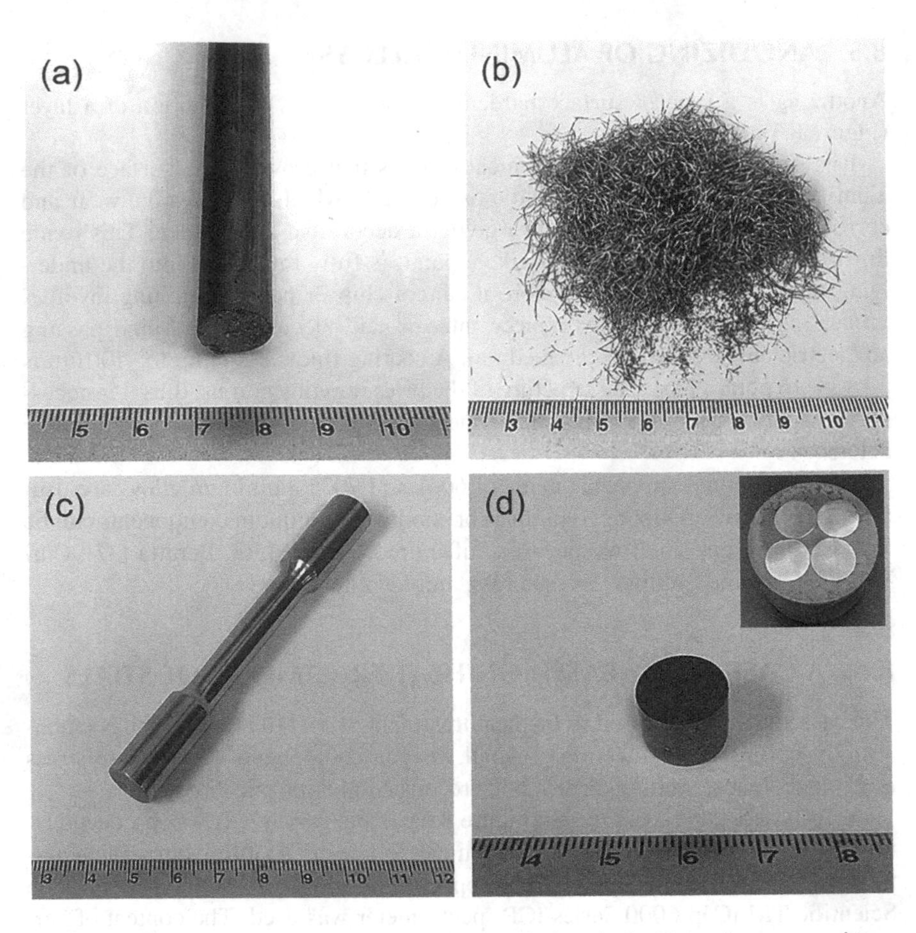


**FIGURE 8.6** 4140 steel. (a) Bar, (b) burr for chemical analysis, (c) tension test specimen, and (d) hardness test specimen (inset corresponds to samples prepared metallographically).

**TABLE 8.6**
**Chemical Composition (wt%) of the 4140 Steel Bar**

| Sample | C | Mn | P | S | Si | Cr | Mo | Fe |
|---|---|---|---|---|---|---|---|---|
| Bar | 0.39 | 0.94 | 0.017 | 0.03 | 0.20 | 1.02 | 0.24 | Bal. |
| AISI/SAE | 0.38–0.43 | 0.75–1.00 | 0.035 | 0.04 | 0.15–0.35 | 0.80–1.10 | 0.15–0.25 | Bal. |

ethanol) reagent to reveal the microstructure (see inset in Figure 8.6d). These samples were used for other analyses such as optical microscopy and scanning electron microscopy (SEM).

Optical microscopy commonly uses visible light and a lens system to generate magnified images of a sample [41]. It is a nondestructive and real-time imaging

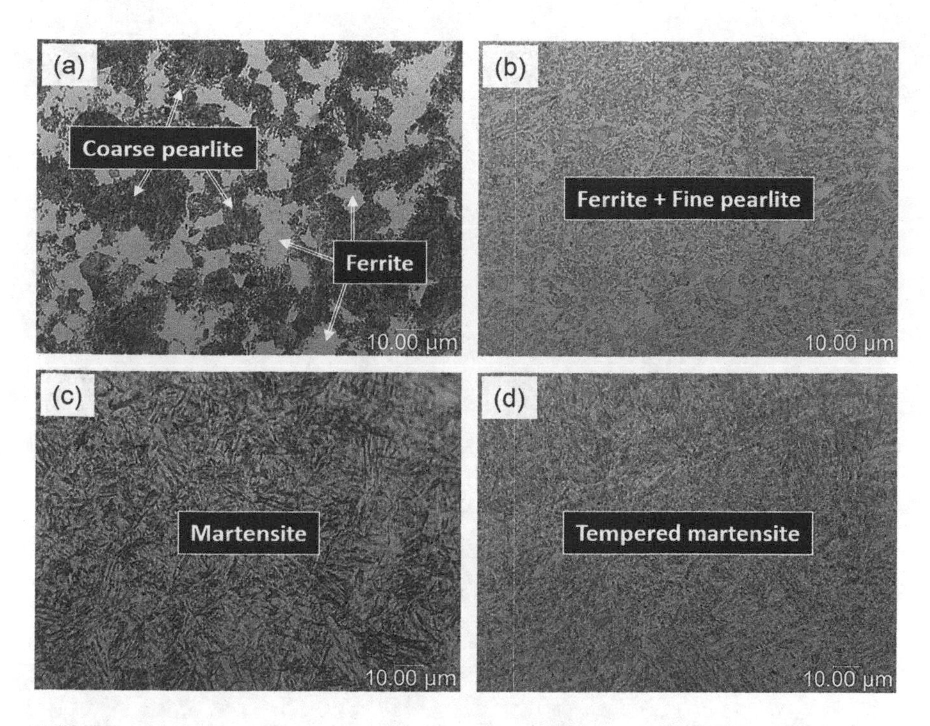


**FIGURE 8.7** Optical microscopy images of the 4140 steel specimens in the condition of (a) full annealed, (b) normalized, (c) hardened, and (d) hardened and tempered.

technique, being one of the most powerful and versatile investigation techniques in material sciences. It is the simplest technique for the superficial and morphological characterization, which allows determining fairly quickly several features about the microstructure of analyzed samples.

Figure 8.7 shows optical microscopy images of the resulting microstructures; an inverted metallurgical microscope Olympus GX71 and a PAX-it image analysis software module were used. The microstructure of the fully annealed samples (Figure 8.7a) consists of coarse pearlite (dark phase) with excess ferrite (light phase), characteristic of a hypoeutectoid steel (<0.76 wt% C); full annealing results in a soft and ductile steel, as will be seen below with the mechanical test results. The microstructure of the normalized samples (Figure 8.7b) presents ferrite and fine pearlite colonies; the mechanical properties of these specimens should be higher than those of the fully annealed specimens. The microstructure of the quenched samples (Figure 8.7c) consists of martensite, the hardest microconstituent in a steel. The quenched and tempered samples (Figure 8.7d) show tempered martensite, which may contain less residual stresses than those of the quenched specimens; the mechanical properties of these samples are expected to be lower than those of the hardened samples.

SEM is a technique for the observation and characterization of materials from the micro (μm)- to the nano (nm)-scales [41]. It is one of the most versatile

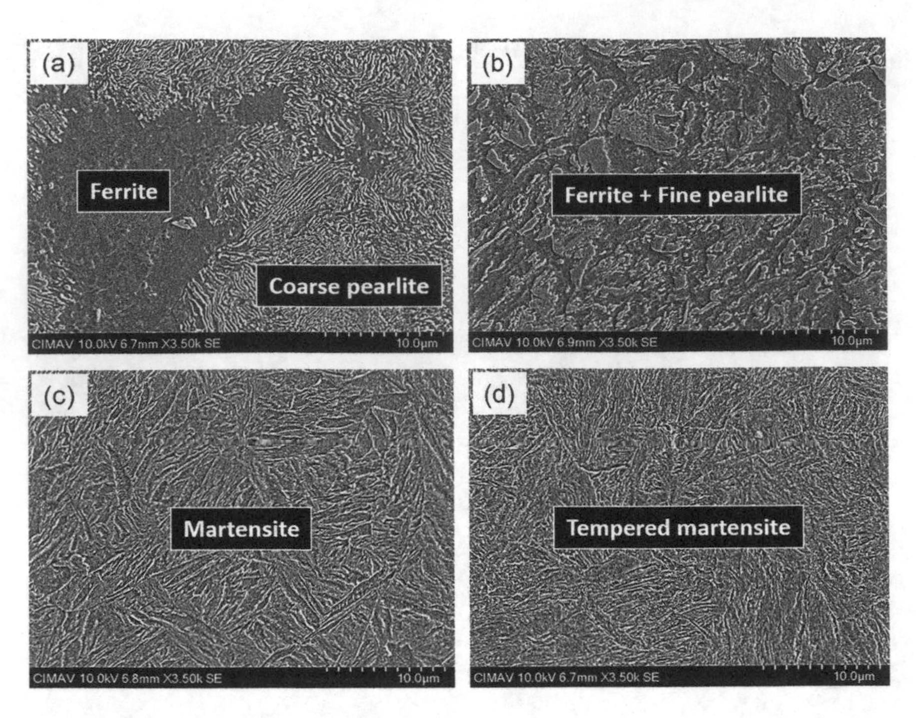


**FIGURE 8.8** SEM images of the 4140 steel specimens in the condition of (a) full annealed, (b) normalized, (c) hardened, and (d) hardened and tempered.

techniques available for the examination and analysis of the microstructural features of solid objects, which takes advantage of the use of electrons to form high-resolution images with highly detailed information about the sample. The re-emitted particles produced by the electron beam can be analyzed by different detectors, allowing the acquisition of information related to the morphology, surface defects, phase distribution, depth of field, and chemical composition.

Figure 8.8 shows some of the capabilities of SEM by providing higher resolution in the microstructure analysis of heat-treated 4140 steel samples; a Hitachi SU3500 scanning electron microscope with a secondary electron signal was used. As can be seen, SEM provided more detailed topographical information of the microstructure, compared to that of optical microscopy (Figure 8.7).

In addition to the morphological observations provided by SEM, a chemical characterization of samples is possible with the use of energy-dispersive spectroscopy (EDS) coupled to the scanning electron microscope [41]. The EDS technique detects X-rays emitted from the material during the beam collision with its surface to determine the elemental composition of the analyzed volume. The screened X-ray energy is characteristic of the element from which it was emitted, since each element possesses a unique atomic structure, obtaining a distinctive set of peaks on its X-ray spectrum. This technique helps to identify the elemental composition of the phases formed during some process. As an example,

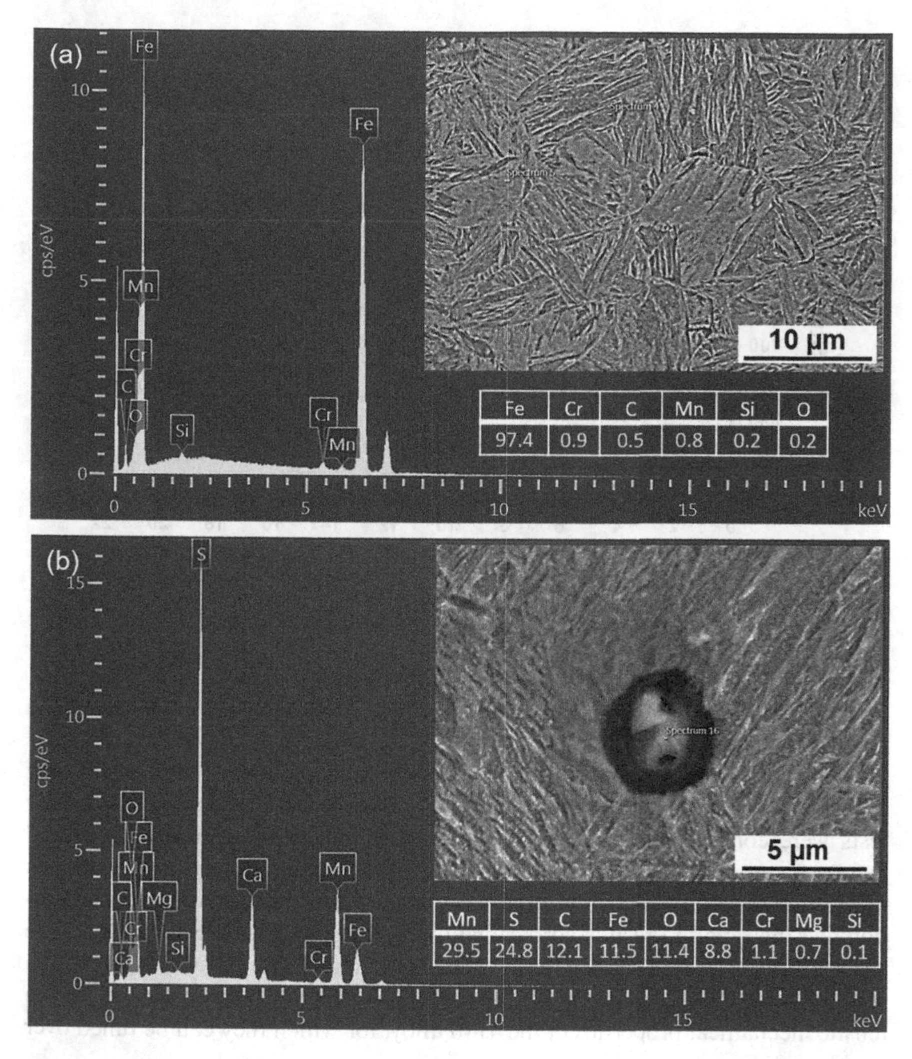


**FIGURE 8.9** SEM-EDS results of the hardened and tempered sample. (a) Analysis of the overall visualized area and (b) punctual analysis of a precipitate.

Figure 8.9 shows the SEM-EDS results of the hardened and tempered 4140 steel sample. Spectrum in Figure 8.9a presents the expected elements (Fe, Cr, C, Mn, and Si) of the alloy along with oxygen as a contamination, which can come from the environment during sample handling; it is worth mentioning that the technique has a detection limit; therefore, the rest of the alloy elements could not be identified. Spectrum in Figure 8.9b corresponds to a precipitate rich in Mn, S, and C, which could have been formed during the heat treatment processes.

The evaluation of the mechanical behavior of a sample under conditions of tension can be achieved to provide basic material property data, which are critical

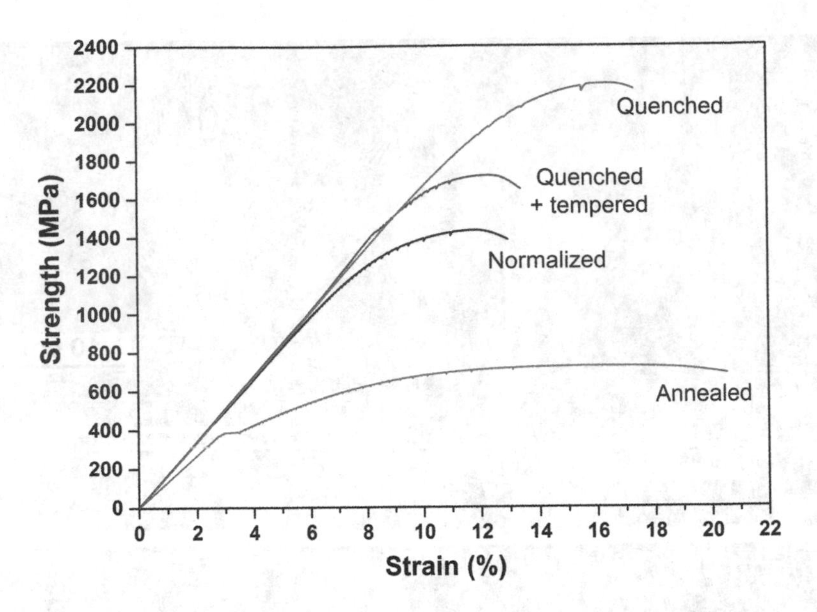


**FIGURE 8.10** Stress-strain curves of heat-treated 4140 steel samples.

for the component design and service performance assessment. A tensile test is a method for determining the behavior of materials under axial tensile loading. The test is conducted by fixing the specimen into the test apparatus and then applying a load by separating the testing machine crossheads. In this case study, tensile tests were conducted for samples with the different heat treatments described before. A universal testing machine Instron 3382 with a capacity of 100 kN was used at a strain rate of 0.35 mm/min; three specimens were tested for each condition and the reported results correspond to the average. Figure 8.10 presents the stress-strain curves generated from these tests, and Table 8.7 summarizes the resulting mechanical properties. It is evident the effect of heat treatments on the tensile mechanical properties of the 4140 alloy, for which they can be tuned over a wide range.

After the tension tests, the fracture morphology of specimens was analyzed by stereomicroscopy, which is a variant of optical microscopy designed to see a three-dimensional view of a sample at low magnification. A Carl Zeiss Stemi DV4 Series stereomicroscope with LED illumination was used. Figure 8.11 shows the morphology originated by each condition of heat treatment. The formation of a neck is evident in all samples in the rupture zone. Such a neck is formed in a tensile test specimen when the ultimate tensile strength (maximum stress) of the stress-strain curve is reached. However, an analysis of the fracture surfaces indicates some differences. In the case of the fully annealed specimen (Figure 8.11a), a moderately ductile fracture is seen. For the normalized specimen (Figure 8.11b), a combination of ductile fracture and brittle fracture is observed. The fracture surface of the quenched specimen (Figure 8.11c) reveals a mostly brittle fracture;

**TABLE 8.7**
**Mechanical Properties of Heat-Treated 4140 Steel Samples**

| Condition | Yield Strength (MPa) | Tensile Strength (MPa) | Elastic Modulus (GPa) | Elongation at Break (%) | Rockwell C Hardness (HRC) |
|---|---|---|---|---|---|
| Annealed | 385 ± 6 | 720 ± 5 | 195 ± 5 | 21.5 ± 1 | 74 ± 2 (HRB) |
| Normalized | 1184 ± 9 | 1424 ± 13 | 200 ± 8 | 13.0 ± 0 | 31 ± 1 |
| Quenched | 1716 ± 5 | 2195 ± 8 | 198 ± 6 | 17.3 ± 4 | 51 ± 1 |
| Quenched+ tempered | 1579 ± 17 | 1700 ± 28 | 201 ± 8 | 13.4 ± 1 | 44 ± 1 |

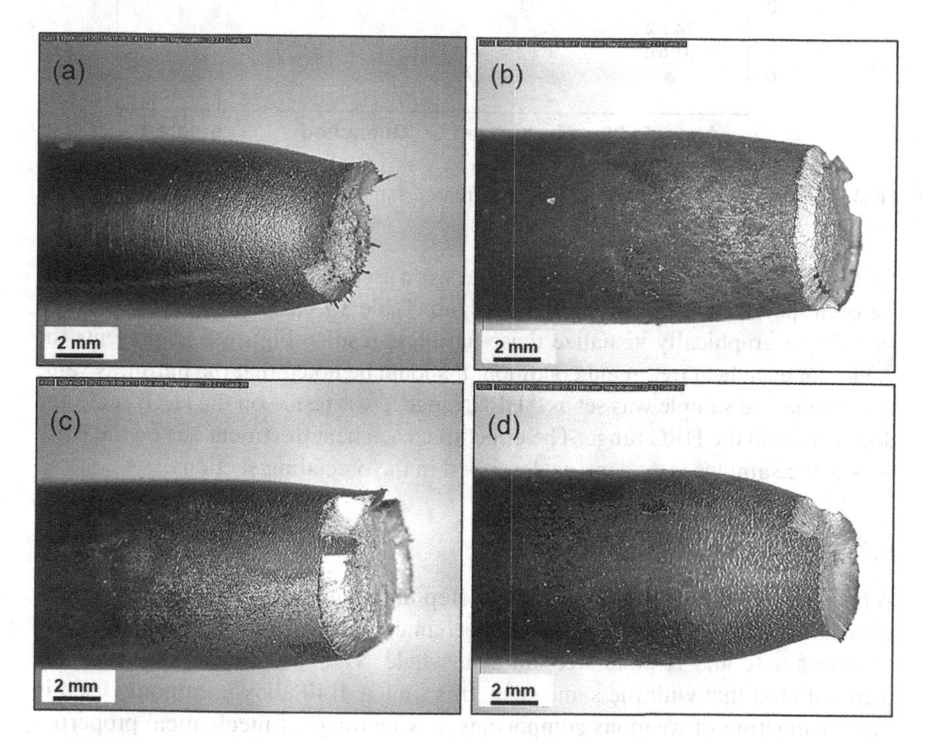


**FIGURE 8.11** Tensile fracture morphology of (a) full annealed, (b) normalized, (c) hardened, and (d) hardened and tempered specimens.

a behavior like this is not desirable on any workpiece, such as weapon components. Finally, the quenched and tempered specimen (Figure 8.11d) presents a fracture surface similar to that of the fully annealed specimen (Figure 8.11a) but with the exception that its mechanical properties are much superior. The key for a workpiece to have a good mechanical behavior is in the control of the microstructure, which is achieved with the application of a suitable heat treatment.

Hardness tests were conducted on a 574 Series Wilson Rockwell hardness tester. A load of 100 kgf for the hardness Rockwell B (HRB) scale and a load of 150 kgf

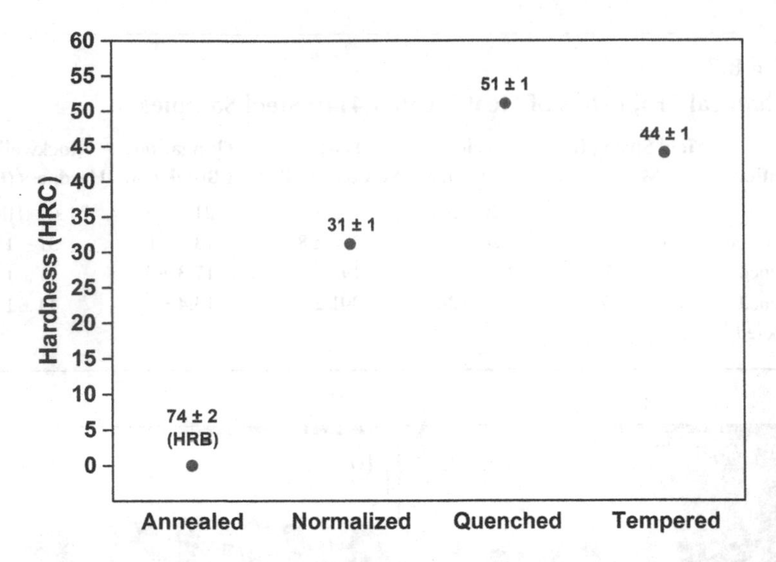


**FIGURE 8.12** Rockwell hardness of heat-treated samples.

for the hardness Rockwell C (HRC) scale were used; five indentations were made for each specimen and the reported results in Table 8.7 correspond to the average. In order to graphically visualize these hardness results, Figure 8.12 presents the values for each heat treatment condition. It should be noted that the hardness value of the annealed sample was set as 0 HRC, since it was tested on the HRB scale and does not reach the HRC range. The effect that each heat treatment has on the hardness of the samples is evident, as described in the preceding sections.

## 8.7 CLOSING REMARKS

A heat treatment process is an essential step in the manufacture of weapon components to provide them adequate mechanical properties and, ultimately, to achieve a safe and reliable weapon as a whole. The case study presented here demonstrated that with the same steel, this time a 4140 alloy commonly used in the manufacture of weapons components, a wide range of mechanical properties and forms of fracture can be attained, which depend on the way the heat treatments are carried out. A weapon component made of the best steel can perform the worst if an inadequate heat treatment is applied.

## REFERENCES

1. ISO 4885:2018(en) Ferrous materials-heat treatments-vocabulary. 2018, International Organization for Standardization: Geneva, Switzerland.
2. Dossett, J. and G.E. Totten, *ASM Handbook Volume 4: Heat Treating*. 1991, ASM International: Materials Park, OH.

3. Chandler, H., *Heat Treater's Guide: Practices and Procedures for Nonferrous Alloys*. 1995, ASM international: Metals Park, OH.
4. Canale, L.C., et al., Hardenability of steel. *Comprehensive Materials Processing*, 2014. **12**: pp. 39–97.
5. Callister, W.D. and D.G. Rethwisch, *Materials Science and Engineering: An Introduction*. 10th ed., 2018, Wiley: New York.
6. AISI 1015 Steel, hot rolled, 19–32 mm (0.75-1.25 in) round. [May 4, 2021]; Available from: http://www.matweb.com/search/DataSheet.aspx?MatGUID=8186049c0bdb42e2a683ae5809a9f9ec.
7. AISI 1015 Steel, cold drawn, 19–32 mm (0.75-1.25 in) round. [May 4, 2021]; Available from: http://www.matweb.com/search/DataSheet.aspx?MatGUID=09a91af8914847398c60e77ba60e7d12.
8. AISI 1045 Steel, hot rolled, 19–32 mm (0.75-1.25 in) round. [May 4, 2021]; Available from: http://www.matweb.com/search/DataSheet.aspx?MatGUID=4b0553daf9c245e684f2199a48179d89.
9. AISI 1045 Steel, Quenched and Tempered to 450 HB. [May 4, 2021]; Available from: http://www.matweb.com/search/DataSheet.aspx?MatGUID=bbc93d88f424449ba85dbc06314080d2.
10. AISI 1060 Steel, hot rolled, 19–32 mm (0.75-1.25 in) round. [May 4, 2021]; Available from: http://www.matweb.com/search/DataSheet.aspx?MatGUID=55cb4415eb704a46b3537627e9619d52.
11. AISI 1060 Steel, oil quenched 845°C (1550°F), 480°C (900°F) temper, 25 mm (1 in.) round. [May 4, 2021]; Available from: http://www.matweb.com/search/DataSheet.aspx?MatGUID=66e9cb816c344ecbb5464f92ff943f30.
12. AISI 1080 Steel, hot rolled, 19–32 mm (0.75-1.25 in) round. [May 4, 2021]; Available from: http://www.matweb.com/search/DataSheet.aspx?MatGUID=5cda6846cdfb4d2383b673abb072e4c0&ckck=1.
13. AISI 1080 Steel, oil quenched from 815°C (1500°F), tempered at 480°C (900°F). [May 4, 2021]; Available from: http://www.matweb.com/search/DataSheet.aspx?MatGUID=075e15cd89214f7ca1b84d62d340db58.
14. AISI 4140 Steel, annealed at 815°C (1500°F) furnace cooled 11°C (20°F)/hour to 665°C (1230°F), air cooled, 25 mm (1 in.) round. [May 31, 2021]; Available from: http://www.matweb.com/search/DataSheet.aspx?MatGUID=7b75475aa1bc41618788f63c6500d36b.
15. AISI 4140 Steel, oil quenched, 25 mm (1 in.) round [845°C (1550°F) quench, 540°C (1000°F) temper]. [May 31, 2021]; Available from: http://www.matweb.com/search/DataSheet.aspx?MatGUID=07d1795c3f034c97b52cccda78ae1409.
16. AISI 4340 Steel, annealed, 25 mm round. [May 31, 2021]; Available from: http://www.matweb.com/search/DataSheet.aspx?MatGUID=fd1b43a97a8a44129b32b9de0d7d6c1a.
17. AISI 4340 Steel, oil quenched 800°C (1470°F), 540°C (1000°F) temper, 25 mm round. [May 31, 2021]; Available from: http://www.matweb.com/search/DataSheet.aspx?MatGUID=1d276ee7ee184c718cdbc1af10671c0b.
18. AISI 8620 Steel, annealed 13 mm (0.5 in.) round. [May 31, 2021]; Available from: http://www.matweb.com/search/DataSheet.aspx?MatGUID=d7f44bdac4514000a0ffe5c983485c1f.
19. 8620 Acero al Cromo-Níquel-Molibdeno. Perfil: Redondo. [May 31, 2021]; Available from: https://www.palme.mx/productos/8620.
20. Davis, J., Process selection guide. in *Surface Hardening of Steels: Understanding the Basic*, J. Davis, Editor. 2002, ASM International: Materials Park, OH, pp. 1–15.

21. Budinski, K.G., *Surface Engineering for Wear Resistance.* 1988, Prentice-Hall, Inc: Englewood Cliffs, NJ.
22. Rudnev, V., D. Loveless, and R.L. Cook, *Handbook of Induction Heating.* 2nd ed., 2017, CRC Press: New York.
23. Gupta, K., N.K. Jain, and R. Laubscher, Surface property enhancement of gears, in *Advanced Gear Manufacturing and Finishing: Classical and Modern Processes.* 2017, Academic Press: Cambridge, MA.
24. ASM International, *ASM Handbook Volume 2: Properties and Selection: Nonferrous Alloys and Special-Purpose Materials.* 1990, ASM International: Materials Park, OH.
25. Mouritz, A.P., *Introduction to Aerospace Materials.* 2012, Woodhead Publishing Limited: Oxford.
26. Aluminum 1100-O. [May 14, 2021]; Available from: http://www.matweb.com/search/DataSheet.aspx?MatGUID=db0307742df14c8f817bd8d62207368e&ckck=1.
27. Aluminum 1100-H12. [July 12, 2021]; Available from: http://www.matweb.com/search/DataSheet.aspx?MatGUID=e98fdee3896f4cfa8b01e232c7fe33b6.
28. Aluminum 1100-H18. [July 12, 2021]; Available from: http://www.matweb.com/search/DataSheet.aspx?MatGUID=4bd86a84512445528eca1d91b7e14c49.
29. Aluminum 6061-O. [May 14, 2021]; Available from: http://www.matweb.com/search/DataSheet.aspx?MatGUID=626ec8cdca604f1994be4fc2bc6f7f63.
30. Aluminum 6061-T4. [July 12, 2021]; Available from: http://www.matweb.com/search/DataSheet.aspx?MatGUID=d5ea75577b1b49e8ad03caf007db5ba8.
31. Aluminum 6061-T6. [Jul 12, 2021]; Available from: http://www.matweb.com/search/DataSheet.aspx?MatGUID=b8d536e0b9b54bd7b69e4124d8f1d20a.
32. Aluminum 6061-T8. [Jul 12, 2021]; Available from: http://www.matweb.com/search/DataSheet.aspx?MatGUID=90404a0c001c4016b2b359a6c19f9127.
33. Aluminum 7075-O. [May 14, 2021]; Available from: http://www.matweb.com/search/DataSheet.aspx?MatGUID=da98aea5e9de44138a7d28782f60a836.
34. Aluminum 7075-T6. [July 12, 2021]; Available from: http://www.matweb.com/search/DataSheet.aspx?MatGUID=4f19a42be94546b686bbf43f79c51b7d.
35. Bononi, M., R. Giovanardi, and A. Bozza, Pulsed current hard anodizing of heat treated aluminum alloys: Frequency and current amplitude influence. *Surface and Coatings Technology*, 2016. **307**: pp. 861–870.
36. Colt-Still making history. [July 15, 2021]; Available from: https://www.colt.com/.
37. Beretta. [July 15, 2021]; Available from: https://www.beretta.com/en/.
38. Walther. [July 15, 2021]; Available from: https://carl-walther.de/.
39. Smith & Wesson. [July 15, 2021]; Available from: https://www.smith-wesson.com/.
40. ASTM E8/E8M-21 Standard test methods for tension testing of metallic materials. 2021, ASTM International: West Conshohocken, PA.
41. Herrera Ramirez, J.M., et al., *Unconventional Techniques for the Production of Light Alloys and Composites.* 2020, Springer: Cham, Switzerland.

# 9 Manufacturing Processes for Small Weapon Components

## 9.1 INTRODUCTION

Mechanical design is closely linked to the choice of materials and the manufacturing processes to achieve the materialization of workparts. Manufacturing is the "art" of transformation of raw materials into products [1]. The word manufacture was coined to describe "made by hand," since originally only manual methods were available [2]; this was the case with implements and weapons, which were produced as handicrafts and trades. Nowadays, most modern manufacturing, including that of weapons, is accomplished by automated and computer numerical control (CNC) machines, which are essential to achieve the tight dimensional tolerances required in the production of weapon components. The manufacture of firearms requires high precision because even if the smallest component does not fit or work properly, the weapon could present from slight to catastrophic failures.

In general, manufacturing processes can be classified into three basic categories: subtractive manufacturing processes, where the desired geometry of a workpart is obtained by removing material; formative manufacturing processes, where the geometry of a workpart is altered by applying external loads or heat; and additive manufacturing (AM) processes, where 3D workparts are created by depositing materials. Figure 9.1 shows a broad classification of the manufacturing processes. Among the numerous existing methods, the selection of the suitable technology and manufacturing process is necessary, depending on the desired features of the workparts to be produced. This chapter is focused on the key processes for making small weapons. Each of these processes is briefly described below.

## 9.2 CASTING

Casting is a process in which molten metal flows by gravity or other force into a mold, where it solidifies in the shape of the mold cavity [2]. Casting can be used to create complex workpart geometries, which may range from a few grams to more than 100 tons. It can be performed on any metal that can be heated to the liquid state. Some casting processes are quite suitable for mass production, and some of them are capable of producing parts to net shape. A variety of casting processes are available, thus making it one of the most versatile of all manufacturing processes. Examples of these processes are sand casting, precision sand casting, high-pressure die casting, investment casting, lost foam, and centrifugal casting.

DOI: 10.1201/9781003196808-9

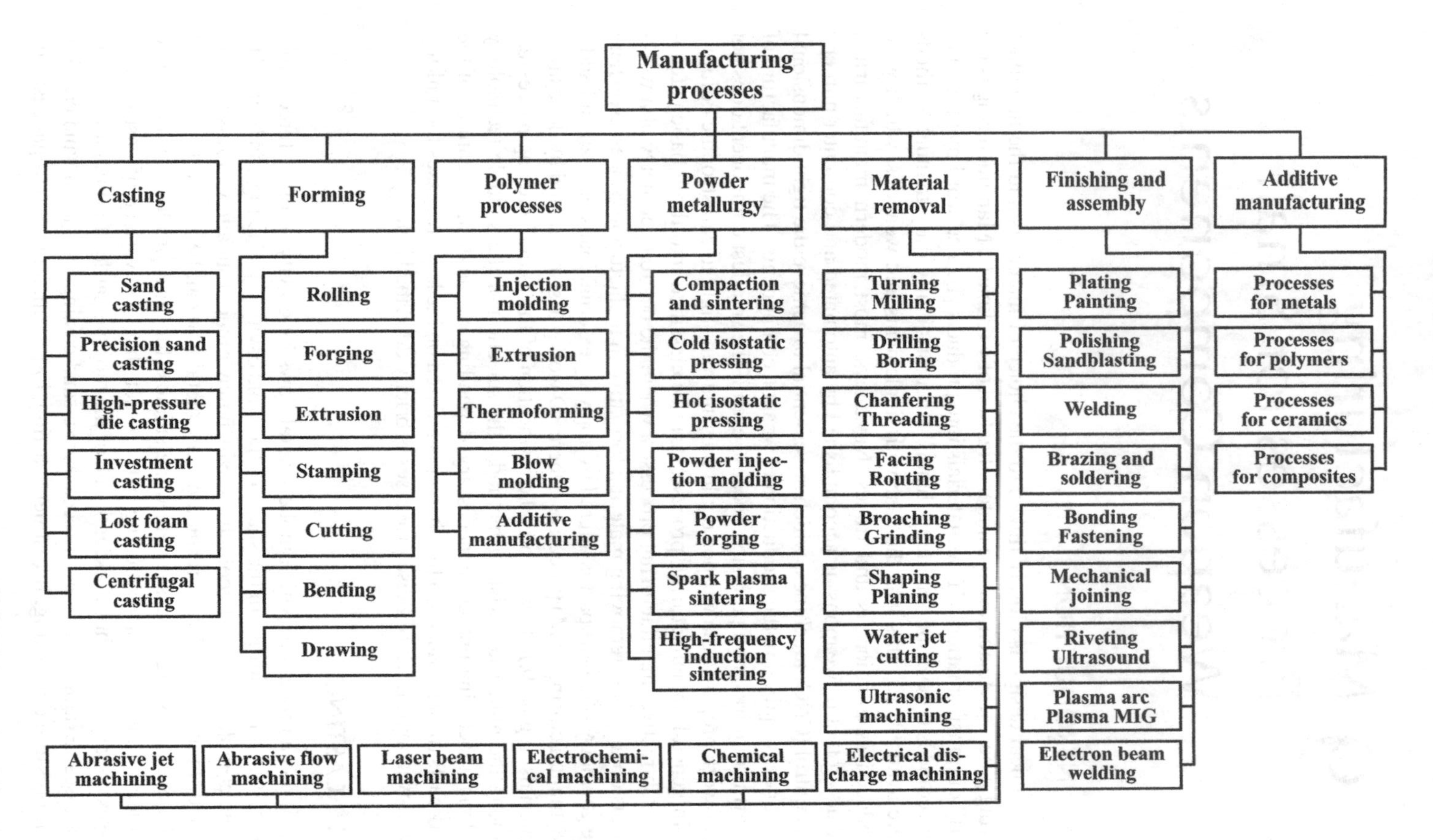


**FIGURE 9.1** Classification of the main manufacturing processes.

Investment casting process, also known as lost-wax casting process, is one of the most used for the manufacture of firearm components. The process basically consists of the following steps (Figure 9.2):

1. The desired workpart pattern is created in wax (Figure 9.2a). Beeswax has been used in this process to form the intended casting, but today more sophisticated reclaimable waxes are used [3]. This step can be performed by injection molding, hand carved from a wax block, or AM.

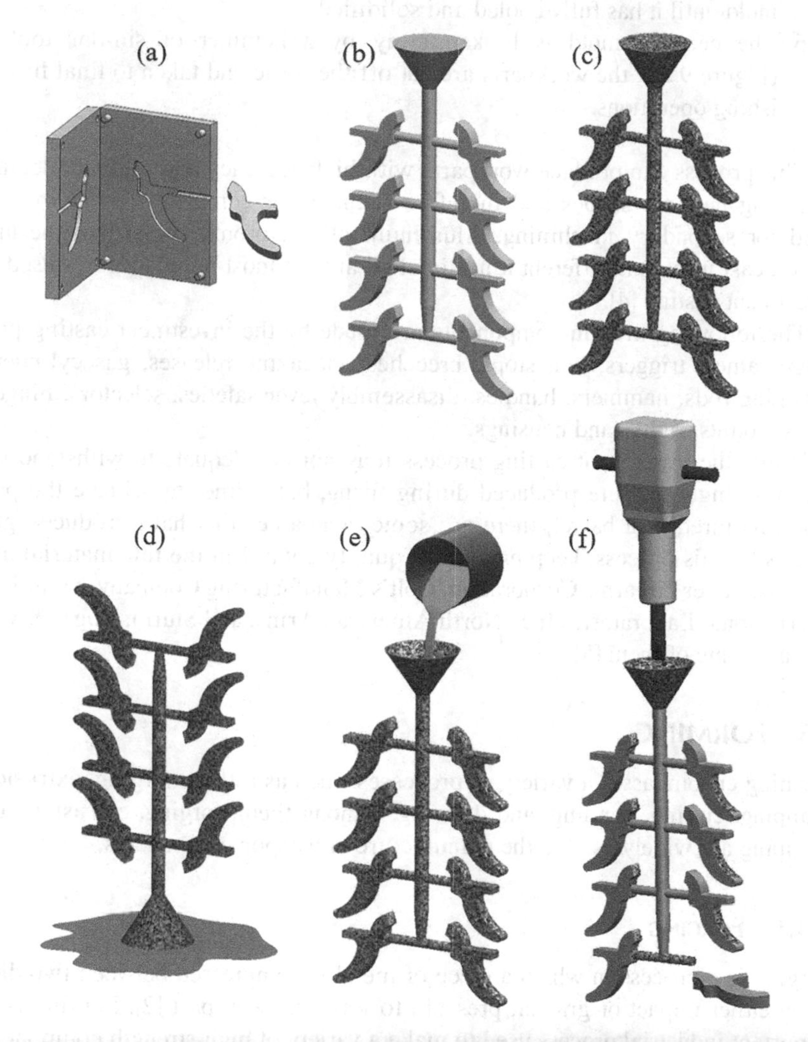


**FIGURE 9.2** Representation of the investment casting process. (a) Creation of the workpart pattern in wax, (b) assembly of multiple wax patterns, (c) immersion in a ceramic suspension and coating with a refractory material, (d) removal of wax inside an oven, (e) pouring of molten metal into the ceramic mold, and (f) breaking of the ceramic mold to obtain the workparts.

2. Multiple wax patterns are assembled onto a sprue (Figure 9.2b).
3. The wax pattern assembly is immersed into a ceramic slurry and coated with a refractory material, which is allowed to dry (Figure 9.2c).
4. The wax is removed from the ceramic mold by heating it inside an oven; the mold is placed upside down and the wax is recovered and reused later (Figure 9.2d).
5. The molten metal is poured into the ceramic mold either by gravity, vacuum, or centrifugal force (Figure 9.2e). The material remains in the mold until it has fully cooled and solidified.
6. The ceramic mold is broken away by a hammer or similar tool (Figure 9.2f); the workparts are cut off the sprue and taken to final finishing operations.

This process can produce workparts with high accuracy and intricate details, obtaining near-net shapes and therefore reducing or completely removing the need for secondary machining. Aluminum alloys, bronze alloys, magnesium alloys, cast iron, and different kinds of steel are the most popular alloys used in investment casting [4].

The following firearm components are made by the investment casting process: frames, triggers, bolt stops, breeches, magazine releases, gas cylinders, operating rods, hammers, handles, disassembly lever safeties, selectors, hinges, locks mounts, sights, and housings.

While the investment casting process may not be adequate to withstand the relatively high pressure produced during firing, but neither to achieve the precision required in a barrel, there are some companies that have produced gun barrels by this process, keeping careful quality control in the raw material and process. Bauer Firearms Corporation, Colt's Manufacturing Company, Munitions International Laboratories Inc., North American Arms, and Sturm, Ruger & Co., Inc. are some of them [5].

## 9.3 FORMING

Forming encompasses a variety of processes, such as rolling, forging, extrusion, stamping, cutting, bending, and drawing. Among them, forging, extrusion, and stamping are widely used in the manufacture of weapon components.

### 9.3.1 Forging

Forging is a process in which a piece of metal is compressed between two dies, using either impact or gradual pressure to form the workpart [2]. Forging is an important industrial process used to make a variety of high-strength components for different applications. An advantage of this process is that it compacts the metal, making the workpart stronger owing to the strain hardening.

Several firearm components are manufactured by this process, such as receivers, bolt mechanisms, and revolver cylinders.

Big forging machines are available for hammering large components. The process is known as hammer forging, and it can be done either by swaging or radial forging. Both processes are used to reduce the diameter of a tube or solid rod [2]. The difference is that in swaging the dies rotate around the workpart, whereas in radial forging the workpart rotates as it feeds into the hammering dies. Hammer forging is most often used to manufacture symmetric parts, such as gun barrels [6]. It can be performed as hot or cold working and the number of hammers to be used can be three, four, or six. The machines used for forging large gun barrels are of a horizontal type and can size the bore of the gun barrel to the exact rifling that is machined on the mandrel. Most mandrels are made from shock-resistant tool steel and high-speed steels; they are hardened, ground and polished, and sometimes plated with chromium to improve wear resistance and surface finish on the inside diameter of the barrel. Tungsten carbide mandrels are used for superior wear resistance when production volume justifies their increased cost; a mandrel can be used to rifle 1000–2000 barrels depending on its quality [5]. Workparts produced by this process often have improved microstructural and mechanical properties compared to those produced by other techniques like broaching, button rifling, or even investment casting.

Specifically, the rifling of firearm barrels by hammer forging begins with a metal barrel blank with a certain length and diameter, depending on the desired final dimensions, in the center of which a finely finished hole has been previously made (Figure 9.3). A mandrel with the pattern of the rifling lands and grooves is then inserted into the blank hole. Next, a forging machine with a series of radially opposed hammers is used to compress the blank inward against the mandrel. As the hammers compress the outer surface, the blank is reduced in diameter and lengthened, at the same time creating the bore and rifling. If desired, the process can form the chamber, throat, and outer profile. The spiral tracks of the hammers can often be seen on the outer surface of barrels; barrels may be subjected to a turning process to remove such tracks. One advantage of hammer forging is that high-quality barrels are produced, both in their surface finish and at a microstructural level; in addition, there is no metal removal, so no waste

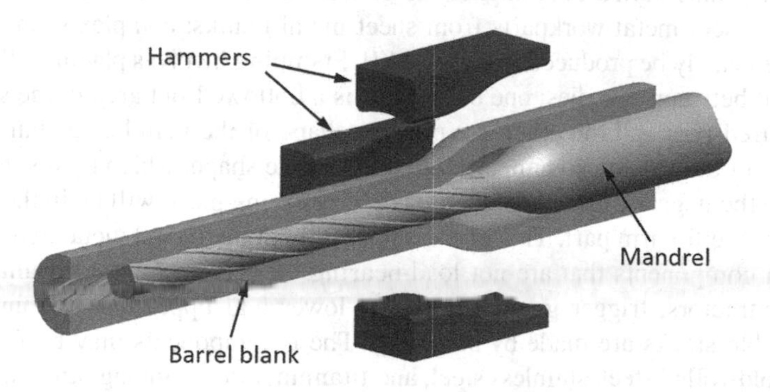


**FIGURE 9.3** Schematic of the hammer forging process.

or chips are generated. Among other manufacturers, the following are using the hammer forging process to produce barrels in some of their gun models [5]: Bushmaster; Fabrique Nationale Herstal; Fratelli Tanfoglio SAS; Glock; Heckler & Koch; Makarov; Pietro Beretta; Remington; Sturm, Ruger & Co., Inc.; Sabatti S.p.a.; SIG Sauer; Springfield armory; Steyr-Mannlicher; U.S. Repeating Arms Company; and Weatherby.

Button rifling, also known as buttoning or button swaging, relies on an elliptical button with the negative pattern of lands and grooves [5]. The tool is usually made of tungsten carbide. The button has a larger diameter than the inside of the barrel blank. Its penetration into the barrel is carried out under hydraulic pressure, either by pulling it through the bore or by pushing it through. When passing through the barrel, the button turns on itself (producing the lands and grooves of the required twist) as the bore expands to take the negative form of the profile manufactured on the button. Bushmaster; Carl Walther; Colt's Manufacturing Company; Eddy; Fratelli Tanfoglio SAS; Iver Johnson & Painfield carbine; John Slough of London; Lothar Walther; Lorcin Engineering; Marlin firearms; Mossberg; Remington; Sturm, Ruger & Co., Inc.; Savage Arms; Shilen; Sphinx systems limited; Springfield armory; Sterling Arms; Wilson Arms Company; and Winchester Repeating Arms are using the button rifling in some of their barrels.

### 9.3.2 Extrusion

Extrusion is a compression process in which material is forced to flow through a die orifice to provide a long continuous product whose cross-sectional shape is determined by the shape of the orifice [2]. It is one of the fundamental shaping processes for metals, polymers, and ceramics. This process is commonly used to make aluminum bases for mounting telescopic sights.

### 9.3.3 Stamping

Stamping, also called blanking, is the general name of sheet metal pressing to produce sheet metal workparts from sheet metal blanks; complex-shaped workparts can easily be produced at low cost [7]. Stamping involves placing a flat piece of metal between two dies; one of the dies is a hollowed-out area in the shape of the desired part and the other is a positive shape of the item being stamped [8]. A large amount of force is imparted to the positive shape, which drives the metal flat into the negative hollowed-out area. The resulting piece will be in the desired shape of the firearm part. This process is usually done to cold metal and used for firearm components that are not load-bearing. Magazines, ammo drums, clips, pins, extractors, trigger guards, brackets, lower and upper receiver units, and collapsible stocks are made by stamping. These components may be fabricated from cold-rolled steel, stainless steel, and titanium alloys, among other materials. Stamping is also used to put serial numbers on firearms.

## 9.4 POLYMER PROCESSES

Polymers are typically processed by injection molding, extrusion, thermoforming, and blow molding, although, in recent years, they are being processed by AM.

In the weapon industry, injection molding is the most widely used process to manufacture different components. Although the process is more suitable for thermoplastics, some thermosets and elastomers are injection molded, with modifications in equipment and operating parameters to allow for cross-linking of these materials. In this process, a polymer is heated to a highly plastic state and forced to flow under high pressure into a preheated mold cavity, where it solidifies by the flow of cold water through channels within the mold [2]. The molded part is then removed from the cavity. The production cycle time is typically in the range of seconds, although for dimension-critical workparts, such as frames, longer times are necessary to allow uniform cooling without deformation. The process can produce complex and intricate-shaped components with almost always net shape. The manufacture of firearm components from polymer is usually cheaper and faster than that of metallic components.

Frames, handguards, grips, stocks, forearms, Picatinny rails, trigger guards, head guards, magazines, sights, some components of the trigger mechanism, and recoil pads, among others, are firearm components produced from polymers by injection molding. For instance, the following pistols have a large number of polymeric components: Glock 17, SIG SAUER P320 Compact, Springfield Armory XD Mod. 2, CZ-USA P-10, HK VP9, Beretta APX, Smith & Wesson Bodyguard 380, Ruger LCP II, Walther PPQ M2, and FN Herstal FNX.

## 9.5 POWDER METALLURGY

Powder metallurgy is defined as the process to consolidate fine powders to produce monolithic or solid components. Powders are compressed into a mold with the desired shape (Figure 9.4) and then heated to cause bonding (sintering) of the particles into a hard, rigid mass [2]. Some powder metallurgy processes allow compression and sintering to be done at the same time. The sintering is performed at a temperature below (typically around 20%–30% lower) the melting point of the metal. There are several consolidation methods: conventional sintering, cold isostatic pressing, hot isostatic pressing, powder injection molding, powder forging, spark plasma sintering, high-frequency induction sintering, dynamic powder compaction, powder roll compaction, powder extrusion, spray techniques, and arc-melting process, among others [3].

Powder metallurgy is attractive due to the superior properties that emerge from the refined and uniform microstructure that evolves during processing. Additional attributes to the method include its diversity in terms of manufacturability and the potential to obtain high-quality, complex, neat net-shaped workparts that may require tight tolerances. The technique is usually cost-effective when compared to the benefits that can be obtained [3]. A disadvantage of this process is that the

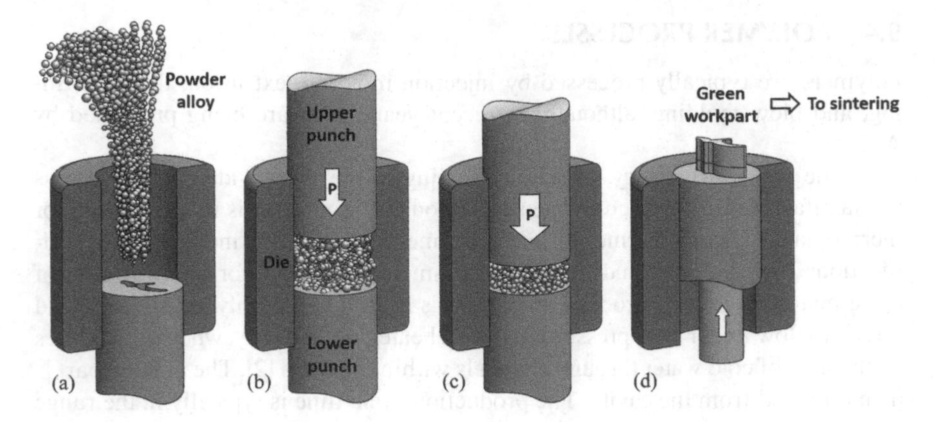


**FIGURE 9.4** Representation of the compaction process: (a) the metal powder is placed in a workpart-shaped tool; (b, c) an increasing load is applied to consolidate the loose powder; (d) a green product is obtained and sent to the sintering process.

workparts may not be as strong as those made by investment casting or forging, mainly due to the remaining porosity, typical of the process.

In combination with other routes, powder metallurgy is able to synthesize advanced materials with attractive physical characteristics for different applications. The versatility of powder metallurgy is such that it is currently used by various industries in the manufacture of workparts that require a calculated control in the porosity, as well as high precision in the design, while simultaneously focusing on mass production and their uniformity. The complexity in the geometries and the accuracy in the control of the alloying elements that constitute an alloy, which can be obtained by this process, are hardly attainable through conventional casting routes [9]. Powder metallurgy allows minimizing the machining process in those workparts where the tolerance required in their application is a critical factor for their use and operation. This reduces in a complementary way the amount of waste material, making this technique an economically viable solution in the weapon industry.

Small weapon industry finds in powder metallurgy a technology that allows the manufacture of weapon components with high precision in their production and minimum waste, and where additional machining processes are not required for large batches in mass production [10]. Hammers, triggers, gas intake blocks, sight components, and breechbolt catching pieces are being produced by powder metallurgy.

## 9.6 MATERIAL REMOVAL PROCESSES

Material removal involves a series of processes through which material is removed from the surface of a workpart until reaching the desired shape. Figure 9.1 presents various material removal processes, which can be broadly classified into three categories [2]: machining, in which excess material is removed by cutting tools (e.g. turning, milling, drilling); abrasive processes, in which excess material is removed by hard, abrasive particles (e.g. grinding); and nontraditional processes where various energy forms other than sharp-cutting tool are used to remove material (mechanical, electrochemical, thermal, chemical energy). It is important to underline the evolution of weapon production from handcrafted processes to productions based on CNC [5]. These automated processes can be used for the entire production of the firearm components or only for some steps of the process.

To address some of the material removal processes, look at the example of the manufacture of a barrel, in which several operations are involved (Figure 9.5) [5]:

1. **Selection of the suitable steel bar:** Different steels are available for barrels, such as carbon steel, chrome-molybdenum steel, and stainless steel.
2. **Cutting:** The steel bar is cut to the required dimension.
3. **Drilling:** The piece of bar is drilled from end to end through the center to produce a long straight hole, forming the bore with a diameter smaller than the desired one and a very coarse inner surface.
4. **Reaming:** A reaming tool is used to smooth the internal contour of the drilled steel bar to improve the inner surface of the bore.
5. **Polishing:** A polishing step can also be performed before rifling.
6. **Rifling:** The rifling process generates the grooves and lands characterizing the barrel of a rifled firearm. There are different rifling forms, such as conventional rifling with lands and grooves, square and polygonal; these last two types of rifling have no lands but rounded profile flanks.

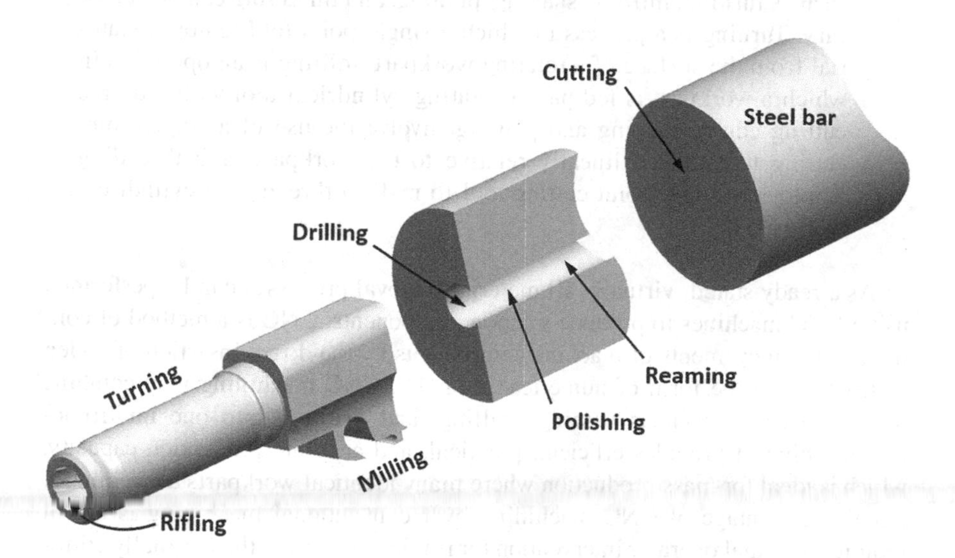


**FIGURE 9.5** Schematic of some material removal processes involved in the manufacture of a barrel.

The process can be performed by hook or scrape cutting, which are the oldest methods of rifling a barrel. A hook cutter consists of a single hook-shaped cutter at its end, with which one groove is dug at a time when it is pushed into the barrel. A scrape cutter consists of multipoint cutters inset into the opposed sides, with which a minimum of two opposed grooves to all grooves can be scraped at a time. Savage (.32-20 Winchester) and S&W Victory were rifled by hook cutting, while .45 ACP calibers were rifled by scrape cutting [5].

Broaching is another process for rifling barrels. A rifling broach is a long rod with a set of progressive cutting discs, also known as cutting blades, perpendicular to its axis, which correspond to the number of grooves. As the broach moves through the barrel, each disk cuts slightly deeper than the previous one until the desired groove depth is reached. The following manufacturers have been using the broaching process to produce barrels in some of their gun models [5]: Bushmaster; Colt's Manufacturing Company; Fabryka Broni "Łucznik" Radom; Iver Johnson & Painfield carbine; Jennings Firearms, Bryco Arms, Jimenez Arms; Mossberg; Pietro Beretta; Raven Arms; Sturm, Ruger & Co., Inc.; Savage Arms; Smith & Wesson; Springfield armory; Star Bonifacio Echeverria, S.A.; Wilson Arms Company; and Winchester Repeating Arms.

There are other processes for rifling which are not based on material removal but on swaging methods, such as hammer forging and button rifling, which were explained in Section 9.3.1.

7. **Other processes:** To form the exterior of a barrel, various operations such as turning, milling, shaping, planing, and threading can be necessary. Turning is a process in which a single-point tool removes material from the surface of a rotating workpart; milling is an operation in which a workpart is fed past a rotating cylindrical tool with multiple cutting edges; shaping and planing involve the use of a single-point cutting tool moved linearly relative to the workpart; and threading employs a single-point cutting tool to make a thread on a cylindrical workpart [2].

As already stated, virtually all material removal processes can be performed using CNC machines to produce weapon components. CNC is a method of controlling the movements of machine components by the direct insertion of coded instructions in the form of numerical data [11]. CNC machining may combine many processes, such as turning, milling, and drilling, into one multifunction machine. It provides efficient, practical, and accurate production capacity, which is ideal for mass production where many identical workparts are required. Another advantage of CNC machining over conventional machining is that it requires minimal operator intervention for routine operation, thus virtually eliminating human error; further, one skilled operator can manage several machines simultaneously. Additionally, there is saving of floor space by having one CNC machine rather than multiple workstations.

There are advanced processes that are based on nonmechanical material removal operations [11]. Some of these processes are (Figure 9.1):

1. **Water jet cutting**, where a jet of water is used to cut materials.
2. **Ultrasonic machining**, where material is removed from a surface by microchipping and erosion with loose, fine abrasive grains in a water slurry.
3. **Electrical discharge machining**, which removes material by melting small portions of the workpart by a spark.
4. **Chemical machining**, which is carried out by chemical dissolution using reagents or etchants.
5. **Electrochemical machining**, where material is removed by the action of an electrical power source and ion transfer inside an electrolytic fluid.
6. **Laser beam machining**, where a high-energy beam is used.
7. **Abrasive flow machining**, which involves the use of abrasive grains, such as silicon carbide or diamond, that are mixed in a putty-like matrix and then forced.
8. **Abrasive jet machining**, where the water jet contains abrasive particles (silicon carbide or aluminum oxide), which increase the material-removal rate above that of water jet machining.

An example of the application of electrochemical machining (ECM) process is in barrel rifling. ECM forms lands and grooves thanks to a controlled dissolution of the barrel metal by a conductive fluid (the electrolyte), which is subjected to a high-density current [5]. The cathode is a mandrel bearing spiraling metal strips, which shape the rifling pattern according to the desired groove shape. The anode is the barrel blank, into which the mandrel is inserted. The system is then immersed in the electrolyte and an electric current is applied. The mandrel is then moved through the barrel and twisted to create the grooves at the desired rate of twists. As the current moves from the steel bar to the electrode, the metal is removed by electrolysis, producing the barrel grooves by duplicating the shape of the metal strips on the mandrel. The electrolyte flows through the barrel under pressure to remove the reaction products. Coonan Arms, SIG Sauer, and Smith & Wesson are using the ECM process for rifling some of their barrels.

## 9.7 ADDITIVE MANUFACTURING PROCESSES

AM is a process of joining materials to make objects from 3D model data, usually layer upon layer, as opposed to subtractive manufacturing methodologies [12]. As stated in Chapter 5, AM is also known as 3D printing, but there are certain differences when it comes to the capabilities of this process in rapid prototyping. The potential of AM has been rapidly disseminated since its appearance, reaching the interest of scientific, technological, and governmental sectors around the world [10]. The diffusion of its capabilities has been reflected in an increasing number of scientific articles year after year and has been considered as an emerging technology, with wide possibilities in fields such as aerospace, automotive, biomedical, electronics, instrumentation, and weapons.

Although AM is currently a technology away from serial production processes, its versatility makes it a valuable tool in prototype production due to the freedom of design and the fine features it offers in the manufacture of, for instance, channels or internal networks. This feature gives AM a competitive advantage over other manufacturing processes used in the production of workparts [10].

AM is capable of producing workparts with complex shapes with acceptable mechanical properties for the development of materials for the weapon industry, among others. The manipulation of metals in the powder form gives AM significant advantages in the production of composite materials that are currently in development and where the imagination of designers is the limit in the architectures that can be built [10].

Kumar [13] conducted a review of existing AM processes, finding that they total 34 main processes: additive friction stir deposition (AFSD), aerosol jetting (AJ), binder jet three-dimensional printing (BJ3DP), big area additive manufacturing (BAAM), ceramic laser fusion (CLF), cold spray additive manufacturing (CSAM), composite extrusion modeling (CEM), continuous liquid interface production (CLIP), digital light processing (DLP), electrochemical additive manufacturing (ECAM), electron beam melting (EBM), electron beam additive manufacturing (EBAM), fused deposition modeling (FDM), fused pellet modeling (FPM), high-speed sintering (HSS), ink jet printing (IJP), laser engineered net shaping (LENS), lithography-based ceramic manufacturing (LCM), localized microwave heating-based additive manufacturing (LMHAM), micro droplet deposition manufacturing (MDDM), microheater array powder sintering (MAPS), photopolymer jetting (PJ), plasma arc additive manufacturing (PAD), powder melt extrusion (PME), rapid freeze prototyping (RFP), selective heat sintering (SHS), selective inhibition sintering (SIS), selective laser melting (SLM), selective laser sintering (SLS), stereolithography (SLA), thermoplastic 3D printing (T3DP), 3D gel printing (3DGP), two-photon polymerization (2PP), and wire arc additive manufacturing (WAAM). It should be noted that other processes can be found in the literature, but they may be those listed here under another name. Such processes may be classified according to the type of material that can be processed with each of them, as can be seen in Figure 9.6. Note that the principle of several processes is applicable to different materials, such as SLS, SLM, and BJ3DP with which it is possible to process metals, polymers, ceramics, and composites.

The workparts obtained through these technologies offer several advantages compared to other manufacturing processes: (1) production of multiple pieces simultaneously, (2) cost reduction during processing, (3) design freedom, (4) use of multiple materials, (5) zero waste, (6) positive impact on the environment, (7) scaling through its use in hybrid systems, and (8) remanufacturing of workparts is possible [10].

Although there is a high availability of technologies associated with additive manufacturing, one of the most practical ways of highlighting its characteristics is through the description of powder bed-based processes [10]. Take as an example SLS, also referenced as direct metal laser sintering (DMLS), which is the most

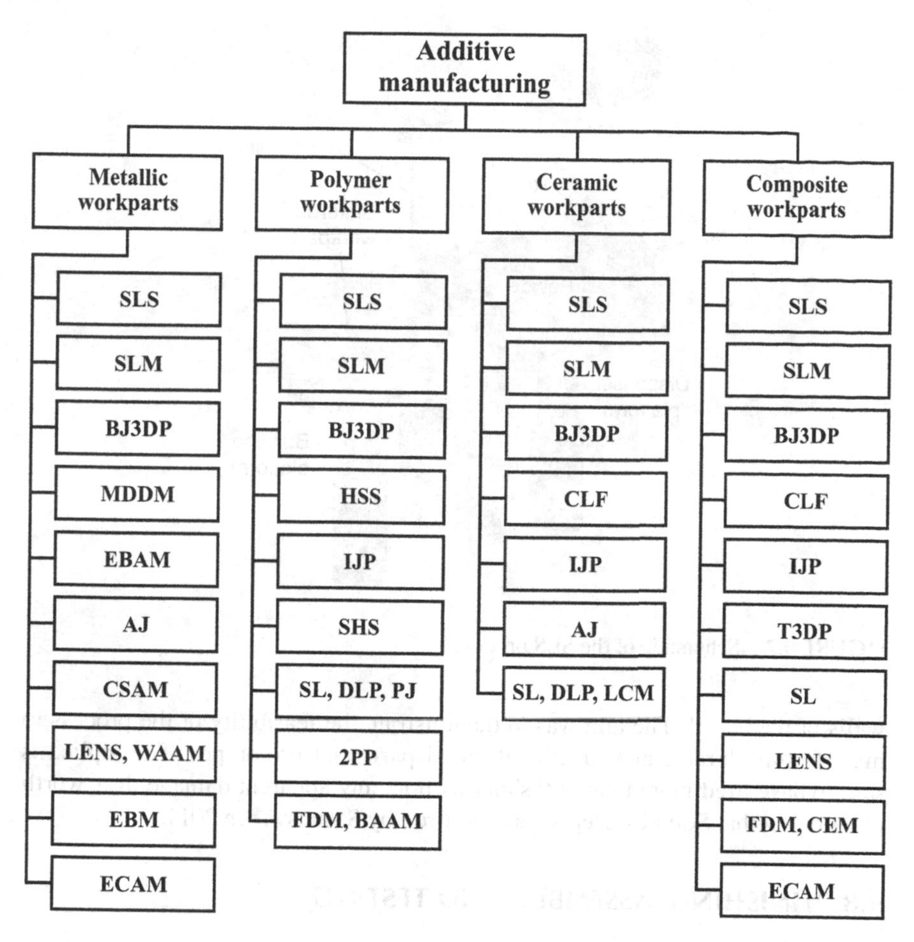


**FIGURE 9.6** Classification of additive manufacturing processes based on materials.

widely used technology in constant research and exploitation by the academic and industrial sectors and seems to be very promising for the manufacture of metallic weapon components. In SLS (Figure 9.7), the formation of metal workparts from powder layers is controlled digitally from 3D-created CAD designs. From a CAD system, a first and thin layer of powder is deposited, then a high-energy laser sinters the selected areas to a high degree of precision. Sintering provides only the energy necessary to cause atomic diffusion among the particles that are in contact with each other, forming metallurgical bonds and providing mechanical stability to the whole. This process is constantly repeated until the entire workpart is formed.

In 2013, Solid Concepts announced that it had successfully 3D-printed a replica of the Browning 1911 pistol in metal [14]. The pistol was made from more than 30 parts printed by SLS (DMLS) using 17-4 stainless steel and Inconel 625 metal powders. The process was capable of printing rifling grooves into the inner

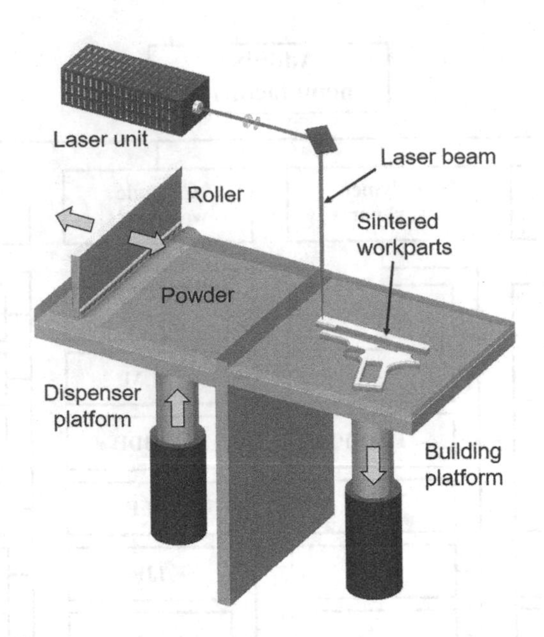


FIGURE 9.7 Schematic of the SLS process.

walls of the barrel. The aim was to demonstrate the feasibility of the process to manufacture durable and functional metal parts and prototypes. The replica is said to have fired more than 600 shots without any apparent damage. It is worth mentioning that Solid Concepts was acquired by Stratasys® in 2014.

## 9.8 FINISHING, ASSEMBLY, AND TESTING

### 9.8.1 Finishing

Secondary and finishing operations are usually performed on firearm components to improve their mechanical properties (strength, toughness, fatigue resistance, hardness), increase their resistance to wear and corrosion, give their dimensional tolerances, remove any surface flaws, give a surface finish, and enhance appearance. These operations include heat treatments and surface hardening (Chapter 8), plating (nickel, chrome, zinc, or cadmium electroplating), powder coating, hot blackening, painting, polishing, and sandblasting, among others.

### 9.8.2 Assembly

Once the components of a firearm have been manufactured, they are assembled to make a functional firearm. This task can involve diverse operations such as welding, tightening threaded fasteners, bolting, riveting, and lubricating, among others. There are different processes for joining two or more workparts into an assembled entity (Figure 9.1). The term joining is generally used for welding, brazing, soldering, and adhesive bonding, which form a permanent joint between the workparts; a joint cannot easily be separated. The term assembly usually refers to mechanical methods of fastening parts together; some of these methods allow for easy disassembly, while others do not [2].

Assembly can be divided into two levels: subassembly and final assembly. The former is the completion of components that require joining several pieces, while the latter consists of creating a finished, functional firearm from the combination of such components. Examples of components that require subassembly are frames, cylinders, bolts, hammers, and triggers. The final assembly includes verification of the proper component fit, which can be done by operator feel or by the use of calibrated gauges. An example of final assembly may include calibrating the headspace of a firearm; the subject of headspace, which is critical to the safety and reliability of a firearm, was addressed in the case study of Chapter 2. While calibrated gauges can be used for this purpose, sometimes the use of a hand reamer is sufficient for the final adjustment.

### 9.8.3 Testing

After final assembly, firearms are tested to make sure they work accurately and reliably and detect any problem. Depending on the country, there may be standard tests that the government enforces, and any firearm must pass before it can be sold. The tests may consist of proof testing and/or test firing. A proof testing consists of firing one or more over-pressured cartridges to verify that the firearm is not defective and that it will not fail on normal firing. If the firearm passes the test successfully, it is stamped with a proof mark to indicate that the proof testing was conducted. A test firing consists of firing the firearm with standard cartridges; the number of cartridges may vary by model and manufacturer.

## 9.9 CLOSING REMARKS

With the growing trend of globalization and strict production demands, the modern weapon industry is experiencing much greater pressure to attain high-quality firearms at a low cost of production. While conventional machining processes continue to be used in the manufacturing of small weapons, new technologies have been emerging over the years. These technologies range from the use of CNC machines, special casting processes, forming processes, injection processes, and powder metallurgy processes, to additive manufacturing processes (3D printing) capable of fabricating a three-dimensional object from an STL format file, which looks like a work of science fiction. Manufacturing processes are an integral part of the design of products, in such a way that they should be taken into account from the early stages of design. The selection of the appropriate technology and manufacturing process is necessary for the firearm components to reach the expected requirements and to ensure that the entire firearm works safely and reliably.

## REFERENCES


1. Davim, J.P., *Modern Manufacturing Engineering*. 2015, Springer International Publishing: AG Switzerland.
2. Groover, M.P., *Fundamentals of Modern Manufacturing: Materials, Processes, and Systems*. 2020, John Wiley & Sons: New York.
3. Robles Hernandez, F.C., J.M. Herrera Ramírez, and R. Mackay, *Al-Si Alloys: Automotive, Aeronautical, and Aerospace Applications*. 2017, Springer: Berlin, Heidelberg.
4. Doan, B.Q., et al., A review on properties and casting technologies of aluminum alloy in the machinery manufacturing. *Journal of Mechanical Engineering Research and Developments*, 2021. **44**(8): pp. 204–217.
5. Werner, D., et al., Manufacturing of firearms parts: Relevant sources of information and contribution in a forensic context. *Wiley Interdisciplinary Reviews: Forensic Science*, 2021. **3**(3): p. e1401.
6. Semiatin, S.L., *ASM Handbook, Volume 14A, Metalworking: Bulk Forming*. 2005, ASM International: Materials Park, OH.
7. Kacar, I. and F. Ozturk, Advances in stamping, in *Modern Manufacturing Processes*, M. Koç and T. Özel, Editors. 2020, John Wiley & Sons: Hoboken, NJ, pp. 3–13.
8. Monturo, C., *Forensic Firearm Examination*. 2019, Academic Press: Cambridge, MA.
9. Akhtar, S., et al., Recent advancements in powder metallurgy: A review. *Materials Today: Proceedings*, 2018. **5**(9): pp. 18649–18655.
10. Herrera Ramirez, J.M., et al., *Unconventional Techniques for the Production of Light Alloys and Composites*. 2020, Springer: Cham, Switzerland.
11. Kalpakjian, S. and S.R. Schmid, *Manufacturing Engineering and Technology*. 6th ed., 2009: Prentice Hall: Hoboken, NJ.
12. ASTM F2792-12a Standard terminology for additive manufacturing technologies. 2012, ASTM International: West Conshohocken, PA.
13. Kumar, S., *Additive Manufacturing Processes*. 2020, Springer: Cham, Switzerland.
14. Tyson, M., Direct metal laser sintering used to 3D-print working metal pistol. 2013 [August 24, 2021]; Available from: https://hexus.net/tech/news/peripherals/62261-direct-metal-laser-sintering-used-3d-print-working-metal-pistol/.

# Index

Note: **Bold** page numbers refer to tables; *italic* page numbers refer to figures.


ABC (atomic, biological, and chemical) weapons 8
abrasive flow machining 215
abrasive jet machining 215
acrylic styrene acrylonitrile (ASA) 118
acrylonitrile butadiene styrene (ABS) 118
additive manufacturing (AM) 205, 215–218, *217*
aging 193–194
Allied Engineering Publication (AEP) 134
Allied Quality Assurance Publications (AQAP) 133
alumina ($Al_2O_3$) 14, 147, 172, 174
aluminum **12,** 13, 162
aluminum alloys
  anodizing 195
  cast 164
  classification 164, **164**
  designation 165, *165*
  heat treatments 191–194, *192, 193*
  influence of alloying elements and impurities 165–167
  phase diagrams 151
  7075 alloy 194
  6061 alloy 194
  temper designation system 191, *192*
  1100 alloy 194
  wrought 164
Aluminum Association (AA) 164
aluminum framing 167
AM *see* additive manufacturing (AM)
American Association for Laboratory Accreditation (A2LA) 134
American Iron and Steel Institute (AISI) 157
American Society for Testing and Materials (ASTM) 151
animation
  CAD 67, *68*
  software 29–30
annealing
  full 182–183
  process 183
  spheroidizing 183
anodizing, of aluminum alloys 195
ANSI-Z87.1 135
Ansys CFX® 83
Ansys Fluent® for polygonal barrel and groove barrel 83, *84*
Ansys® Explicit Dynamics module 100
Anti-attack Materials and Constructions 134
AQAP-2110 NATO 133
aqueous polymer solutions 184
Arcadia Machine & Tool 162
arquebus wheel *see* wheellock system
arsenic (As) 12
arsenical bronze 11–12
artificial intelligence (AI) 108
AS-243 135
AS/NZS-2343 135
assault rifle 10, 33
  bullpup *44*
  product family 42, *43*
assembly 218–219
  assessment and simplification 79, *80*
  mesh configuration 79, *80*
ASTM-E3062M-20 135
ASTM-F1233 134–135
augmented reality (AR) 67
austempering 185–186
austenite 149–150, 181
axiomatic design 16, 54–56, *55, 56*

bainite 182
barrel
  manufacture of 213
  structural analysis of 79, *80*
Bauer Firearms Corporation 208
BCs *see* boundary conditions (BCs)
beeswax 207
Beretta 195
Bergmann, T. 7
Bessemer, H. 13
bias 55
bill of materials (BOM) 16, 21, 59, 67, 113
binary phase diagrams *148,* 148–149
black powder 2
blanking *see* stamping
body-centered cubic (BCC) structure 149
BOM *see* bill of materials (BOM)
Borchardt, H. 7
boron (B), in aluminum alloys 166
boundary conditions (BCs) 76–78
brine 184

Brinell 153, *154*
British Standards Institution (BSI) 152
broaching 214
bronze 11–12, **12**
BRV (ballistic vehicle resistance standard) 135
Bryco Arms 214
BS-5051 135
bullet penetration and perforation, by explicit dynamics 100, 104
  assigning velocity *102*
  meshing of plate and bullet *101*
  plate penetration *102*
    directional deformation *103*
    total deformation *103*
  target plate configuration *101*
bullpup firearm 33
Bureau of Alcohol, Tobacco, and Firearms (ATF) register 132
Bureau of Indian Standards (BIS) 152
burst action 9
Bushmaster 210, 214
button rifling 210

CAD *see* computer-aided design (CAD)
CAE *see* computer-aided engineering (CAE)
CAM *see* computer-aided manufacturing (CAM)
carbine 10
carbon (C), in steels 156
carbonitriding **189,** 190
carburizing **189,** 190
cardboard, corrugated 174
Carl Walther 195, 210
Carter, R.H. 175
case-hardening *see* surface hardening, of steels
cast aluminum alloys 164
casting 205–208, *206*
cast irons 12, 151
CDP *see* critical design parameter (CDP)
cementite 151
centerfire ammunition 6
Centre for Applied Science and Technology (CAST) 134
ceramic-matrix composites (CMCs) 147, 172
ceramics **12,** 14–15, 147, 174–175
chemical machining 215
Chevalier® 124
Christensen, C. 50
chromium (Cr)
  in aluminum alloys 166
  in steels 157
CI *see* commonality index (CI)
Cimatron® 114
clearance verification
  between hammer and disconnector 68, *69*
  hammer and trigger 68, *69*
CMCs *see* ceramic-matrix composites (CMCs)
CNC machines *see* computer numerical control (CNC) machines
Code of Federal Regulations (CFR) 135
Collier, E.H. 6
Colt, S. 6, 195
Colt's Manufacturing Company 208, 210, 214
commonality index (CI) 28–29, 39–40, *40*
compaction process 211, *212*
composites **12,** 14, 147
  ceramic-matrix 147, 172
  classification 171, *171*
  matrices 171–172
  metal-matrix 147, 171
  polymer-matrix 147, 171–172
  reinforcements 172–174, *173*
compound annual growth rate (CAGR) 33
compression test 153
computational fluid dynamics (CFD) 76
computational modeling and simulation 63–64, *64*
computer-aided design (CAD) 16, 22, 29, 30
  animation 67, *68*
  bill of material property manager 67
  digital modeling 65–66
  documentation and report 107
  drawings 74
  interference detection 67–68, *68*
  materials database storage 67
  models, types of 65, *66*
  for pistol *66*
  software evolution **17**
  tolerance stack-up analysis 70–74, *70–73,* **74**
computer-aided engineering (CAE) 22, 29, 30
  bullet penetration and perforation by explicit dynamics 100, 104
  CAD model treatment 79
  computational fluid dynamics 83–89
  co-simulation 100
  documentation and report 107
  emulation 105–107
  finite element analysis 79–81
  framework *75,* 77
  interference detection of trigger mechanism 82, *82*
  motion analysis 81–82, *82*
  multibody dynamics 96
  multi-domain 100
  structural analysis of barrel 81, *81*
  thermal simulation by FEA, CFD, and FSI 105
  workflow of optimization study 79, *79*
computer-aided manufacturing (CAM) 16, 22, 29
  assessment 112–115
    for high-volume manufacturing 113, *113*
    for low-volume manufacturing *112,* 112–113

CNC machining 115–116, *116*
industry 4.0 122–124, *123*
laser cutting 116–117, *117*
manufacturing devices and rapid tooling *121,* 121–122
prototyping 120, *120*
rapid prototyping 120–121, **121**
3D printing 117–118, *119*
computer numerical control (CNC) machines 115–116, *116,* 205, 214
conservation of momentum 36, 37
constraints
external 28, 39, **40,** 50
internal 28, 39, **40**
conventional weapons 8
coordinate measuring machine (CMM) 113
copper (Cu), in aluminum alloys 165
co-simulation 100
critical annealing *see* full annealing
critical design parameter (CDP) 56–58, *57*
Cura® 114, 124
customer experience (CX) 132
customer journey mapping (CJM) 50
Cusumano 131
cutter location (CL) file 114
cutting process 213

data acquisition cards (DACs) 133
Daw, G.H. 6
DECKRA 134
declaration of conformity (DoC) 130
δ-ferrite 151
Department of Defense (DOD) 131
design for assembly (DFA) index 29, 40–41, **41, 42**
design for excellence (DFX) 16, **17,** 29
Design for Inspection 16
design for test (DFT) 16
design parameters (DPs) 51, 54–56, *55,* 56, 130
design structure matrix (DSM) 16
Design Thinking 16
design tools, evolution of 15–18
first period 15–16
second period 16
third period 16–17
DFA index *see* design for assembly (DFA) index
DFX *see* design for excellence (DFX)
digital modeling 65–66
digital twin (DT) technology 122
DimXpert® 71, *71,* 72, *72*
DIN-52-290 134
DIN-EN-1063 134
DIN-EN-1522 134
direct metal laser sintering (DMLS) 216
direct numerical control (DNC) 115
DPs *see* design parameters (DPs)
Dreyse needle system 5
drilling process 213

Eddy 210
Edgecam® 114
eDrawings® 67
elastomers 14, 147, 169–170, 211
electrical discharge machining (EDM) 120, 215
electrochemical machining (ECM) 215
electronic card 28
emulation, CAE 105–107
end user, firearm design 130, *130*
energy-dispersive spectroscopy (EDS) 198–199, *199*
engineering resources product (ERP) software 21
EN ISO-9001 (quality management systems) 133
enterprise resource planning (ERP) 16
EPMs *see* experimental physical models (EPMs)
European Committee for Standardization (CES) 152
eutectic alloy 149
evolution
of design tools 15–18
of firearm manufacturing process 15
of firearms 3–7
of weapon materials *11,* 11–15, **12**
expectation for innovation
market pull 24–25, 33–35, *35*
technology push 25–26
experimental physical models (EPMs) 29, 30, 112, 132–133, 137
explicit method 76
extended reality tool 67, *68*
external constraints 28, 39, **40,** 50
extrusion 210

Fabrique Nationale Herstal 210
Fabryka Broni "Łucznik" Radom 210, 214
face-centered cubic (FCC) structure 149
failure modes and effects analysis (FMEA) 16, 44
FEA *see* finite element analysis (FEA)
Fe-C phase diagram 149–151, *150*
ferrite 149
ferrous alloys 12, 149
fiber-optic laser cutting technology 124
fine pearlite 181
finishing operations 218
finite element analysis (FEA) 76, 79–81
finite element method (FEM) 79
firearm manufacturing process, evolution of 15
firearm performance case studies (FPCS) 138–140, *139*

firearms
ALPHA 132–133
assault rifle 10
automatic action 9
BETA 133
burst action 9
calibration 132
classification by *7*, 7–11
level of harm 7
physical characteristics, size, and support 8
portability 8
traditional structure 7–8
type of firearm 9–11
weapon action 9
conventional weapons 8
design, requirements
critical design parameter 56–58, *57*
design matrix 54, *54*, *55*
DP determination using axiomatic design 54–56, *55*, *56*
FR determination and assessment 51–54, *52*, *53*
functional architecture 50, *51*
identification 49–51
interaction of tools 50, *50*
process variable *58*, 58–59
to reach readiness of system 48–49, *49*
technology transfer 59, *60*
test bench feature 56–58, *57*
types 48, *48*
evolution of *3*, 3–7
experimental physical models 132–133
headspace 35, *36*
heavy weapons 8
lethal 7
light weapons 8
long or shoulder-fired firearms 8
machine gun 10
manufacturing processes 15
non-conventional weapons 8
non-lethal 7
other types 10–11
performance case studies 138–140, *139*
pistol 10
polymer 137–138
product validation 129–132
prototypes 132–133, 136–137
qualification or certification 130
rapid prototyping 120–121, **121**
recoil 35
repeating action 9
revolver 9
rifle or carbine 10
semi-automatic action 9
short or hand-held firearms 8
shotgun 10
single-shot action 9
small arms 8
sub-machine gun 10
test benches and standards 133–136, *134*, *136*
3D printed prototypes 137–138
timeline of technologies for manufacturing 137, *137*
usability testing 131, 132
user experience testing 131, 132
Firearm Safety Certificate (FSC) 132
firearm shooting 33, *34*
fit 70
flame hardening 190–191
flintlock system 4–5
Flobert, L.-N. 6
Fluent® module
cumulative moment coefficient 85, *86*
flow simulation of bullet across fluid
cell Reynolds number *95*
internal energy *93*
match number *94*
meshing 86, *87*
particle pathlines *88*
radial velocity *96*
skin friction coefficient *90*
sound speed *89*
static temperature *92*
strain rate *89*
tangential velocity *95*
total energy *93*
total temperature *92*
turbulent dissipation *91*
turbulent intensity *90*
turbulent kinetic energy *91*
velocity *87*
velocity magnitude *87*, *88*
vorticity magnitude *94*
mesh configuration 83, *85*
velocity, simulation of 85, *86*
Fordism 16
forging 208–210, *209*
forming
extrusion 210
forging 208–210, *209*
stamping 210
Forsyth, A.J. 5
4D printing 124
FPCS *see* firearm performance case studies (FPCS)
Fratelli Tanfoglio SAS 210
French Standardization Association (AFNOR) 152
FRs *see* functional requirements (FRs)
full annealing 182–183

functional requirements (FRs) 27–28, 38, **39,** 51, 54–56, 131
functional tree 51
fused deposition modeling (FDM) 117, 118
FX05 rifle 138

G-code 114, 115
geometric dimensioning and tolerancing (GD&T) 116
German Institute for Standardization (DIN) 152
Glock 210
Goal Seek of Excel® 71
Governmental Standard (GOST) 134
graphical user interface (GUI) 29
graphics processing units (GPUs) 108
Grujicic, M. 175
gunpowder 1, 3

hammer acceleration 82, *83*
hammer angular displacement 82–82, *82*
hammer forging 209, *209*
hand cannon 4
hardening 184
hardness Rockwell B (HRB) 201
hardness Rockwell C (HRC) 160–162, 202
hardness-testing techniques 153
headspace, firearm 35, *36*
heat treatments
  of aluminum alloys 191–194, *192, 193*
    aging 193–194
    quenching 193
    solution treatment 193
  4140 steel bar
    chemical composition 195, **196**
    full annealing 197
    hardness tests 201–202
    mechanical properties of heat-treated steel samples 200, **201**
    microstructures 197, *197*
    optical microscopy 196–197, *197*
    quenching 197
    Rockwell hardness 201–202, *202*
    scanning electron microscopy 196–198, *198*
    SEM-EDS 198–199, *199*
    stress-strain curves 200, *200*
    tensile fracture morphology 200–201, *201*
    tension test specimen 195, *196*
  of steels 179–186
    austempering 185–186
    case study 195–202
    full annealing 182–183
    hardening 184
    martempering 185
    normalizing 183
    process annealing 183
    spheroidizing annealing 183
    tempering 185
    TTT diagram 181–182
heavy weapons 8
Heckler & Koch 210
history, of weapons
  design tools, evolution of 15–17
  firearms
    classification *7,* 7–11
    evolution of *3,* 3–7
    manufacturing processes 15
  first period 1
  fourth period 2
  second period 1
  third period 2
  weapon materials, evolution of 11–, *11,* **12**
Home Office Scientific Development Branch (HOSDB) 133, 134
homogeneous transformation graphical (HTG) 24, 29
Hooke's law 152
house of quality (HOQ) 27, 28, 51, 52, *52, 53*
Howard, E.C. 5
Human Factors Department 131
Human Factors Society 131
human-machine interface (HMI) 114, 122
hypothetical binary phase diagram *148,* 149

implicit method 76
incremental innovation 47
induction hardening 190
industrial internet of things (IIoT) 22, 122
industry 4.0 122–124, *123*
industry 5.0 122–124
injection molding 172, 211
*The Innovator's Dilemma* (Christensen, C.) 50
intellectual property 59
interference detection, on trigger mechanism 67–68, *68*
internal constraints 28, 39, **40**
International Alloy Designation System (IADS) 164
International Organization for Standardization (ISO) 152
investment casting process *207,* 207–208
investment readiness level (IRL) 30, *31,* 48, 131
IPG® fiber laser 117
IRL *see* investment readiness level (IRL)
iron (Fe) 12, **12,** 12–13, 166
iron carbide ($Fe_3C$) 151
Ishikawa diagram 16
ISO-16935 135
ISO/IEC-17020 133
ISO/IEC-17025 133
Iver Johnson & Painfield carbine 210, 214

Janissary corps 4
Japanese Industrial Standards (JIS) 152
Jennings Firearms 214
Jimenez Arms 214
job mapping framework 50
Jobs To Be Done (JTBD) 50
John Slough of London 210
just in time (JIT) 16

Kaizen 16
Kano model 51, 131
Kevlar® 169
Kiefuss, J. 4
Knoop 153, *154*
Kotter, A. 6
Kumar, S. 216

LabVIEW® 133
laminated glass windshield 174
laser beam machining 215
laser cutting 116–117, *117*
lean manufacturing 16
Lefaucheux, C. 5
Lenherd 131
lethal firearms 7
light weapons 8
long or shoulder-fired firearms 8
Lorcin Engineering 210
lost-wax casting process *see* investment casting process
Lothar Walther 210

machine gun 10
magnesium (Mg), in aluminum alloys 165
magnitude
  of hammer acceleration 82, *83*
  of hammer angular displacement 82, *82*
Makarov 210
manganese (Mn)
  in aluminum alloys 166
  in steels 156
Mankins, J.C. 131
manufacturing readiness level (MRL) 30, *31*, 48, 131
Marin le Bourgeoys 4
market pull analysis 24–25, 33–35, *35*
Marlin firearms 210
martempering 185
martensite 182
Mastercam® 114
matchlock system 4
material library 67
material removal process 212–215
Math model 30
MATLAB® 9, 96
Maya® 67
MBD *see* multibody dynamics (MBD)
M-code 114
mercury fulminate 5
meshing
  of bullet and fluid 83, *84*
  of fluid and bullet refining bullet meshing 83, *84*
metal-matrix composites (MMCs) 147, 171, 173
metals 146
Meyer 131
Military Standard (MIL-STD) 132, 134
milling 214
MIL-SAMIT (ballistic resistance) 135
mineral oil 184
MMCs *see* metal-matrix composites (MMCs)
modal mass 74
Model 1900 Colt 0.38″ automatic 7
Model 1902 Colt Military Automatic Pistol 7
Model 1911 in 0.45″ ACP 7
Model 60 revolver 162
molybdenum (Mo), in steels 157
Monte Carlo method (MCM) 70
Mossberg 210, 214
MRL *see* manufacturing readiness level (MRL)
multibody dynamics (MBD) 76, 96
multi-domain 100
multi-operation machines (MMs) 115
multiphysics *see* multi-domain
Munitions International Laboratories Inc. 208

National Aeronautics and Space Administration (NASA) 26
National Association of Testing Authorities (NATA) 134
National Firearms Act (NFA) 132
National Institute of Justice (NIJ) 134
National Rifle Association (NRA) 132
National Voluntary Laboratory Accreditation Program (NVLAP) 134
NATO Commercial and Government Entity (NCAGE) codes 133
natural polymers 147
new product development (NPD) 22, *23*
Newton's third law of motion 36
nickel (Ni)
  in aluminum alloys 166
  in steels 157
nitriding **189,** 190
nitrile-butadiene rubber (NBR) 170
noise, vibration and harshness (NVH) analysis 74
non-conventional weapons 8
nonlethal weapon 33–34
nonlinear FEA 76
non-uniform rational B-spline (NURBS) 65, 66
normalizing 183

North American Arms 208
NRBC (nuclear, radiological, biological, chemical) weapons 8
numerical control (NC) 115
NX Nastran® 107
Nylon 118, 147

Official Mexican Standards (NOM) 152
open architecture control (OAC) 115, 116
optical microscopy 196–197

partial differential equations (PDEs) 77
Pauly, J.S. 5
PDMs toolkit *see* product design methods (PDMs) toolkit
pearlite 180
percussion system 5
Permanent International Commission (CIP) 135
phosphorus (P), in steels 156
Pietro Beretta 210, 214
pinfire system 5–6
pistol 10
  Browning 1911 217
  CAD model for *66*
  product family 42, *43*
  simulation 96, *97, 98*
  SW22 Victory 162
plastics 13–14, 169
plate perforation, by bullet 100, 104, *104, 105*
PLC *see* programmable logic controller (PLC)
PLM *see* product lifecycle management (PLM)
PMCs *see* polymer-matrix composites (PMCs)
Poiseuille flow 76
poka-joke 16
polishing 213
polyamide 14
polycarbonate (PC) 118
polyethylene terephthalate glycol (PETG) 118
polygon mesh 65
polylactic acid (PLA) 118
polymer-matrix composites (PMCs) 147, 171–173
polymers **12,** 13–14
  elastomers 147
  firearms 137–138
  natural 147
  process 211
  synthetic 147, 168–170
    classification 169–170
    glass transition temperature 168–169
    nomenclature 169
  thermoplastics 147
  thermosets 147
powder metallurgy 211–212
PowerPoint® 67
prEN-ISO-14876 135
process annealing 183
process variable (PV) *58,* 58–59
product approval process 130, 131, *131*
product design methods (PDMs) toolkit **24,** 24–33
  case study 33–43
  commonality index 28–29, 39–40, *40*
  constraints 28, 39, **40**
  design attributes 26–27
  design criteria 26, 35–37, **36,** *36–38*
  design parameters 28, 39, **39**
  DFA index 29, 40–41, **41, 42**
  frameworks 29–33
  functional requirements 27–28, 38, **39**
  market pull, expectation for innovation using 24–25, 33–35, *35*
  modeling and simulation 29
  product portfolio 41–42, *42–44*
  requirements 27, 38
  technology push, expectation for innovation using 25–26
product lifecycle management (PLM) 16, 21–24, *22, 23*
  cycle of products development 22, *23*
  framework 22, *22*
  new product development 22, *23*
product platform 131
product portfolio 41–42, *42–44*
product validation 129–132
programmable logic controller (PLC) 120, 133
prototypes 132–133, 136–137
prototyping 22–23, 30
pure iron 149

quality function deploy (QFD) 16, 24, 27, 28, 51
quenching 193

radical innovation 47
rapid prototyping (RP) 112, 120–121, **121**
  types 120–121
  workflow 113, *114*
rapid tooling (RT) *121,* 121–122
Raven Arms 214
Raycus® 117
RCM (Canadian ballistic standard) 135
reaming process 213
recoil 35–36
  components in 36, *37*
  penetration force and impulse, calculation of 37, *38*
  velocity, calculation of 37, *37*
reinforcements 172–174
  fibers *173,* 173–174
  particles 173, *173*
  structural *173,* 174
Remington 210

repeating action 9
revolver 6, 9
rifle 10
rifling process 213
rimfire ammunition 6
root sum squared (RSS) method 70, 71
RP *see* rapid prototyping (RP)
RSS *see* root sum squared (RSS) method
RT *see* rapid tooling (RT)

SAAMI *see* Sporting Arms and Ammunition Manufacturers' Institute (SAAMI)
Sabatti S.p.a. 210
SAFE-MCF-10012 135
Sanderson 131
Savage Arms 210, 214
scanning electron microscopy (SEM) 196–198, *198*
SD-Std-01.01 135
selective laser sintering (SLS) 118, 216, 217, *218*
self-loading firearm 6–7
semi-automatic action 9
semi-automatic pistol 33
serviceable available market (SAM) 30
shape-morphing systems 124
Shaw, J. 5
Shilen 210
Shore 153
short or hand-held firearms 8
shotgun 10
"shots" 10
Siemens® technology 115
SIG Sauer 210
silicon (Si)
  in aluminum alloys 165
  in steels 157
silicon aluminum oxy-nitride (SiAlON) 174, 175
silicon carbide (SiC) 174
silicon nitride ($Si_3N_4$) 174, 175
Simplify3D® 114
Simscape® 76
  pistol simulation 96, *97, 98*
  trigger mechanism block diagram 96, *99*
  trigger mechanism simulation 96, *99*
single-shot action 9
sintering 211, 218
Six Sigma 16
Sketchpad system 16
Slic3r® 114
SLS *see* selective laser sintering (SLS)
"slug" 10
small arms 8
Smith & Wesson 162, 195, 214
Smith, W.H.B. 6
snaphaunce system 4
sniper rifle 33
Society of Automotive Engineers (SAE) 157
solid 65
SolidWorks® 67, 107, 113, 114
solution treatment 193
spheroidizing annealing 183
Sphinx systems limited 210
Sporting Arms and Ammunition Manufacturers' Institute (SAAMI) 35, 135, 141
sports guns 33–35, *35*
spring-elastomer mechanism 54, 56
Springfield armory 210, 214
stamping 210
Standardization Agreement (STANAG) 35, 134
Star Bonifacio Echeverria, S.A. 214
steel bar, selection of 213
steels **12,** 13
  AISI steels
    416 alloy 161–162, **163**
    1015 alloy 160, **163**
    1020 alloy 160, **163**
    1045 alloy 160, **163**
    1060 alloy 160–161, **163**
    1080 alloy 161, **163**
    4140 alloy 161, **163**
    4340 alloy 161, **163**
    8620 alloy 161, **163**
  carbon and alloy steels
    composition 157, **158**
    mechanical properties 159, **159**
  classification 154–156, *155*
  designation 157–162
  heat treatments 179–186
    austempering 185–186
    Fe-$F_3C$ phase diagram 180, *180*
    full annealing 182–183
    hardening 184
    martempering 185
    mechanical properties before and after **187**
    normalizing 183
    process annealing 183
    spheroidizing annealing 183
    tempering 185
    TTT diagram *181,* 181–182
  influence of alloying and residual elements 156–157
  microstructures in **186**
  surface hardening 187–191, *188*
    carbonitriding **189,** 190
    carburizing **189,** 190
    flame hardening 190–191
    induction hardening 190
    nitriding **189,** 190
STEP-NC 115, 116
stereolithography (SLA) 118
Sterling Arms 210

Steyr-Mannlicher 210
Stratasys® 124
stress-corrosion cracking (SCC) 166
Sturm, Ruger & Co., Inc.208, 210, 214
styrene-butadiene rubber (SBR) 170
sub-machine gun (SMG) 10
sulfur (S), in steels 157
surface hardening, of steels 187–191, *188*
SURFCAM® 114
SW22 Victory pistol 162
synthetic polymers 147, 168–170
  classification 169–170
  glass transition temperature 168–169
  nomenclature 169
system usability scale (SUS) 131

target market 30
TBF *see* test bench feature (TBF)
technical importance rating (TIR) 52
technology push 25–26
technology readiness level (TRL) 16, 30, 31, *31,* 48, 112, 131, 136
  deployment phases *32,* 32–33
  development phases 32, *32*
technology transfer 59, *60*
tensile strength 152, 153
tensile stress-strain curve, of ductile metal 152, *153*
tension test 151
test benches (TBs) 129, 140–141
  design 133
  and standards 133–136, *134, 136*
test bench feature (TBF) 56–58, *57*
test firing 219
theory of inventive problem solving (TIPS) 24
thermal conductivity 105, *106*
thermodynamics 78
thermoplastic copolyester elastomers (TPCs) 170
thermoplastic elastomers (TPEs) 170
thermoplastic polyether block amides (TPAs) 170
thermoplastic polyolefins (TPOs) 170
thermoplastic polyurethane elastomers (TPUs) 118, 170
thermoplastics 147, 169, 172, 211
thermoplastic styrenic block copolymers (TPSs) 170
thermoplastic vulcanizate elastomers (TPVs) 170
Thermo Scientific FlashSmart CHNS-O elemental analyzer 195
Thermo Scientific TM iCap 6000 Series ICP spectrometer 195
thermosets 147, 170, 211
3D Max® 67
3D printed prototypes 137–138
3D printing 117–118, *119,* 124, 215
3D scanning 66
3D Sprint® 114
3D Systems® 124
3DXpert® 114
timeline of technologies 137, *137*
time-temperature-transformation (TTT) 180–182, *181*
titanium (Ti)
  alloys 167–168
  in aluminum alloys 166
  corrosion resistance of 168
TolAnalyst® 71, 74
tolerance 70, 71
tolerance stack-up analysis 70–74, *70–73,* **74**
total available market (TAM) 30, 33
Toyota design techniques (TDTs) 16
Toyota production system (TPS) 16
TRL *see* technology readiness level (TRL)
tungsten carbide mandrels 209
Twaron® 169

UL-752 135
Ulrich 131
ultrasonic machining 215
United Nations Office on Drugs and Crime (UNODC) 7, 10
Unreal Engine® 67
usability testing (UT) 131, 132, 138
user-centered design (UCD) 16, 43
user experience (UX) testing 131, 132, 138
user interface (UI) 63
U.S. Repeating Arms Company 210
Utterback 131
Uzumeri 131

valuable minimum product (VMP) 30
vanadium (V), in steels 157
V50 ballistic limit 134
Verowhite 118
Vickers 153, *154*
virtual reality (VR) software 64, *65*
Visiclear 118
voice of the customer (VOC) 27–28, **39,** 51, 52
Von Dreyse, N. 5
Von Mannlicher, F. 7

water 184
waterfall Toyota 16
water jet cutting 215
weapon(s)
  ADC 8
  action 9
  conventional 8
  heavy 8
  history of

weapon(s) (*cont.*)
design tools, evolution of 15–17
firearms *7,* 7–11, 15
first period 1
fourth period 2
second period 1
third period 2
weapon materials, evolution of 11–, *11,* **12**
light 8
non-conventional 8
NRBC 8
weapon action systems **17**
weapon materials, evolution of 11–, *11,* **12**
ceramics **12,** 14–15
composites **12,** 14
metals
aluminum **12,** 13
bronze 11–12, **12**
iron **12,** 12–13
steel **12,** 13
polymers **12,** 13–14
stone, wood, and bone 11, **12**
Weatherby 210
Wesson, D.B. 6
wheellock system 4
White, R. 6
Wilson Arms Company 210, 214
Winchester Model 1866 12
Winchester Repeating Arms 210, 214
wireframe 65
wootz 13
World War I 1
World War II 15, 24
worst-case tolerance analysis 70
wrought aluminum alloys 164
composition 166, **166**
designation 165, *165*
heat-treatable 166, **167**
mechanical properties 194, **194**
7075 alloy 167
6061 alloy 167
1100 alloy 167
wrought iron 12
Wuwei Bronze Cannon 12

"yellow boy" 12
Young's modulus 152

zinc (Zn), in aluminum alloys 166
zirconia ($ZrO_2$) 174

[illegible]inted in the United States
by B[illegible]er & Taylor Publisher Serv[illegible]

Printed in the United States
by Baker & Taylor Publisher Services